Media and Mapping Practices in the Middle East and North Africa

Media and Mapping Practices in the Middle East and North Africa

Producing Space

Edited by
Alena Strohmaier and Angela Krewani

Routledge
Taylor & Francis Group
LONDON AND NEW YORK

First published in 2021 by Amsterdam University Press Ltd.

Published 2025 by Routledge
4 Park Square, Milton Park, Abingdon, Oxon OX14 4RN
605 Third Avenue, New York, NY 10158

Routledge is an imprint of the Taylor & Francis Group, an informa business

ISBN: 9789462989092 (hbk)
ISBN: 9781041182627 (pbk)
ISBN: 9781003699491 (ebk)
NUR 670

Cover illustration: VJ Um Amel (2017) Body of Cyborg 8.8/ سايبورغ مجسّم من البيانات Self-portrait. Dye sublimation on canvas. One original print. 30" x 20"

Cover design: Coördesign, Leiden

DOI: 10.5117/9789462989092

Acknowledgements

This book would not have been possible without the valuable contributions of our authors, whom we would like to thank here. Furthermore, we are very grateful for the great support of the research network *Re-Configurations History, Remembrance and Transformation Processes in the Middle East and North Africa* (Center for Near and Middle Eastern Studies, Philipps-Universität Marburg) which facilitated this edited volume both conceptually and financially. The Philipps-Universität Marburg founded the research network in spring 2013 with funding from the German Federal Ministry for Education and Research (Bundesministerium für Bildung und Forschung, BMBF). Its founding was triggered by the developments in the Middle East and North Africa (MENA) region since 2009, which highlighted the need for a partial reassessment of scholarship on MENA countries, for a development of new interdisciplinary perspectives, and for a deepening of our understanding of the events unfolding in the region: their underlying reasons, historic roots, and future perspectives. Our gratitude extends also to Gregory Fisk for style editing as well as to Leslie Karina Debus and Isabelle Wientzek for copy editing this manuscript.

Alena Strohmaier and Angela Krewani

Table of Contents

Introduction: About Space as a Media Product

Alena Strohmaier and Angela Krewani

Keywords: Media practices, mapping, cartography, movements, agencies, Middle East, North Africa

Bourj Al Shamali, South of Lebanon, red balloons in the sky. In 1948, 7000 refugees fled or were expelled from their homes in Tiberias and Safad in historic Palestine, now Israel; second- and third-generation refugees (22,000 registered) currently form the majority of the population in the camp. The ancestors of Bourj Al Shamali's population led an agricultural existence that has now been completely lost; the camp residents have increasingly grown detached from the land. Al Houla Association, one of the local NGOs working in the camp, which also serves as the base for the local camp committee working to improve conditions in the camp, began exploring the possibility of launching an urban agriculture pilot project and creating a green space in the camp.[1] For this initiative, a map of the camp was needed to discuss potential locations and to visualize potential water sources. However, it turned out to be difficult to find a map of Bourj Al Shamali, even though it has been in existence for over 60 years. With the complex politics of the region, the maps that do exist are withheld by international organizations that justify their discretion in the name of security and do not share them with the camp inhabitants or with the local camp committee. On internet maps, only the main street is marked, and on Google Earth, the very low-resolution images of the area obscure the space, the narrow streets, and the buildings. Therefore, in 2015, the inhabitants themselves launched an initiative in cooperation with the local camp committee to map the area.

1 See: *Greening Bourj Al Shamali. An Initiative that Aims to Green and Improve the Living Conditions in a Refugee Camp in South Lebanon*: http://bourjalshamali.org/

Strohmaier, A. and A. Krewani (eds.), *Media and Mapping Practices in the Middle East and North Africa: Producing Space.* Taylor & Francis Group, 2021
DOI 10.5117/9789462989092_INTRO

The solution was a reusable latex/chloroprene balloon measuring at one and half metres wide, a 300-metre-long line, swivel clips for attaching the balloon and the camera, rubber bands for making a camera cradle, reusable Velcro for closing the balloon, some carabiners to attach things together, and a camera that can be set on an automated mode to take images every few seconds. Everything was tied up, the helium-filled balloon rose up in the air, and after a flight of 10–20 minutes, it could be brought down again.[2]

Technology, digital media, and activism brought this project into being. However, the balloon mapping alludes to more enduring concerns that arose from the need to capture one's own space as a map. In its use of digital media, bottom-up cartography, and citizen science, the balloon mapping of Bourj Al Shamali offers a significant point of departure for any discussion of contemporary media and mapping practices in the Middle East and North Africa (MENA) region. Aspects of media, mapping practices, and the construction of spaces are interrelated and reflected into each other. Space is not a given, but produced in activities such as the balloon mapping of Bourj Al Shamali. Without media, the dimensions of space can hardly be experienced. Bruno Latour claimed, in reference to central perspective in painting, space could be a mobile medium in itself: 'in linear perspective, no matter from what distance and angle an object is seen, it is always possible to transfer it and to obtain the same object at a different size as seen from another position' (1990, p. 27). Besides this European linear perspective, diverging combinations of space and media can be considered valid, such as the song lines of the Australian Aborigines, which combine the constitution of spaces and places with their song lines – in this manner, orally constructing their cultural-geographical spaces (Winkler 1997).

When viewed against this background, media could always have been conceptualized as spatial agents, since most media – traditional and analogue ones as well as new and digital ones – inscribe themselves into spaces or help to construct and communicate these spaces. The history of the letter connects to the history of post, combining media with the crossing of spaces (Siegert 1993). The Hollywood genre of the western or the road movie recounts the conquest of the vast spaces of the American West with the media of the stagecoach, Bible, and law while completely ignoring the native inhabitants of the region. Some media carry their conceptualized spaces in their name: viewing into the distant spaces – 'tele' 'visioning' – addresses television as a window opening up into distant spaces. Other media have

2 See: *Balloon Mapping Guides*: http://bourjalshamali.org/2016/10/12/balloon-mapping-guides-in-arabic-and-german/

brought about new constellations of spaces and places, particularly the digital mobilization that offers new access to the spatial dimension, since it recombines spatial and medial aspects. Locative media – smartphones, GPS devices, tablets, and others – combine local and virtual elements. Adriana de Souza e Silva conceives spaces as inherently mobile, relating to the definition of 'augmented space' (Manovich 2005) as a connection of virtual and material aspects, 'mobile spaces are networked spaces defined by the use of portable interfaces as the nodes of the network' (De Souza e Silva 2006, p. 266). The idea of networked spaces offers a theoretical framework for the 'nomad existence and the spatio-geographic aspect' (De Souza e Silva 2006, p. 267). Accordingly, a nomad moves within predefined spaces and routes. De Souza e Silva connects to this concept of the nomad and rekindles it in the light of mobile media:

> Mobile technology users take the nomadic concept one step further, because not only their paths are mobile but also the nodes. With the fixed Internet, and fixed landlines, computers and telephones were primarily connected to places. Conversely, cell phones represent movable connection points, accompanying their users' movement in physical spaces. (2006, p. 267)

Technologically, the application of mobile phones is supported by 3G to 5G mobile networks, which are often not operated by governments, but by commercial companies. The prevalence of mobile media demands a modified conceptualization of space. De Souza e Silva concludes: 'I regard space as a concept produced and embedded by social practices, in which the support infrastructure is composed of a network of mobile technologies' (2006, p. 271). Following De Souza e Silva's references to the nomad as the new inhabitant of networked spaces, the features of these new spaces have to be reconsidered.

An additional consequence resulting from mobile media is a different aspect of media participation.[3] The idea of participatory culture is also recasting the way in which citizens, like the inhabitants of Bourj Al Shamali, engage with their surroundings and in political life. This is especially the case with the media upheaval away from broadcast mass media towards more individualized mobile media practices. This generates a shift in the connection between political power relationships and regimes of knowledge. Thus, mobile media practices constitute reconfigured spaces that provide

3 See also: Giaccardi 2012; Jenkins, Ford, and Green 2013.

both knowledge *about* spaces as well as knowledge *through* spaces. Hence, this book is centrally concerned with how space is not, as Henri Lefebvre famously stated, a social but rather a media product. In the following, we shall elaborate to what extent cartographies, movements, and agencies play a decisive role in both media and mapping practices in perpetually producing and creating space in and of the MENA region.

Media Practices

A few months into the popular uprisings in the MENA region, the promises of social media, including its ability to influence a participatory governance model, grassroots civic engagement, new social dynamics, inclusive societies, and new opportunities for businesses and entrepreneurs, became more evident than ever. Arguably, social media tools and their influence on online and offline identities played a critical role in the MENA region before and after the popular uprisings, which tested the conceptual suitability of classical and traditional ideas of space. At the same time, digital and especially mobile media were a crucial factor in the reconfigurations of the production and circulation of the images of these popular uprisings:

> to the extent that there would appear to be an ineluctable logic to these developments, it is all the more crucial to observe how the rhetoric of revolution effects a subservience of the aesthetic to the spectacle of conflict, not to mention the claims and counter-claims of politics and the often ideological expectations of historicization. (Downey 2014, p. 18)

The Green Movement in Iran in 2009 and the simultaneous waves of popular uprisings across the Arab world in 2011 were strongly documented by so-called citizen journalists; in other words, private persons or amateurs with digital or mobile cameras who disseminated their images via social media. These videos soon circulated beyond social media: between 2010 and 2015, at least 20 documentary films were made both inside and outside the MENA region, consisting partly or entirely of this kind of internet footage. The documentary films reassemble the videos by shortening, recontextualizing, and adding intertitles or voice-overs. The origin of the source material featured in these documentaries is often partly or sometimes completely unknown; the original clips had often gone viral and become iconic in the social media landscape. The videos are presented in a new framing and with novel meanings through the revised narrative and aesthetic contexts of the film.[4] These videos have also appeared in art installations, such as Rabih Mroué's lecture performance *The Pixelated Revolution* from 2012 and Birgit Hein's *Abstract Film* from 2014. Non-profit media collectives, such as the Egyptian Mosireen and Syrian Abounaddara and Bidayyat collectives have used them as well. Finally yet importantly, they have been stored in digital open-access and open-source archives, such as the Syrian Archive, the Yemeni Archive, 858 – An Archive of Resistance, and the Public Access Digital Media Archive (Dang and Strohmaier 2018).

What these different appropriation processes have in common is that they pursue the visualization of crises in documentary and in post-cinematographic forms. In this context, post-cinema does not mean the pure accumulation of new media forms triggered by the changes of film as a material object of photochemical recordings. Nor is it limited to the cinema as a cultural and architectural institution in the digital age. Rather, the term enables a view that perceives media reconfigurations as interlocking cultural phenomena (Hagener, Hediger, and Strohmaier 2016, pp. 1–12). The theoretical field of post-cinema thus makes it possible to pursue such interactions on three levels: first, at the level of the specific technical and aesthetic possibilities of the audiovisual of so-called citizen journalists; second, at the level of the representation of norms and conventions that emerge in their media processing; and third, at the level of geopolitical aspects and positionalities, such as the Western view on popular uprisings. Artist and theorist Hito Steyerl proposed, with what she calls 'circulationism' (Steyerl 2014), that the importance of images lies no longer in the art of making them, but in the art of their post-production, distribution, and acceleration. On another note, media scholars Ramon Lobato and James Meese use the concept of 'circumvention' (Lobato and Meese 2016, p. 9) to analyse the geo-blocking of videos to emphasize the subversive potential of users of streaming platforms. Therefore, media practices, this book argues, mean more than the triad of production, distribution, and reception; instead, media practices designate not only forms of technical and economic dissemination but also a holistic view of various media practices that build on each other, are interwoven, networked, and, at times, contradictory.

In this context, it is important to note that as other world regions, the MENA region is not a well-defined, closed, and homogeneous area. On the contrary, it

4 For recent research on cinematic appropriation processes of protest videos in the MENA region since 2009, see: https://www.uni-marburg.de/de/fb09/medienwissenschaft/forschung/forschungsprojekte/mena-film

is actually a region that is subject to continuous reconfigurations and for which numerous designations exist, such as the Orient, the Levant, and the Near or Middle East.[5] At the same time, economic, cultural, social, and political interrelations and processes of exchange unfold in very different regional contexts, which intersect with established spatial ideas. Political scientist Claudia Derichs equates such a shift to what she calls a 'new areas studies current' (2015, p. 33). It demarcates an open concept of the notion of 'area' as itself not hermetically sealed or self-contained. Her main argument is that how 'area' is understood, how it is defined, and what it designates is subject to historical and cultural processes. She refers in this context to Arjun Appadurai, who claimed that 'we need to recognize that histories produce geographies and not vice versa' (2010, p. 9). Likewise, social and regional scientists Anna-Katharina Hornidge and Katja Mielke argue that 'area' should not be understood as a singular or hermetic unit per se, but as a dynamic category (2015, p. 14). This development in area studies itself, and the accompanying change from an inflexible to an open understanding of 'area', therefore emphasizes structures that conceive of the MENA region not as a homogeneous and unified space but rather as a set of interdependent and interrelated places which are in a constant process of reconfiguration (Ouaissa, Pannewick, and Strohmaier 2021, pp. 1-21).

Following these approaches, the underlying assumption of this book is that mobility and movement increasingly challenge static political, socio-cultural, ethnic, and religious boundaries. An open and flexible concept of 'area' is crucial for our understanding of what we would like to call *mobile media spaces*, which bring about varied and amorphous modes of producing space. Therefore, this book promotes a transregional perspective on the circulation of ideas, narratives, and images, without being limited to fixed metageographies. Thus, the second argument is that media do not provide simple technological tools or equipment, but *mobile media spaces* which work over time and beyond different world regions, including diaspora formations (Strohmaier 2019). Mobile media spaces undermine, defy, and blur state-centred notions of territory as well as notions of nation and political communities based on territorially fixed states. Therefore, the starting point for an examination of media as a factor of spatial production, which is based on social conditions and practices in the broadest sense, needs to include precisely the conditions of media and their transmission between academia and the public, between centres and peripheries.

'Mapping' has become a buzzword in this context. Digital cartography modifies conventional cartography by conflating it with forms of digital

5 See also: Culcasi 2010; Culcasi 2012.

communication practices. Web spaces combine conventional cartography with the communication structures of social networks and the archive functions of computer hard drives. These new hybrid forms of digital spaces merge with mobile mapping practices.

Mapping Practices

With the start of the Cold War and the first attempts at space travel, efforts were undertaken to study the Earth from outer space, following from the military tradition of using either moored balloons or airplanes to produce photographs of landscapes from an elevated vantage point. Older satellites worked partly with photographic stills or moving pictures, and the resulting films were physically transported down to Earth in capsules. More modern technology worked in a manner similar to television. A television camera was used, and its images were stored on magnetic tape until the satellite had passed by a receiving station and the images could be passed as electric signals. This kind of technology was used in satellite transmissions until the end of the 1970s. At that point, photography was replaced by a new technology, which recorded and stored light using a sensitive electrooptical technique – just as in the video camera, which was to be developed later (Dyring 1992, p. 29). In cartography, there has been a long-standing tradition of collaboration between media and the military. In *War and Cinema*, French media philosopher Paul Virilio uses media theory to highlight the parallels between war and cinema, working with a tightly woven technological analogy between the apparatus of war and the film camera:

> It was in 1861, whilst travelling on a paddle-steamer and watching its wheel, that the future Colonel Gatling hit upon the idea of a cylindrical, crank-driven machine-gun. In 1874 the Frenchman Jules Janssen took inspiration from the multi-chambered Colt (patented in 1832) to invent an astronomical revolving unit that could take a series of photographs. On the basis of this idea, Etienne-Jules Marey then perfected his chrono-photographic rifle, which allowed its user to aim at and photograph an object moving through space. (1986, p. 19)

Nonetheless, Paul Virilio's argument fails to address some important aspects of the origins of visual traditions and of the discursive attributions of media apparatuses. In addition to the technical inscriptions of images, formative visual traditions and routinizations of images exist, which function as more than just technical inscriptions. In this way, his ideas confirm the persuasive power of images, which has emerged in the course of a long historical process.

Over the last few years, cartography has received a considerable amount of renewed interest. Notably, with the introduction of Geographical Information Systems (GIS), cartography has merged with social media platforms and communication structures. Geobrowsing has emerged as a new name for this conflation. Apart from being an innovative aspect of communication in social networks, geobrowsing includes a variety of traditional media forms, particularly with regard to the transformation of the imagery of the Earth. Traditional cartography, specific forms of film and photography, and the satellite image are important predecessors of geobrowsing applications. Traditional cartography and digital satellite imagery were produced to document the shape of the Earth and the features of landscapes; these images nonetheless carry within them fixed visual conventions and meanings which, as Denis Cosgrove puts it, 'have drawn upon and reconstituted a repertoire of sacred and secular, colonial and imperial meanings, and [...] these representations have played an especially significant role in the self-representation of colonial and post-colonial nations' (1985, p. 46). In this sense, cartography and the possession of the landscape occurred in parallel, while the application of geometry often made the acquisition of actual space easier.

In this context, the visual zoom function provided in Google Earth is based on the visual and epistemic history of the images of the Earth in cartography and satellite photography. With the introduction of Google Maps and Google Earth, both released in 2005, cartographic images of the world have experienced a surge on social media platforms. Although a variety of GIS had previously been in use, Google Maps and Google Earth brought about a new quality and a broader popularization within GIS systems, which can be described with respect to their popular proliferation, as Michael Jones, Chief Executive Officer of Google Earth, stated in 2007:

> What's happening now [...] is that instead of just GIS experts talking to each other, or experts making maps for regular people, regular people are talking to each other, and they are making maps for each other. And that's very important [...] the story of the where is very important. (cited in Crampton 2010, p. 25)

Contrary to classical GIS, Google Maps and Google Earth, and later OpenStreetMap and Wikimapia, allow for the combination of social media with cartographic systems and thus take cartographic competence away from the experts. Contemporary cartographic systems can be installed on personal computers or on smartphones. Data can be stored on a server and the client computer merely requires a browser to enter cartographic systems. Data and modules are loaded onto the client server on demand (Abend 2013, p. 150). These innovations made the general use of cartographic systems even easier, but not less controversial.

Image-related conventions and structures of mapping practices transform, so this book's third argument, means to enable them to function in adjacent discourses beyond technologies. Accordingly, images become nodal points for a multitude of different discourses. Taking for granted that geographical and national spaces are always mediatized spaces, the turn towards social media has nevertheless changed the understanding of space, which has become fragmented and individualized. This is reflected in the manifold approaches in this book towards space as a media product, which encloses a variety of practices such as digital cartography, gamification, video activism, cinema, parkour, data mining, oral transmission, and audiovisual representations.

Cartographies, Movements, Agencies

The following contributions are informed by the idea that the fragmentation of spaces brings about new forms of media and mapping practices and are, at the same time, a product of these practices. Therefore, the book features multiple voices that share collective and individual stories within larger contexts of political and social challenges, enabling a plurality of interdisciplinary perspectives. This displaces binaries such as the division between the public and private sphere or the nation and the citizen in a collage structure that embraces disjuncture, heterogeneity, plurality, and dialogue. The contributions in this book provide compelling perspectives that privilege context over text. They engage different levels of creative abilities, participation, and viewpoints in dynamic iterative relationships. They circulate and reorganize scattered media remnants across different platforms and within different communities. In so doing, they create a more nuanced and shared construct to reimagine how we might understand space as a media product in the MENA region today. The book mirrors three different aspects of media-generated spaces. The first part, labelled 'Cartographies' relates to historic and contemporary practices; the second part, 'Movements', addresses the media and spatial activities of cinema, migration, and parkour; and the third part, 'Agencies', discusses the viable interest of global and local groups in specific areas.

The first part starts with a discussion of historic cartographic practices in the last century of the Ottoman Empire. Nour Nicole Dados focuses on the early period of cartographic practices in the MENA region (up to the 20th century) to consider how surviving maps and notes, in conjunction with other textual sources, can be used to develop a historical account of subjectivity and the production of social knowledge. She uses textual and cartographic material about Beirut to consider the ways in which the practice of mapping not only expands our geographical 'memory' of the city but also allows us to reflect upon specific urban histories.

Aspects of the cartographic organization of social groups also play a role in Christian Bittner and Georg Glasze's exploration of patterns of exclusion and inclusion in OpenStreetMap and Wikimapia, especially in the examples of the Gaza Strip and Jerusalem. These mapping platforms have proven that geographical information and maps can be generated via crowdsourcing by cartography amateurs. Yet several studies have excavated powerful exclusion mechanisms in such Web 2.0 cartographies, showing that privileged demographic groups often produce user-generated geographical content. Besides these strategies, the cartographic systems themselves are not neutral actors within these activities.

Frederik von Reumont's examination of cartographies within the Western Sahara takes a closer look at the connection between identity, territory, and borders as well as their construction via mapping processes. His analysis of different web maps voices the rootedness in dominant political world views and contrasts them with the perspectives of the parties of the opposition (Sahrawi Arab Democratic Republic and Morocco). The delineation of borders is here directly linked to territorial identity and nation-building. This demonstrates the important role of web maps and other digitally disseminated maps in the definition of identity and (self-) governance.

Besides cartography's social and political partiality, its gamification represents a further step towards its becoming virtual. Janina Schupp analyses the evolution of digital training cartographies and critically investigates the growing gamification ensuing in these representations of Middle Eastern battlefields. The mapping capability – offered by the resulting simulation game *73 Easting* – to visualize the battlefield from any position and point in time revolutionized military exercises. With ongoing conflicts in the MENA region, these digital training cartographies have matured over the years and are now linked to real bodies and vehicles through digital and mobile technologies during live training in artificially constructed Iraqi and Afghan villages.

The following part, on 'Movements', focuses on the media impact of political movements and events, which are informed by audiovisual media and/or are represented throughout manifold media practices. In this part, space is not exclusively related to mapping practices and cartography, but it is understood as being brought about through cinematic strategies on the one hand and bodily as well as representational strategies on the other hand. In this manner, media practices interact with real locations and geopolitical discourses. Interestingly, these processes are not restricted to specific media but can occur in a variety of media configurations. Cinema seems to be the most obvious form of constituting social and political spaces, but these processes extend to other forms such as print media and parkour.

This section starts with Hamid Nacify's examination of the emergence of 'internet cinema' by focusing on Iran in the late 2000s, particularly since the 2009 disputed re-election of Mahmoud Ahmadinejad as president. Ahmadinejad's re-election unleashed the opposition Green Movement, undergirded by this internet cinema. Iran provides an example of the use of the new internet-driven digital global media in support of democratic ideals that, soon after the Green Movement, rocked many authoritarian regimes of the MENA region.

In her contribution, Katarzyna Ruchel-Stockmans investigates the possibilities of documenting historical events through mobile video footage. The so-called Arab Spring appeared on the news in countless video testimonials shot by participants and bystanders. Viewers all over the world could watch distant events immediately after they were recorded and uploaded on social media. Yet the sheer amount of such visual data, combined with the untranslatability and poor quality of the footage, often delivered opaque and ambiguous material. She compares Peter Snowdon's *The Uprising* (2013) to Harun Farocki and Andrei Ujică's *Videograms of a Revolution* (1992), and questions the possible shift in knowledge produced and performed by the amateur video-makers.

Whereas Naficy and Ruchel-Stockmans deal with documentary forms, Alena Strohmaier provides a close reading of Mohamed Diab's feature film *Clash* (2016), through which she foregrounds the idea of the truck as a cinematic space predicated on its ability to accommodate movement, both in a literal and metaphorical sense. The analysis allows for a discussion of cinematic spaces of the so-called 'Arab street' created by both mise en scène and cinematography that go against the more prevalent images of street fights and mass demonstrations as seen in documentaries. Her main argument is that cinema challenges and inverts traditional spaces of social upheavals, such as streets and squares, in their capacity to be spaces of knowledge and solidarity, in negotiating them as enhanced media-sensible spaces.

The last two chapters of this section turn away from the realm of moving images towards print and everyday media practices in urban spaces. Ines Braune follows the idea of a mediated constitution of spaces, placing her considerations within the context of everyday urban practices of parkour in Morocco. Furthering de Certeau's concept of reclaiming urban spaces, she introduces this movement as a corporeal practice of claiming urban spaces.

Regarding print media, Ayça Tunç Cox analyses the varied ideological affiliations subscribed to differing interpretations of the coverage on Syrian refugees in Turkey. She examines the coverage in *Hürriyet*, a mainstream conservative paper; *Cumhuriyet*, the nationalist Kemalist representative of the Turkish press; *Yeni Şafak*, a right-wing Islamist paper; and *Birgün*, a left-wing paper. Moreover, she focuses on certain periods that raised momentous public interest: the three days following the outbreak of the Syrian crisis; the three days following the discovery of baby Aylan's body washed ashore; and the three days following the European Union's three-billion-euro fund for Turkey.

The third and last part of the book, 'Agencies', aims to broaden the concept of space to agencies and artistic interventions into political structures and processes. These agencies can relate to visual media representations such as the media representation on YouTube, but it can also cover political unrest or be applied to define a state or nation.

The book's third part starts with Angela Krewani's exploration of the media coverage of the Arab Spring and the ensuing reactions of Western media communities. Focusing on interactive documentaries and websites, her contribution clearly demonstrates the extent to which media featured individualized coverage of major events. Digital media in particular have merged with cartographic competencies to provide topical information. Compared to the informational range of classic print media and television, these digital platforms and digitally distributed art forms create new and interactive forms of media participation.

Whereas Angela Krewani focuses on the site of visual representation of images, Laila Shereen Sakr looks at the digital practice of data mining. She presents a basic genealogy of the existing literature on the 2010–2012 era of the Arab Spring within Middle East and media studies. The aim of her analysis is to map the plurality of patterns, stories, ideas, and analyses networked across media forms. Her contribution is designed to refute the possibility of a monolithic narrative about contemporary Egypt and situates the study within the growing scholarship on social media and political change, introducing key concepts in design research methodology.

Annabelle Sreberny and Gholam Khiabany explore in their contribution how contemporary Iranian politics exist through media and how these politics are played out through political geographies. They investigate three powerful spatial dynamics at work in the analysis of Iran's political orbit. The first is the classic remit of the international relations between states. The second is the mainstream remit of political analysis, the national dialogue – sometimes open, often constrained – between the state and its inhabitants. The third is the cross-border space between the state and its citizens who – as diaspora, exiles, and migrants – live in other countries.

Also concentrating on the visual politics of an ethnic community, Sebastian Maisel underlines an increased exposure to sacred knowledge as well as a re-intervention of religious dogma in Yezidi culture by unorthodox scholars, intellectuals, and politicians. He argues that their use of technology to spread deliberate knowledge has contributed to the development of new forms of identity and loyalty among Yezidi groups in Syria and Iraq.

Finally, Andrea Fischer-Tahir shows that colonial images of Kurdish bodies and of Kurdistan's geography still have use value in identity discourses shaped by the aid of social media, which witnessed the simultaneity of modes of representation, multiplied the possibilities of translocating knowledge, and promoted Kurdish national aspirations for a nation state.

Across the MENA region today, these varied, empirical, theoretical and interdisciplinary contributions show there is not only a development of a critical field of digital media, but also a proposition of alternative platforms for social and political engagement. In an attempt to rearticulate the relationship between media and mapping practices, while also addressing new and social media, this book nevertheless abides by one relatively clear point: space is a media product. The overall focus of this book is accordingly not so much on the role of new technologies and social networks as it is on how media and mapping practices expand the very notion of cultural engagement, political activism, popular protest, and social participation. To this end, space is produced, created, re-configured, reproduced, and recreated, which testifies to its status and role as a cultural media and mapping practice in itself.

References

Pablo Abend, *Geobrowsing. Google Earth & Co. Nutzungspraktiken einer digitalen Erde* (Bielefeld: transcript, 2013).

Arjun Appadurai, 2010, 'How Histories Make Geographies', Unpublished manuscript. Online Resource: http://heiup.uni-heidelberg.de/journals/index.php/transcultural/article/view/6129 (accessed 12 November 2020).

Denis E. Cosgrove, 'Prospect, Perspective and the Evolution of the Landscape Idea', *Trans. Inst. Geogr. N.S.*, 10 (1985), pp. 45–62.

Jeremy W. Crampton, *Mapping: A Critical Introduction to Cartography and GIS* (Malden: Wiley-Blackwell, 2010).

Sarah-Mai Dang and Alena Strohmaier, 'Collective Collecting: The Syrian Archive and the New Challenges of Historiography,' *View – Journal of European Television History and Culture*, 7(14) (2018), pp. 88–102. Online Resource: https://www.viewjournal.eu/articles/abstract/201/ (accessed 12 November 2020).

Claudia Derichs, 'Shifting Epistemologies in Area Studies: From Space to Scale', *Middle East – Topics & Arguments*, 4 (2015), pp. 29–36. Online Resource: https://meta-journal.net/article/view/2981 (accessed 12 November 2020).

Karen Culcasi, 'Constructing and Naturalizing the Middle East', *Geographical Review*, 100(4) (2010), pp. 583–597. Online Resource: https://doi.org/10.1111/j.1931-0846.2010.00059.x (accessed 12 November 2020).

—, 'Mapping the Middle East from Within: (Counter-)Cartographies of an Imperialist Construction', *Antipode*, 44(4) (2012), pp. 1099–1118. Online Resource: https://doi.org/10.1111/j.1467-8330.2011.00941.x (accessed 12 November 2020).

Anthony Downey, *Uncommon Grounds. New Media and Critical Practices in North Africa and the Middle East* (London: I.B. Tauris, 2014).

Eric Dyring, 'Wie die Erde entblößt wird', in *Erdsicht – Global Change*, ed. by Bert Bolin, Eric Dyring, Annagreta Dyring (Stuttgart: Hatje Cantz, 1992).

Elisa Giaccardi, 'Reframing Heritage in a Participatory Culture', in *Heritage and Social Media: Understanding Heritage in a Participatory Culture*, ed. by Elisa Giaccardi (New York: Routledge, 2012), pp. 5–10.

Malte Hagener, Vinzenz Hediger, and Alena Strohmaier, ed., *The State of Post-Cinema: Tracing the Moving Image in the Age of Digital Dissemination* (London: Palgrave Macmillan, 2016).

Anna-Katharina Hornidge and Katja Mielke, 'How to Conceptualize Area Studies? Thesis 1 Crossroads Studies: From Spatial Containers to Studying the Mobile', *Middle East – Topics & Arguments*, 4 (2015), pp. 13–19. Online Resource: https://meta-journal.net/article/view/3577 (accessed 12 November 2020).

Henry Jenkins, Sam Ford, and Joshua Green, *Spreadable Media: Creating Value and Meaning in a Networked Culture* (New York: New York University Press, 2013).

Bruno Latour, 'Drawing Things Together', in *Representation in Scientific Practice*, ed. by Steve Woolgar and Michael Lynch (Cambridge: MIT Press, 1990), pp. 16–68.

Henri Lefebvre, *The Production of Space* (London: Blackwell, 1991).

Ramon Lobato and James Meese, ed., *Geoblocking and Global Video Culture* (Amsterdam: Institute of Network Cultures, 2016).

Lev Manovich, 'Die Poetik des erweiterten Raums. Von Prada lernen', in *Black Box White – Cube*, ed. by Lev Manovich (Berlin: Merve, 2005), pp. 105–144.

Rachid Ouaissa, Friederike Pannewick, and Alena Strohmaier. 2021. 'Introduction: Re-Configurations: A New Conceptual Framework for Research on the MENA Region', in *Re-Configurations: Contextualizing Transformation Processes and Lasting Crises in the Middle East and North Africa*, ed. by Rachid Ouaissa, Friederike Pannewick, and Alena Strohmaier. Wiesbaden: Springer International Publishing. Online Resource: https://www.springer.com/gp/book/9783658311599 (accessed 12 November 2020).

Bernhard Siegert, *Relais. Geschicke der Literatur als Epoche der Post 1751–1913*. (Berlin: Brinkmann & Bose, 1993).

Adriana de Souza e Silva, 'From Cyber to Hybrid. Mobile Technologies as Interfaces of Hybrid Spaces', *Space and Culture*, 9.3 (2006), pp. 261–278. Online Resource: https://doi.org/10.1177/1206331206289022 (accessed 12 November 2020).

Hito Steyerl, *VAM Symposium Circulationism*, 24 May 2014, Online Resource: https://www.youtube.com/watch?time_continue=2&v=TA2uNR-6Lr4&feature=emb_logo (accessed 12 November 2020).

Alena Strohmaier, *Medienraum Diaspora: Verortungen zeitgenössischer iranischer Diasporafilme* (Wiesbaden: Springer, 2019). Online Resource: https://www.springer.com/de/book/9783658246051 (accessed 12 November 2020).

Paul Virilio, *Krieg und Kino. Logistik der Wahrnehmung* (Munich and Vienna: Fischer, 1986).

Hartmut Winkler, 'Landschaft und Architektur als Modell für den Datenraum', *HyperKult. Geschichte, Theorie und Kontext digitaler Medien*, ed. by Martin Warnke, Wolfgang Coy, and Georg Christoph Tholen (Frankfurt: Stroemfeld, 1997), pp. 227–239.

About the Authors

Angela Krewani is a full professor of media studies at Philipps-Universität Marburg, with a special focus on the impact of digital media on social practices and art. She is interested in the epistemic, aesthetic, and cultural configurations of digital media. She has widely published on these topics; her publication *Medienkunst. Theorie, Praxis, Ästhetik* (2016) examines the practices and the developments within media art. She recently co-edited a book on *Canadian Ecologies* (2020) (with Alessandra Boller and Martin Kuester) and another one on contemporary authorship *Constructions of Media Authorship* (2020) (with Christiane Heibach and Irene Schuetze). Currently she is preparing a publication on *Corona and Media* (with Peter Zimmermann).

Alena Strohmaier is project leader (postdoc) of the research project "'But I'm not filming, I'm just doing a bit of video'. Cinematic appropriation processes of protest videos from the Middle East and North Africa since 2009" at Philipps-Universität Marburg (funded by the German Federal Ministry for Education and Research, 2019–2023). She did her PhD on films of the Iranian diaspora and initiated the scientific Open Media Studies Blog (Zeitschrift für Medienwissenschaft, with Sarah-Mai Dang). Her recent publications include the edited volume *Re-Configurations. Contextualizing Transformation Processes and Lasting Crises in the Middle East and North Africa*, edited with Rachid Ouaissa and Friederike Pannewick (2021) and her first monograph *Medienraum Diaspora. Verortungen zeitgenössischer iransicher Diasporafilme* (2019). Currently she is editing a book on racism and film (*Rassismus und Film*, 2022) together with Ömer Alkin and Rauf Ceylan).

Part I

Cartographies

1. Mapping Empire: Knowledge Production and Government in the Nineteenth-Century Ottoman Empire

Nour Nicole Dados

Abstract
Many studies of the nineteenth century Middle East and North Africa (MENA) region have been concerned with the economic, social and political influence exerted by European colonial governments through the accumulation of knowledge about the region and its subsequent military domination. The case of the Ottoman Empire in the nineteenth century demonstrates that European techniques of knowledge production were also strategically adopted by ruling elites outside the colonial metropolises. Ottoman adoption of European technologies and techniques were politically entwined with the empire's territorial claims against nascent nationalisms and a calculated move towards knowledge-based forms of government administration in the quest to hold onto power. Cartographic and demographic methods used by the Ottomans produced new assemblages of territory and population that profoundly reshaped the objective of government and the conduct of imperial administration. Statistics and geography became the choice tools of social progress and advancement, underpinning the numerous reforms of the nineteenth century aimed at rationalisation and centralisation.

Keywords: Ottoman Empire, maps, historical knowledge, Beirut

Mapping Empire

Studies of the nineteenth-century Middle East and North Africa (MENA) region have focused on the economic, social, and political influence exerted by European colonial administrations through the accumulation

Strohmaier, A. and A. Krewani (eds.), *Media and Mapping Practices in the Middle East and North Africa: Producing Space.* Taylor & Francis Group, 2021
DOI 10.5117/9789462989092_CH01

of knowledge about the region and its subsequent military domination. However, the case of the nineteenth-century Ottoman Empire demonstrates that emergent techniques of knowledge production were not geographically bound in their origin and development, and were strategically adopted by ruling elites outside the colonial metropolises. The inter-imperial knowledge networks between the Ottoman and European empires, essential to these developments, remain underexplored. Writing this history requires thinking about 'the connected world of empires' (Çelik 2008), and the 'overlapping territories' and 'intertwined histories' (Said 1995), in ways that may not be 'unidimensionally hierarchical' or 'reducible to core-periphery relationships' (Go 2013; 2016). This would allow for the recuperation of a 'multidirectional communication pattern that dismantles the one-way vector of European influence on Ottoman modernizing reforms' (Çelik 2008, p. 5).

The epistemic shifts in the social sciences of the nineteenth century were the product of a convergence of technologies and techniques developed over more than two centuries of colonial expansion (Said 1995; Chakrabarty 2000; Connell 2007; Go 2014). The dominant tendency to see these developments as originating within European society overlooks the centrality of the dynamics of empire (Connell 2007) and the contribution of non-Europeans (Chakrabarty 2000; El-Shakry 2007). Raewyn Connell's study of the history of sociology demonstrates the significance of the imperial relationship to the development of the discipline (Connell 2007). Connell's database study of early scholarship reveals that the relationship between metropolis and colony was far more important to the development of the early social sciences than internal changes within European society (Connell 2007). The representation of modern knowledge formations within the social sciences as a strictly European phenomenon has been extensively critiqued for systematically omitting the contribution of non-Europeans in the universalization of modern disciplines (Chakrabarty 2000; El-Shakry 2007). In the case of inter-imperial knowledge, it also means that the 'unexpected circulation of influences' between the Ottoman and European empires has been overlooked (Çelik 2008, p. 5).

European methods for surveying territory and population in the nineteenth century were entangled in the instrumentalization of knowledge as a technique of government (Foucault 1975; 2007). Similarly, the Ottoman adoption of these technologies and techniques was no less politically entwined with the empire's territorial claims against nascent nationalisms and a calculated move towards knowledge-based forms of government administration in the quest to hold on to power. The cartographic and demographic methods adopted by the Ottomans in the nineteenth century produced new assemblages of territory and population that began to profoundly reshape the objectives of government and the conduct of imperial administration. When Beirut's status as a provincial capital was abolished in 1865 as a result of the new Provincial Law, for example, the bureaucrats in Istanbul argued that the decision was based on statistics and geography (Hanssen 2005).

Statistics and geography were central to the numerous reforms of the nineteenth century aimed at centralization, rationalization, and bureaucratization. New methods of survey and analysis in geography and demography led to greater precision in the representation of territory and made possible detailed inquiries and statistical analyses about the empire's population. The new maps were not just important representational tools, but also critical planning instruments for building the empire's infrastructure of roads and rail and port networks, and for linking the administrative provincial units. Similarly, methods of social survey introduced after 1831 systematically captured far greater detail about the population than earlier census data (down to eye colour, complexion, and physical marks). This was combined with the expansion of the state registration system across three institutions and the expansion of the census workforce of population officials (Karpat 1985, pp. 28–29). Ottoman bureaucrats and imperial subjects participated in the production of knowledge about the empire's territory and population by adapting new social science techniques to local conditions. These exercises were part of a broader (though not necessarily equal) transactional global circulation of knowledge increasingly shaped by emerging discursive formations across the social sciences and the creation of discrete disciplinary units.

Disciplinary Formations and Cartographic Knowledge Production

The authority of modern disciplines, and their claim to positive, scientific, and objective knowledge, is founded upon the erasure of the complex and heterogeneous history of their emergence (Foucault 2002; 1972). Cartography, like other modern disciplines, began shedding its pre-modern roots in the seventeenth century and, by the end of the nineteenth century, had established itself as a systematized knowledge formation (Foucault 1980; de Certeau 1988). Prior to the seventeenth century, geography was always 'something *more*' than 'merely positive knowledge' (Said 1995, p. 55). Premodern maps were a performative record of the journey and documentation of an itinerary. Cartographic representations were secondary to their accompanying texts. In a passage from *Walking in the City*, Michel de Certeau laments the moment of the discipline's disengagement from 'the itineraries that were the conditions of its possibility' (de Certeau 1988, p. 120):

> Each of the [medieval] maps is a memorandum prescribing actions. The tour to be made is predominant in them [...]. This drawing outlines not the 'route' (there wasn't one) but the 'log' of their journey on foot – an outline marked out by footprints with regular gaps between them and by pictures of the successive events that took place in the course of the journey (meals, battles, crossings of rivers or mountains, etc.): not a 'geographical map' but 'history book.' (de Certeau 2001, p. 120)

The treatise of Moroccan-born Sicilian geographer Muhammad al-Idrisi, *Nuzhat al-mushtaq fi ikhtiraq al-afaq* (*Diversions for him who longs to travel the world*), offers one example of mapping as a history book of itineraries. In the treatise, commissioned by the Sicilian King Roger II in 1154, Idrisi incorporates geographical knowledge from the Ancient Greeks with his own observations, as well as accounts of merchants and travellers (Edson and Savage-Smith 2004). Idrisi's grand map of the world – the *mappae mundi* referred to as the *Tabula Rogeriana* (Figure 1.1) – is considered one of the most important surviving pre-modern maps. Like other cartographic records from the Middle Ages, the map was not intended to be used without its accompanying itineraries, textual commentaries, and detailed regional maps, which included a smaller map of Sicily (Figure 1.2).

While borders on pre-modern maps were 'designated in terms of rivers, littorals and some vague notion of where one people ends and another begins' (Brummett 2007, p. 28), from the seventeenth century, dotted demarcation lines and shaded areas denoting ownership became common features as expanding colonial empires produced new forms of knowledge with territory as their strategic objective (Brummett 2007, pp. 32–33; Fortna 2002). Early modern maps were increasingly restricted to trade and military concerns as the itinerant features of pre-modern maps gave way to demands for scientific rigour and topographical detail (Brummett 2007, p. 50). The itineraries of Idrisi and other medieval cartographers vanished. The map, in its modern sense, became a privileged site for 'scientific' knowledge that was distinctly separate from topographical city or town views. In the same historical moment of geography's establishment as a science, drawings and engravings of city views proliferated. Extensive collections of landscapes and panoramas made in this period presage the emergence of monochrome photography (Brummett 2007, pp. 32–33).

City views, as a distinct and separate category of visual representation, were made popular by European travellers. The panoramic engravings and sketches of Constantinople by Antoine Ignace Melling are one example of this break in technique with earlier itinerary-type maps (Figure 1.3). European by birth, Melling worked as an imperial architect in the palace

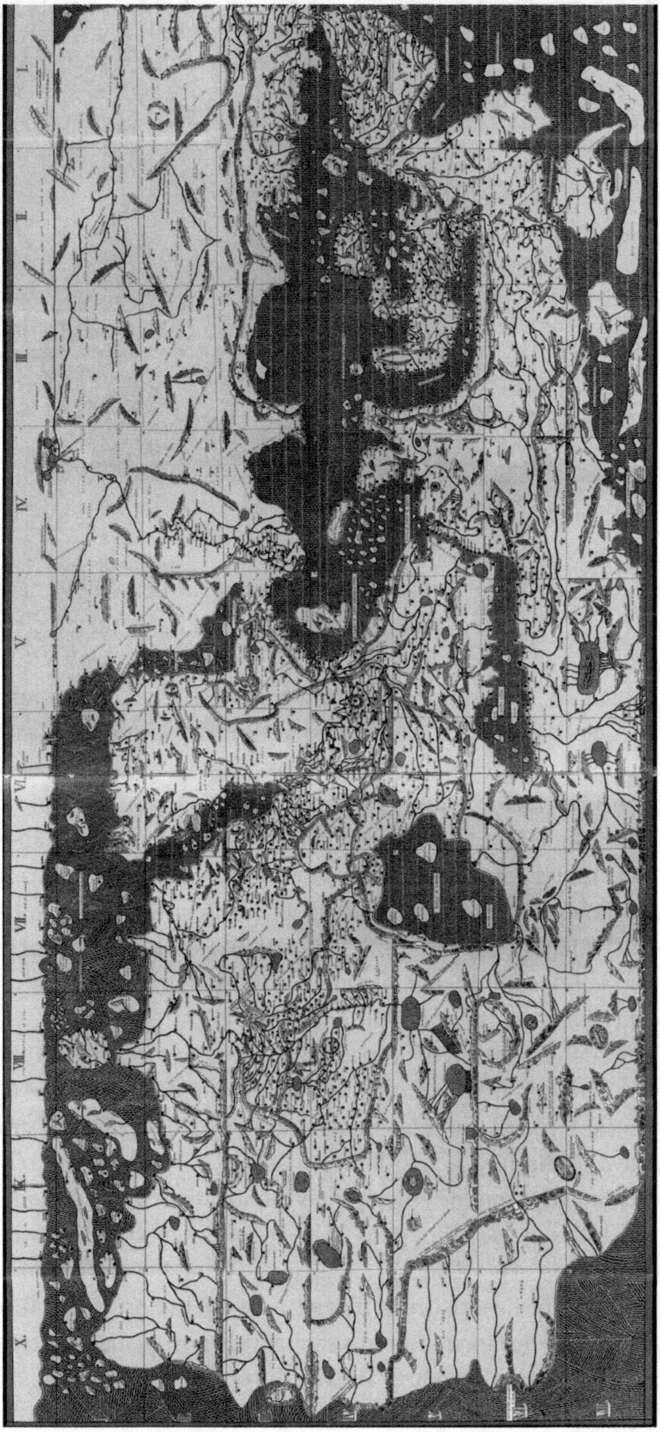

Figure 1.1: Idrisi, Tabula Rogeriana, Source: Wikipedia.

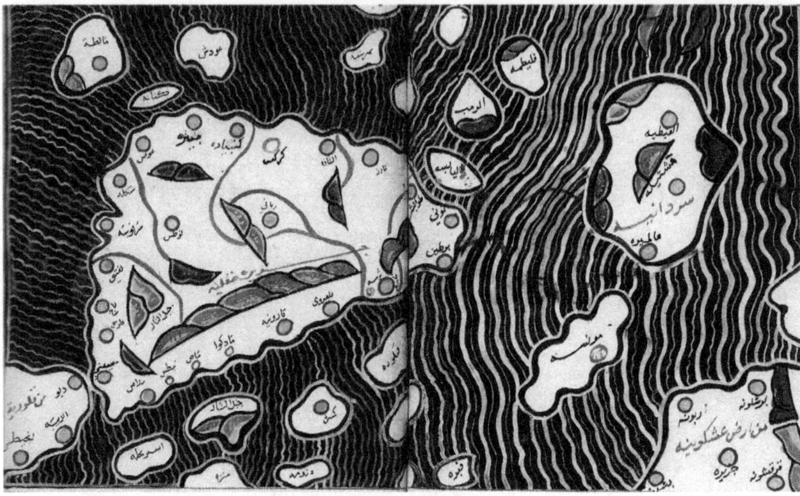

Figure 1.2: Idrisi, Map of Sicily, Source: Edson and Savage-Smith 2004, p. 89.

of Sultan Selim III for almost eighteen years. His drawings and engravings attest to the creative elements that came to characterize city and town views, now firmly separated from the science of geography.

However, the separation was never complete. The production of maps in the nineteenth century included many narrative-driven 'educational' tools that not only surveyed the space of empire, but also visually communicated the imperial project. Visual mappings from earlier periods also continued to be reproduced and consumed as instrumental sources of geographical evidence. City views of Beirut drawn by W.H. Bartlett and referred to in the travel memoirs of John Carne (1836) reappear in a geographical survey of Beirut's ancient gates in 1921. The author, Comte du Mesnil du Buisson, makes a sketch of an earlier city view of Beirut and uses it to study changes to the city's spatial organization over time (Figure 1.4).

While the separation of map and view was never complete, the division was part of a broader reorganization of 'the sciences of nature (the inquiry and the table)' and 'the human sciences (examination, discipline)' into separate and distinct fields (Foucault 2002). As Foucault's archaeological excavation of the modern disciplines demonstrates, the emergence of systematized knowledge formations – 'disciplines' – entailed the erasure of their pre-modern histories (Foucault 1980; 1972). What Foucault does not address however, is that the erasure of disciplinary prehistories also reconstituted the new knowledge formations as exclusively European (Chakrabarty 2005; Connell 2007; El-Shakry 2007). The North became the geographical reference point for the production of all knowledge (Connell 2007; Connell et al. 2017;

Figure 1.3: Melling's Bosphorous c. 1800, Source: Pamuk 2005, pp. 58–59.

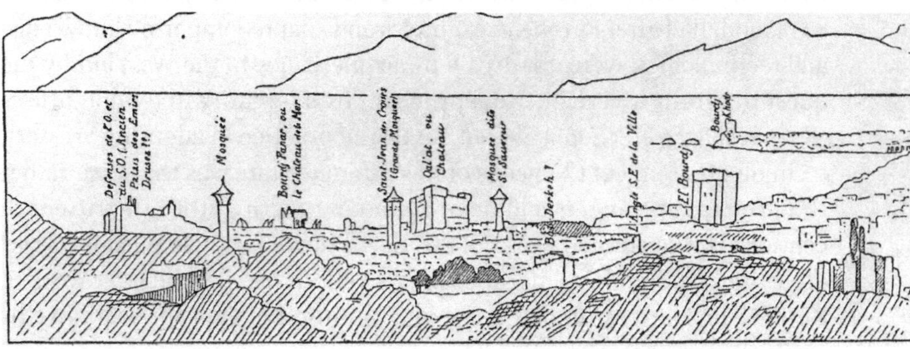

Figure 1.4: Panorama of Beirut in 1835, drawn after W.H. Bartlett, Source: Mesnil du Buisson 1921, p. 236.

Connell and Dados 2014). Idrisi's treatise and other works produced in the pre-modern era attesting to a global circuit of knowledge production are either co-opted into the Northern canon, dialectically positioned to illustrate the superiority of European knowledge, or assigned to the dustbin of history.

Techniques of Science and Government in the Late Ottoman Period

When the Danish Vice-Consul Julius Löytved presented a map of Beirut to the last Ottoman Sultan Abdulhamid II in 1876, the European military presence in the Ottoman territories was growing, nationalist sentiment was brewing all the way from the Balkans to Mount Lebanon, and Beirut had already been surveyed and mapped several times by the French and the British. Therefore, the Löytved map was more than a gift; it was highly sought-after empirical knowledge that communicated power and sovereignty. The techniques and technologies utilized in the Löytved map had been developing over decades. These rapid advances in knowledge production in the social sciences had been quickly patronized by the Ottoman state several decades earlier as the imperial administration worked actively to contain the growth of geography as a pedagogical tool and to monopolize the production of maps.

The late Ottoman period attests to a variety of these complex cultural exchanges that continued despite the shifts in disciplinary boundaries that were instrumentalized in the consolidation of European imperial power. The hegemonic diffusion of Western cultural and political power can be readily observed in the MENA region well before the Mandate period (Said 1995; Makdisi 2000), but this did not mean that the Ottomans were passively coerced into adopting Western techniques of government. The Ottoman reform programmes of the nineteenth century, many of which were modelled after European examples and adapted imported knowledge and technologies, were not driven by acquiescence to the West but by the quest to retain sovereignty over the empire's territory and populations (Karpat 1985; Çelik 2008; Evered 2012). The practice of adopting Western technologies to meet the needs of the Ottoman state was the norm rather than the exception – it had been common practice since the fifteenth century. The celebrated works of Piri Reis, known for their 'portolan style' after the chart-making methods of the Venetians, like the later work of Katib Celebi, are examples of the long tradition of appropriation in this arena (Rogers 1992; Soucek 1992).

With the establishment of the War College (*Harb Okulu*) in 1834, the military took responsibility for map production, and maps depicting the empire in its entirety were displayed in every school classroom (Fortna 2002, pp. 182–186). Military cartographers were trained in Western map-making techniques both locally and in England, France, and Prussia, and experts and map-making equipment were imported from Europe (Fortna 2002, p. 182). However, while cartographic and representational techniques had long been arenas of cultural exchange, in the nineteenth century, they began to reshape how territory was administered and even conceived. It was not only the techniques of spatial representation that were learned but also their application as techniques of government.

European rivalry for territory in the 1830s and 1840s would have added to the urgency to update the empire's administrative organization. Between 1839 and 1842, the British Army's Royal Engineers produced precise topographical maps of Beirut that could be used for military purposes (Davie 1987). These were not the semi-detached maps of previous decades nor the scholarly or worldly sketches of Western travellers, which would have been produced from memory, but military maps surveyed by 'military engineers, with state of the art instruments' (Davie 1987, p. 3).

Along with many other War Office documents of that period, these maps can now be viewed at the National Archives in Kew, but in 1841, they would have been classified military documents, and the convergence of technologies and techniques that produced them would not have been widely available. 'Containing all the information required for military operations in the area and for policing an occupied town' (Davie 1987, p. 3), the military surveys of the city were instrumental in the 1841 British bombardment that ended the reign of Egyptian Ibrahim Pasha in Beirut and facilitated its return to the Sultan (Salibi 2003, p. 157; Davie 1987; Chevallier 1971). With Beirut in their grip again thanks to the British, the Ottomans would have been acutely aware of the power vested in new European technologies and techniques of survey, and the danger of the empire's continued reliance on old-world representations.

While Ottoman cartography pre-dates the nineteenth century, geographical representation and territorial administration had always remained two separate things. Unlike the French and British practices of using the army to administer colonial provinces, Ottoman imperial administration was almost exclusively done by bureaucrats (Çelik 2008). Up until the Tanzimat Reforms (1839–1876), the organization of localities was equally bureaucratic: long tabled lists rather than cartographic representations documented the empire's territories (Karpat 1985; Hanssen 2005; Çelik 2008). In the 1830s, this began to change as European advances in technology, administration, and education were carefully studied. This was most keenly observed during the reform periods in the nineteenth century: the *Tanzimat* era (1839–1876) and the *Hamidian* era (1876–1901) (Çelik 2008, pp. 13–14):

> Prior to the *tanzimat*, a place's demarcations and administrative belonging were ranked in pyramid-shaped lists from the capital down to the smallest hamlet unit, the *nahiye*. While Ottoman cartography had a long legacy, *tanzimat* maps of the Ottoman Empire represented a break in technique. (Hanssen 2005b, Ft74, p. 41).

The impact of European techniques of government on the Ottomans was partly a product of the power of cartographic representation in the eighteenth and nineteenth centuries (Çelik 2008, p. 10; Hanssen 2005b, pp. 40–42; Fortna 2002, pp. 165–201). Following the post-1789 programme of the French state engineered by Armand Mattelart, the organization, classification, and documentation of imperial space were strategically realigned (Çelik 2008, p. 7). Mattelart's project 'of mastering space' centred on maps (Çelik 2008, p. 6), which were 'a projection of a rational system in which everything should communicate' (Çelik 2008, p. 7). The map as a scientific survey of space was reimagined as a tool of territorial administration. Zeynip Çelik writes:

> In 1790, with the goal of replacing the former provinces of the Old Regime, characterized by intersecting administrative districts and judicial and religious units, with a unified set of administrative units, the French National Assembly divided France into eighty-three departments, which in turn were divided into districts and communes. Eighty years later, the 1871 Ottoman Law on the General Administration of Provinces (Idare-yi Umumiye-yi Vilayet Nizamnamesi, based on an 1864 'vilayet law', or 'law of provinces') established administrative divisions and uniform principles of management and reform that would be pursued throughout the empire. Modeled after the *Code-Napoléon*, it reorganised the imperial territories into twenty-seven *vilayets* (provinces), each governed by a *vali* (governer-general) appointed by the capital. (Çelik 2008, pp. 7–8; emphasis and translations in original)

The Provincial Law (*Vilayet* Law) of 1864, which led to the 1871 reforms, was similarly a uniform, empire-wide administrative reorganization based on 'the sum of twenty years of provincial inspection tours, petitions, rule by councils and model provinces' (Hanssen 2005, pp. 39–40). Provincial capitals became the centre of administrative power, and their governors were directly appointed by the Sultan. The law had a direct impact on Beirut, which lost its provincial capital status. The announcement of Damascus as the new capital of the *vilayet* led to an outcry amongst Beirut's notables, and a petition was sent to the Sultan (Hanssen 2005b, pp. 40–41). To justify his reform proposal to the imperial government, the mastermind of the initial *Vilayet* Law of 1864, Rushdi Pasha, declared that his proposal 'was based on statistics' and 'on a map of Syria, and on recourse to those with an understanding of true information about geography' (Hanssen 2005b, pp. 40–41).

As the British and French military presence in the region grew, so, too, did the Western cultural influence through education and the influence of missionary schools (Fortna 2002, pp. 50–60; Hanssen 2005, pp. 149–151; Salibi 2003, pp. 161–162). By the 1880s, missionary and *millet* schools[1] catered to almost the entire Christian and Jewish population of the empire, leaving the

1 A *millet* school was an educational institution that was under the authority of its respective confessional community, or *millet*. These institutions came to prominence in the nineteenth century as minority communities of the Ottoman Empire had greater means and motivations to establish them.

Ottoman state school system to be perceived as entirely Muslim (Evered 2012; Fortna 2002, p. 76). The influence of foreign schools created problems for the Ottomans. In addition to the growing European presence, there was concern about the 'dangerous ideas' discussed behind the walls of these institutions, particularly about geography (Fortna 2002). In response to this, and as part of the process of state modernization, the Ottomans introduced the Education Act of 1869. The state was to create a centralized and compulsory education system modelled after the French example (Evered 2012). Many minority communities, with long-established and high-quality private schools, found themselves in direct conflict with the state.

Education and pedagogy remained central concerns through the following decades as the empire struggled to maintain legitimacy and control of its territories. Geography remained a key site of contestation: it communicated both the power of the state and the challenges to it. In response to the Russo-Turkish War (1877–1878), a large three-dimensional topographical map depicting the empire's war with Russia was produced to educate the public (Fortna 2002, pp. 166–167). The empire's defeat in that war, and the outcomes of the Congress of Berlin in 1878, started chipping away at its outer edges. Borders were increasingly politicized as geography became a defining principle of group identity among nationalist movements (Anderson 2003). Nationalist organizing within the empire's borders was met with increasing repression. The Armenian and Greek population suffered the worst of this repression, which culminated in the Hamidian massacres perpetrated by Ottoman troops and Kurdish irregulars between 1894 and 1897.

Statistics and the Birth of the Population

In the late nineteenth century, the production of knowledge about the empire's subjects rapidly expanded with the introduction of modern statistics that completely overhauled older census methods employed primarily for taxation purposes. The development of new methods in the social sciences redefined the boundaries between the known and the unknown as the object of government shifted from territory to population. Significantly weakened through a series of military defeats and the rising influence of European military power and scientific knowledge, the rulers of the Ottoman Empire were particularly interested in developments in government administration. European resources were widely translated and circulated. They informed the numerous Ottoman reforms of the nineteenth century aimed at rationalization, centralization, bureaucratization, and secularization.

The *Tanzimat* Reforms, enacted through a series of statutes following the major decrees of 1839 (Edict of Gülhane) and 1856 (Reform Edict on Religious Freedom) had attempted, with varying degrees of success, to move towards centralization and rationalization (Evered 2012; Karpat 1992). The most significant reforms of this period were those designed to bolster a single national identity by applying the law equally to all citizens regardless of religion or ethnicity. The so-called *millet* system, which had organized people into subordinate religious communities with degrees of autonomy, was abolished. Ottomanism was to become the unifying identity of the empire. An anthem, flag, and state banknotes were introduced. Centralization was bolstered with numerous public works, the establishment of bodies such as the Ministries for Health and Education, and the foundation of state-run training academies and codes of conduct for civil servants and some professions.

When the *Tanzimat* Reforms came to an end with the appointment of Abdulhamid II in 1876, they had not fully realized their objectives. The declaration of equality among Ottoman subjects was greeted with varying degrees of enthusiasm, but its application was limited. Neither the declaration of equality nor the shift in state education policy (following the Education Act of 1869) could fully mitigate the influence of foreign and *millet* schools or the rise of nationalism. After decades of learning separately, and learning to be governed separately, the *millet* system had become entrenched in the social psyche of the empire's subjects. The state's consignment of its minorities to 'benign neglect on the fringes of an imperial Ottoman imaginary' (Makdisi 2000, p. 37) was no longer possible.

As cartographic techniques adopted for the organization of territory merged with new statistical and demographic methods for the government of population, the administration of territory became inseparable from the demographics of population. The original Ottoman population counts, prior to the first modern government census conducted in 1831, were designed to gather information for tax purposes. Census data was gathered only for the male population; the government was not interested in collecting data on the female population or on the geographical distribution of religious or ethnic groups (Karpat 1985, p. 9).

Under the *Tanzimat* Reforms, centralization and rationalization placed the larger portion of responsibility for population health and the provision of education and other services with the imperial government. Management replaced organization as the dominant tendency, and social science methods were the tools of choice to create the 'bureaucratic blueprints for the restructuring of society' (Karpat 1992, p. 134).

By 1868, the Ottoman government had established the Office of Statistics. Population bureaus measured birth and death rates in different localities (Karpat 1992, p. 137) as the Provincial Law of 1864 and the later 1871 reforms charged local provincial administrations with the welfare of the population. By 1874, the Office of Statistics had expanded to become the General Directorate of Statistics, and the central government pondered the systemization of data collection and the creation of a new population registration system (Karpat 1992, p. 137). Increasingly, in the official view of the imperial government, statistics were the choice instrument of progress (Karpat 1992, p. 138). Conversations between Sultan Abdulhamid II and the United States ambassador Samuel Sullivan Cox saw bags full of surveys and reviews collected from the American embassy and translated.

The content of the data collected in the later nineteenth century was significantly different from what had been captured by earlier state censuses. As the reform period wore on, deeply variegated demographic statistics were sought to capture the numerical density of non-Muslim populations:

> The Porte began to embrace the European concept of functional government and service to society. Consequently, it felt the need to reassess its human and natural resources. Moreover, the rising national consciousness among the Christian groups had manifested itself in the form of a variety of demands eg. for the establishment of national churches and schools – as well as in claims for independence and territory, often supported by data ignoring other groups: hence there arose the necessity of learning the exact number of each major Christian ethnic group. These humanist-functionalist-nationalist considerations were manifest in the census and the registration system implemented in 1881/82–1893. This census used several ethnic-confessional categories for the Christian population, but all the Muslims continued to be counted as one homogenous group, despite the ethnic and linguistic differences among them. (Karpat 1985, p. 9)

The concern with organizational matters in the early part of the nineteenth century was fast giving way to inescapable administrative dilemmas, as the signing of the Treaty of Berlin in 1878 saw the establishment of several independent principalities within the empire. The independent principalities in the Balkans were established following the intervention of the Great Powers at the Congress of Berlin that effectively reversed the Russian gains of Ottoman territory from 1877. However, the claims of opposing groups to territory in the Balkans were not fully resolved. This was particularly the case in Macedonia, which had been returned to the Ottomans under the Treaty of Berlin (1878). Statistical campaigns by competing groups made strategic use of numbers to bolster their claims. The empire was a key proponent of these campaigns, constructing population statistics that showed 'a clear Muslim (i.e. Ottoman) majority in Macedonia' and that could be used to bolster 'governance through demographic discourse' (Evered 2012, pp. 72–73). In the case of Beirut, the loss of the city's provincial capital status to Damascus following the implementation of the Provincial Law of 1865 was attributed in no small part to the statistical data about the sectarian composition of the region (Hanssen 2005, p. 45).

Although intended as a tool for exerting control, power, and sovereignty, population surveys in the multi-ethnic and multi-religious Ottoman Empire also undermined the claims of the central government. Since statistics could be used 'to justify or negate territorial claims and ambitions' (Evered 2012, p. 81) by any group, and not just the central government, the science of demography was also bitterly contested. By the late nineteenth century, the use of population data had become an important component of imperial and nationalist claims. The Hamidian massacres of Armenians between 1894 and 1896, following decades of refusal by the imperial government to implement concessions granted to the Armenians under the Treaty of Berlin, also followed decades of competing petitions regarding population.

Conclusion

The rise of Ottoman translations of Western knowledge peaked at the same time as the expansion and patronage of the social sciences. While the new knowledge was intended to augment economic growth and bring social prosperity, neither followed. Historian Kemal Karpat, who has written extensively about Ottoman statistics, attributes this to the 'misappropriation of western scientific products' (Karpat 1992, p. 144). According to Karpat, the Ottomans did not fail in the practical application of statistics, but in developing a theoretical framework to justify their use – in short, failing to see that the organizational methods of the European social sciences were not simply tools for the management of society, but 'actually provided scientific bases for the political ideologies that challenged the established order and called (implicitly) for new social organization' (Karpat 1992, pp. 142–144).

According to Foucault, statistics began to develop in the seventeenth century as an apparatus of knowledge:

> Etymologically, statistics is knowledge of the state, of the forces and resources that characterise a state at a given moment. For example:

> knowledge of the population, the measure of its quantity, mortality, natality; reckoning of the different categories of individuals in a state and of their wealth; assessment of the potential wealth in circulation, of the balance of trade, and measure of the effects of taxes and duties, all this data, and more besides, now constitute the essential content of the sovereign's knowledge. (Foucault 2007, p. 274)

Nevertheless, this knowledge raises difficulties, including the technical issues of administrative size and the logistics of collection. More than this, it is the 'problem of the secret' and the 'problem of the public' that mean that this knowledge has to be managed in particular ways so that it is not fully in public view, but is public enough to appear to be the data on which the sovereign acts (Foucault 2007, pp. 275–277). As disciplinary institutions such as the police, prisons, hospitals, and asylums gradually replace the power of the sovereign, statistics as an apparatus of knowledge also changes in function, becoming at once more purposeful as a technique of power and more transparent as a scientific method (Foucault 2007; 1975). Rather than being directly located within the state, statistical methods are localized within institutions that make up the state. While Karpat's conclusion regarding Ottoman oversight about how Western scientific methods could be used to challenge the state is important, the different trajectory in the development of state institutions as conduits of social and scientific knowledge is also a factor.

The late Ottoman period attests to many complex cultural exchanges that continued despite the consolidation of European imperial power in the MENA region. European techniques of knowledge production in cartography and demography were strategically adopted by the Ottomans and used to bolster the empire's territorial claims against nascent nationalisms. The practice of adopting Western technologies to meet the needs of the Ottoman state was the norm rather than the exception. Statistics and geography became the choice tools of the government in its quest for power, economic prosperity, and social stability. In statistics and geography, the reforms of the nineteenth century found social science models that could be used for the rationalization and centralization of the state. Although intended as a tool for exerting control and safeguarding sovereignty, maps and population surveys in the multi-ethnic and multi-religious Ottoman Empire also undermined the claims of the central government and led to open contestation and confrontation.

References

Marcus Abraham, *The Middle East on the Eve of Modernity: Aleppo in the Eighteenth Century* (New York: Columbia University Press, 1989).

Benedict Anderson, *Imagined Communities. Reflections on the Origin and Spread of Nationalism* (London and New York: Verso, 2016).

Michel De Certeau, *The Practice of Everyday Life*, trans. S. Rendall (Berkeley and Los Angeles: University of California Press, 2003).

Zeynep Çelik, *Empire, Architecture, and the City: French-Ottoman Encounters, 1830–1914* (Seattle and London: University of Washington Press, 2008).

Dipesh Chakrabarty, *Provincializing Europe: Postcolonial Thought and Historical Difference* (Princeton and Oxford: Princeton University Press, 2000).

Raewyn Connell, *Southern Theory: The Global Dynamics of Knowledge in Social Science,* (Sydney: Allen & Unwin, 2007).

Raewyn Connell, Fran Collyer, João Maia, and Robert Morrell, 'Towards a Global Sociology of Knowledge: Post-Colonial Realities and Intellectual Practices', *International Sociology*, 32(1) (2016), pp. 22–37.

Raewyn Connell and Nour Dados, 'Where in the World Does Neoliberalism Come from? The Market Agenda in Southern Perspective', *Theory and Society*, 43(2) (2014), pp. 117–138.

Michael F. Davie, 'Trois cartes inédites de Beyrouth: Eléments cartographiques pour une histoire urbaine de la ville', *Annales de Géographie (Beirut)*, 5 (1984), pp. 37–82.

—, 'Maps and Historical Topography of Beirut', Scanned from: *Berytus*, 35 (1987), pp. 141–164, by Elie Wardini, Online Resource: http://almashriq.hiof.no/lebanon/900/930/930.1/beirut/reconstruction/davie/Davie-text.html (accessed 12 November 2020).

Evelyn Edson and Emilie Savage-Smith, *Medieval Views of the Cosmos: Picturing the Universe in the Christian and Islamic Middle Age* (Oxford: Bodleian Library and University of Oxford, 2004).

Omnia El-Shakry, *The Great Social Laboratory: Subjects of Knowledge in Colonial and Post-Colonial Egypt* (Stanford: Stanford University Press, 2007).

Emine O. Evered, *Empire and Education Under the Ottomans: Politics, Reform and Resistance from the Tanzimat to the Young Turks* (London and New York: I.B. Taurus, 2012).

Benjamin C. Fortna, *Imperial Classroom: Islam, the State, and Education in the Late Ottoman Empire* (Oxford and New York: Oxford University Press, 2002).

Michel Foucault, *The Archaeology of Knowledge* (New York: Tavistock Publications, 1972).

—, *Discipline and Punish* (London and New York: Penguin, 1975).

—, *The Order of Things: An Archaeology of the Human Sciences* (London and New York: Routledge Classics, 2002).

—, *Security, Territory, Population: Lectures at the College de France 1977–1978* (New York: Picador, 2007).

Julian Go, *Post-Colonial Sociologies: A Reader* (Bingley: Emerald, 2016).

—, 'For a postcolonial sociology', *Theoretical Sociology*, 42 (2013), pp. 25–55.

Jens Hanssen, *Fin de Siècle Beirut: The Making of an Ottoman Provincial Capital* (Oxford: Clarendon Press, 2005).

Ahmet T. Karamustafa, 'Introduction to Ottoman Cartography', in *Cartography in the Traditional Islamic and South Asian Societies*, ed. by John Brian Harley and David Woodward (Chicago: Chicago University Press, 1992).

Kemal H. Karpat, *Ottoman Population 1830–1914: Demographic and Social Characteristics* (Madison and London: University of Wisconsin, 1985).

—, *Studies on Ottoman Social and Political History: Selected Articles and Essays* (Leiden, Boston, and Cologne: Brill, 2002).

Henri Lefebvre, *The Production of Space*, trans. by Donald Nicholson-Smith (Oxford and Malden: Blackwell, 1991).

Ussama Makdisi, *The Culture of Sectarianism: Community, History and Violence in Nineteenth-Century Ottoman Lebanon* (Berkeley and Los Angeles: University of California Press, 2000).

Bruce Masters, *Christians and Jews in the Ottoman Arab World: The Roots of Sectarianism* (Cambridge: Cambridge University Press, 2004).

Robert du Mesnil du Buisson, 'Les anciennes défenses de Beyrouth', *Syria (Paris)*, 2 (1921), pp. 235–257 and 317–327.

John Michael Rogers, 'Itineraries and Town Views in Ottoman Histories', in *Cartography in the Traditional Islamic and South Asian Societies*, ed. by John Brian Harley and David Woodward (Chicago: Chicago University Press, 1992).

Edward Said, *Orientalism: Western Conceptions of The Orient* (London and New York: Penguin Books, 1995).

Kamal Salibi, *A House of Many Mansions: The History of Lebanon Reconsidered* (London and New York: I.B. Taurus, 1990).

Svat Soucek, 'Islamic Charting in the Mediterranean', in *Cartography in the Traditional Islamic and South Asian Societies*, ed. by John Brian Harley and David Woodward (Chicago: Chicago University Press, 1992).

About the Author

Nour Nicole Dados is a postdoctoral researcher working in sociology and social history. Her PhD thesis examined the impact of colonial modernity on the social and cultural history of Beirut between 1890 and 1975 (University of Technology Sydney 2011). She has written (with Raewyn Connell) on the theories and origins of neoliberalism, with a focus on the intellectual work and historical experience of the Global South. She is currently working on a project about the impact of labour insecurity on knowledge production and knowledge workers, and a series of studies about the development of intellectual cultures and disciplinary formations in the nineteenth century Ottoman Empire.

2. Who Maps Middle Eastern Geographies in the Digital Age? Inequalities in Web 2.0 Cartographies in Israel/Palestine

Christian Bittner and Georg Glasze

Abstract
Cartographic practices have long been the domain of professionalized and centralized actors from governmental or commercial spheres. Inspired by the emergence of mapping practices in the age of Web 2.0, however, new forms of mapping may offer a bottom-up alternative. Indeed, the mapping platforms Wikimapia and especially OpenStreetMap have proven that geographical information and maps can be generated via crowdsourcing by cartographic amateurs. Yet, several studies have excavated powerful exclusion mechanisms in such Web 2.0 cartographies, showing that user-generated geographical content is often produced by privileged demographic groups. This chapter will explore these patterns of exclusion and inclusion in OpenStreetMap and Wikimapia in the examples of the Gaza Strip and Jerusalem.

Keywords: Cartography, Israel, OpenStreetMap, Palestine, volunteered geographical information, Wikimapia

Introduction

The idea of a neutral and objective presentation of geographical features on a map has been the underlying and constitutive myth of modern cartography from its formation in Renaissance Europe and largely until today. However, the writings of the so-called critical cartography approach have challenged this myth since the late 1980s. They have shown that all mapping practices are embedded in specific sociotechnical contexts and thus are always

Strohmaier, A. and A. Krewani (eds.), *Media and Mapping Practices in the Middle East and North Africa: Producing Space*. Taylor & Francis Group, 2021
DOI 10.5117/9789462989092_CH02

related to questions of social inequalities, politics, and power. The practices of modern cartography, critical cartography writing argues, have been dominated by governmental and commercial actors and thus shaped by the interests of these privileged actors. Against the backdrop of this critique, the emergence of new Web 2.0 mapping practices, which enable the collaboration of many cartographic amateurs, has been welcomed as a game changer – an opening and democratization of cartography.

In Israel and Palestine, the nexus of elitist cartographies and open, collaborative mappings is especially controversial, because the manifold dimensions of the conflict between two opposing nationalist ideologies heavily affect cartographies in Israel/Palestine. This chapter analyses and discusses how the dimensions of this conflict are being dealt with in the open, collaborative, and international online mapping projects OpenStreetMap (OSM) and Wikimapia.[1]

Modern Cartography and Its Critics

While humans from all parts of the world have made maps since prehistoric times (Delano-Smith 1994), the roots of modern cartography lie in Renaissance Europe, when maps became constitutive elements of the emerging nation states (Branch 2014; Kagan and Schmidt 2007). Deeply entangled with political power, maps were not only used for territorial governance, tax collection, or demographic monitoring (Black 2008); they also contributed to the formation of national identity (Winichakul 1994). Yet the application of maps was not limited to the domestic sphere of those emerging European powers. To the contrary, maps played an important role for imperialism and colonialism, as extremely valuable tools for exploring, conquering, and exploiting distant lands and peoples (Stone 1988). As such, cartography has also contributed to the ways the West imagines other regions of the world (Roberts 1985, pp. 194–202, cited in Hall). One example is the region we know today as the 'Orient', 'Arab World', 'MENA', or 'Middle East' (Culcasi 2010).

Of course, this region (as far as we want to think of it as such) has not only been constructed from the outside, but is also a strong point of reference from the inside, for anti-imperialist counter-narratives such as the 'Arab Homeland', which are likewise accompanied by respective cartographic representations (Culcasi 2012). As we will see below, the case of Israel and

1 The empirical data this chapter is based on reflects the status of 2017.

Palestine can also be conceived of as such a contested cartographic 'battlefield' (Leuenberger and Schnell 2010, p. 833), though on a smaller scale.

Thus, it is fair to say that maps have long been entangled with political power in many dimensions. Ironically, even today, the most common stance towards cartography is that of an unpolitical technology for representing geographical features of the Earth's surface. Maps are widely accepted as depicting the world in an objective manner through geometric accuracy and scientifically approved symbologies. Accordingly, a positivist and scientistic perspective on maps has been – and largely still is – the predominant paradigm in cartographic literature since the 1950s, when cartography became a technical academic discipline (Robinson 1952; Robinson 1953), guided by the theoretical framework of the map communication model (Robinson and Petchenik 1976; Morrison 1977).

Since the late 1980s, however, several authors who emphasized questions of power, politics and ethics inherent in all maps and map-making practices have shaken this positivist understanding of cartography. The historical geographer and cartographer Brian Harley was highly influential among those critics. In a series of publications, Harley developed a social constructivist perspective on cartography that constitutes a radical departure from the cartographic mainstream (Harley 1988a; 1988b; 1989; 1990). Drawing on Michel Foucault and Jacques Derrida, Harley conceptualizes maps as rhetorical and power-laden texts (Harley 1989). Power in map production, Harley states, is exerted not only externally, through the interests of funding governmental or commercial institutions, but also internally, through design decisions led by unquestioned assumptions of the cartographer. Harley demonstrates how maps are imbued with societal norms, values, and hierarchies, by selecting and emphasizing certain (powerful) aspects of the social world while omitting and silencing (marginalized) others. Thus, maps do not merely represent physical structures of the geographical world, but rather reproduce and naturalize specific spatio-social orders and hence serve the interests of powerful elites. Consequently, Harley calls for critical map analyses to unmask the hidden agendas and interests behind them – an agenda he terms the 'deconstruction' of the map (Harley 1989).

Harley has often been characterized as the starting point of an academic movement that was later labelled critical cartography (Crampton and Krygier 2005; Glasze 2009; Wood 2002). However, critical cartography is more than just an academic exercise and can be understood as a combination of both theoretical critique and the promotion of alternative mapping practices (Crampton and Krygier 2005). Thus, beside interrogating political dimensions of maps, critical cartography also pursues at least three general avenues of practical intervention (Bittner and Michel 2014): (1) counter-mapping, which aims to map the world from marginalized perspectives (Rundstrom 2009); (2) map art, which breaks from and experiments with established cartographic conventions and rules (Wood 2010, pp. 189–230); and (3) participatory mapping, which aspires to open map-making practices for all kinds of excluded actors (Bittner and Michel 2018).

Lately, these critical mapping practices have benefitted from technological disruptions towards digitized and internet-based forms of cartography. Since the 1970s, the craft of map-making has undergone several dramatic changes. The emergence of the computer, digitization, and remote sensing satellites have enabled the development of GIS and map design software. Cartography has become a data-driven practice and classic paper-and-ink map production has been largely replaced. Since the early 2000s, the spread of Web 2.0 technologies as well as the open-source and open-data movements have brought about another alteration of cartographic practices. Professional GIS and mapping software, which was only available at a high cost a few years ago, can now be downloaded for free under open licences. Mostly non-cartographic web developers and software engineers have created new mapping technologies, such as web mapping services or data visualization libraries. All of this has enabled new ways to create, process, and visualize geographical information. Citizens themselves, with the help of crowd-sourced mapping platforms, are now generating further geographical data, which has long been withheld by governmental land surveying offices. While all three dimensions of practical critical cartography (counter-mapping, map art, and participative cartography) have certainly benefitted from these developments, it is probably the domain of participatory mapping that has undergone the most radical innovations, thanks to the advent of online and user-generated forms of cartography.

User-Generated Cartographies

In recent years, ever more sophisticated online production modes of maps and geographical information have reconfigured traditional actor constellations and power distributions in the realm of cartography. Increasingly, internet users, instead of governmental bodies or commercial map publishing houses, generate maps and their underlying geographical data via crowdsourcing. This development can also be framed by the more general shift towards user-generated digital content in the digital sphere. Traditional mass media and the early internet of the 1990s were characterized by rather static content that were provided centrally for more or less passive recipients. Such a clear distinction between producers and consumers of information became increasingly difficult in the so-called Web 2.0 (O'Reilly 2005), which is reflected in neologisms such as 'prosumer' (Ritzer and Jurgenson 2010, following Toffler) or 'produser' (Budhathoki, Bruce, and Nedovic-Budic 2008). Today, 'consumers' without much prior knowledge or prerequisites generate and modify all kinds of media on innumerable web platforms. As a result, the 'cult of the amateur' (Keen 2007) put pressure on established models of income generation and division of labour in many media industries. Examples encompass the blogosphere vis-à-vis professional journalism, video-sharing platforms vis-à-vis the television industry, or Wikipedia vis-à-vis printed encyclopedias. From this angle, the field of cartography is just another setting of these reconfigurations. A vast and growing number of mapping platforms have radically lowered the entry barriers to participate in the creation, modification, and usage of maps and geographical information. Among them are not only commercial enterprises such as Google Maps or HERE WeGo, which pursue crowdsourcing strategies to lower their expenses for updating and maintaining map data, but also initiatives that operate fully on a non-profit basis, or at least offer their products under open licences.

Probably the most successful example of such a non-profit initiative is OpenStreetMap. At first glance, OSM appears to be a map, or rather an interactive digital map service with functionalities similar to Google Maps or Bing Maps. All these online map services have in common that every geometric feature on the maps is actually a mere visualization of a corresponding object in a geographical database. In the case of OSM, this database contains, at the time of writing, more than 440 million lines ('ways', in OSM terminology) and over four billion point objects ('nodes', in OSM terminology). However, what makes OSM radically different from its commercial competitors is that this mass of data is freely accessible to anyone and that an online community of more than 700,000 mappers[2] created it on a voluntary basis. Accordingly, OSM describes itself as 'a map of the world, created by people like you and free to use under an open license' (OpenStreetMap 2017). Founded in London in 2004, OSM's goal was to provide freely available geographical data, which were hardly released

2 This figure is actually debatable: 700,000 users have made at least one contribution to the OSM database. While the number of assigned user accounts is much higher (over 4 million), the number of active long-term contributors is much lower, between approximately 15,000 and 20,000 (according to statistics from Neis 2016).

by British (or other national) map agencies at the time. In 2006, the newly established OSM Foundation became the project's legal representation, coordination centre, and infrastructure provider. Yet membership is neither restricted nor mandatory for an OSM user account, and contributors are rather self-organized through a tiered system of local and regional mapping communities and with the help of mailing lists, forum threads, and documentary wiki pages.

The mapping agenda of OSM follows the ethos of a ground truth, which has a twofold meaning. First, it favours mappers operating on the ground: local experts who contribute spatial data about places and regions they are familiar with. Second, ground truth indicates that things should be mapped according to their (physical or political) appearance on-site (Glasze and Perkins 2015). This effectively means that OSM reproduces the rationale of topographic maps. It strives for a high level of geometric precision and a strong correspondence between the map and the physical landscape. The data therefore fulfils high standards of consistency and positional accuracy, and an increasing number of geo-services and navigation tools are built upon OSM. However, the ground-truth ethos also precludes other possible mapping purposes, such as fictional mapping or political contestations. If a feature is disputed, such as an ambiguous place name, then OSM generally prefers the name that is either present on signs on the ground or that is most commonly used among the local population. Due to this ground-truth approach, OSM generally tends to reproduce the status quo, to naturalize current orders, and not to subvert them in the spirit of counter-mappings. Furthermore, far from being a form of map art, OSM largely follows common cartographic conventions for aspects such as map orientation, projection, or symbologies. However, even though OSM can hardly be seen as a form of counter-mapping or map art, the project has received much sympathy from critical cartographers (e.g. Crampton 2009; Dodge and Kitchin 2009). This is only comprehensible in the light of its radical participatory approach, and OSM stands for nothing less than revolutionized actor constellations of map-making practices. It is not only the most participatory mapping enterprise that has ever been undertaken; it is also enormously successful. At the time of writing, OSM's map data outstrips the products of commercial or governmental providers in levels of detail or richness for many parts of the world. Most of the more densely populated regions in Germany, for example, contain not only obvious map features such as roads, street names, or prominent landmarks but also footpaths, bicycle lanes, and all kinds of amenities like shops, playgrounds, benches, waste bins, or postboxes, often even with collection times attached.

Despite the undeniable strength of OSM's community-based approach for building a global map database, the above reference to dense regions in Germany also points to pronounced inequalities in OSM, in respect to both data coverage and participation patterns, broadly following the structural cleavages of a digitally divided world (Wessels 2013). Several studies have uncovered socio-demographic biases in the user base of the project (Budhathoki and Haythornthwaite 2012; Steinmann et al. 2013), resembling those of other user-generated data platforms by skewing 'toward a more wealthy, more educated, more Western, more white and more male demographic' (Crampton et al. 2013, p. 132). Since many OSM users focus on mapping their local surroundings, and since many of them live in privileged urban neighbourhoods of the Global North, there are also strong geographical biases in the data. Such regional inequalities of data density have been studied on all scales, whether between different city neighbourhoods (Bittner 2017a), urban and rural areas (Haklay and Ellul 2010), wealthier and less prosperous countries (Quattrone, Mashhadi, and Capra 2014), and different continents (Neis and Zipf 2012) or world regions (Neis, Zielstra, and Zipf 2013; Graham, Sabbata, and Zook 2015). Thus, in contrast to homogeneous and standardized map series that are usually produced by governmental surveys, OSM data is characterized by stark contrasts and inequalities. In some regions, the seemingly continuous map of OSM obscures the fragmented, patchwork-like mosaic of the heterogeneous stories, actors, and data sets it consists of. Although OSM is in many respects the largest and most comprehensive online mapping project, it is not the only one.

Another prominent platform for generating map data collaboratively is Wikimapia. Wikimapia was founded in Moscow in 2006 and has, at the time of writing, more than 2.5 million registered user accounts, although, like on OSM, only a minority of those belong to active contributors (Wikimapia statistics 2017a). Unlike OSM's ground-truth ethos, Wikimapia is committed to the agenda 'let's describe the world', which aims less at accurate geometric representations of the material world and more at rich semantic descriptions of places. Accordingly, the place is the central object in Wikimapia's data structure, represented by hand-drawn polygons of all sizes, from single houses to entire regions. Textual descriptions, category assignments, pictures, or comments can enrich these polygonal places. Thus, Wikimapia is generally more about semantics than about geometry. To date, almost 25 million such places have been added to Wikimapia, making the platform one of the richest sources of spatial descriptions on the web (Wikimapia statistics 2017b). Moreover, Wikimapia's epistemological approach is much more open and adaptable to alternative forms of mapping than the ground-truth mantra of OSM.

Referring to the three avenues of intervention aspired to by practical critical cartographies, we can find numerous examples of both counter-mappings and map art within Wikimapia (Bittner 2017c). Surprisingly though, unlike OSM, Wikimapia has received little academic attention so far, and not much is known about the biases, power distributions, or inequalities of the project.

In general, the participatory bottom-up approach of user-generated cartographies like OSM or Wikimapia bears enormous potential for local empowerment, articulations of marginalized voices, and overcoming power imbalances between those who map and those who are being mapped. This is particularly relevant in the postcolonial context of the Middle East, and especially in the contested region of Israel and Palestine. In the following, we will therefore explore to what extent social inequalities in the region of Israel and Palestine are reproduced or subverted by collaborative cartographies such as OSM and Wikimapia.

Web 2.0 Cartographies in Israel and Palestine

In Israel and Palestine, any cartographic product is deeply intertwined with the manifold dimensions of the conflict between Israelis and Palestinians. Although identities and factionalisms are highly complex and inconsistent among Israelis as well as among Palestinians, the fact remains that the territorial aspirations of the two opposing nationalist ideologies (Zionism and Palestinian nationalism) are rather incompatible (Wallach 2011). Long before Israel was founded, maps had been utilized to assist the Zionist aspirations for a Jewish national state in Palestine. Those maps nurtured a geographical imagination of the region as an uninhabited and deserted land, awaiting and welcoming the return of the Jewish People (Long 2008; Bar-Gal 2003). After the First Arab-Israeli War in 1948–1949 and the founding of the State of Israel, the young state invested large sums into a Hebrew map of the country, erasing and replacing thousands of Arab toponyms (Azaryahu and Golan 2001). After the Second Arab-Israeli War and the Israeli occupation of East Jerusalem, the West Bank, the Gaza Strip, and the Golan Heights in 1967, borders on Israeli maps started to be redrawn, incorporating occupied areas into the state territory (Bar-Gal 2004; Collins-Kreiner, Mansfeld, and Kliot 2006).

The geography of the region today, especially of the Palestinian territories, is highly fragmented and marked by numerous processes of border drawing, disengagement, and segregation. In this setting, political facets of map-making become more than obvious. Which borders are delineated? Which places are depicted; which are omitted? In what language are they labelled? Which font styles are used? Aspects of map design that seem to be rather aesthetic questions in many parts of the world turn out to be strong political statements in Israel/Palestine. Unsurprisingly, maps are repeatedly scandalized and at the centre of heated public disputes, such as over the recent labelling of Palestine on Google Maps (Dewey 2016) or about the usage of misleading historical maps by the US television network MSNBC (Wootliff 2015).

How, then, are these conflictive dimensions dealt with in open, international mapping projects such as OSM and Wikimapia? Do Palestinian and Israeli mappers combat for authority in 'edit wars' such as happened on Wikipedia (Rettig Gur 2010)? Alternatively, does the integrative approach of collaborative mapping platforms bear a chance for Israelis and Palestinians to cooperate in a joint mapping effort? In the following, these questions will be discussed with the help of two exemplary case studies of the Gaza Strip and Jerusalem.

OSM in the Gaza Strip

As described above, OSM prefers data to be contributed by local communities; for example, by people on-site who are familiar with the area and its specific local context. While this model has been successful in large parts of the Global North, including Israel, Palestine can be considered an example of the opposite: to date, no local OSM community has emerged, and contributions to the map have mostly been the result of three external interventions. The first occurred during the First Gaza War between the Israeli army and the Palestinian group Hamas in 2008–2009. During the fighting, international aid organizations demanded accurate geographical information on the Gaza Strip in order to coordinate their relief efforts, because such data simply was not available at the time. Thereupon, a campaign was initiated among international OSM members to map the Gaza Strip remotely from satellite imagery, providing timely geodata to the humanitarian sector. While the OSM map of the Gaza Strip had been almost empty before the war, most of the larger roads were added to the map after this initiative was finished in March 2009. Secondly, in September 2009 the map of the Gaza Strip underwent another sudden leap when the US NGO JumpStart International generated an open-licence geodata set of the Palestinian territories. They had hired local employees (mostly students from Palestinian universities) who recorded GPS data from field surveys in the West Bank and the Gaza Strip. This data was then cleared and processed by staff members and subsequently donated to OSM as one bulky data import. While this data donation literally brought the previously orphaned West Bank onto the OSM map overnight, it also entailed significant improvements to the Gaza Strip map. After this data import, however, there was not much activity on OSM in the Palestinian territories. On the West Bank, most activity occurred in the Jewish settlements, which were edited by Israeli mappers. In the Gaza Strip, there was hardly any activity on the map for years, until the third external mapping intervention took place in the summer of 2014 during the Second Gaza War. In the course of ongoing air strikes by the Israeli military, the aid organizations Global Shelter Cluster Committee and Terre des Hommes Lausanne kick-started an OSM campaign to map all the buildings of the Gaza Strip to generate a database for damage assessment and rebuilding efforts. A great number of international mappers (predominantly from Europe and North America) participated in this initiative, and in terms of data quantity, this third mapping intervention was by far the largest one. As a result, the Gaza Strip is one of the most densely mapped places on OSM in the whole Middle East. However, in contrast to the general policy of OSM, this is not the outcome of a long-term effort by an eager local OSM community. Instead of local empowerment and cartographic emancipation, the OSM map of the Gaza Strip tells a story of development aid and humanitarian relief activities. It is bitterly ironic that the OSM map representation of Gaza seems to become more nuanced and sophisticated with every round of violence and devastation on the ground. In some respects, we can also think of OSM in Palestine as a continuation of postcolonial cartographic traditions, not only because the OSM standard map is, aesthetically, a Western representation, following the principles of the topographic map but also because the map, as we have seen, is largely created from a distance and mostly by Western contributors, with the help of the 'view from nowhere' of satellite imagery (Dodge and Perkins 2009). OSM in the Gaza Strip then, has not subverted the postcolonial relationship between those who map and those who are being mapped; rather, it has reproduced it.

OSM Jerusalem

The city of Jerusalem is a focal point of the Israeli-Palestinian Conflict. After the First Arab-Israeli War in 1948–1949, the city was divided into a Western, Israeli part and an Eastern, Jordanian part. In the Second Arab-Israeli War of 1967, the Israeli army conquered and occupied East Jerusalem as well as the West Bank (then in Jordan). In 1980, Israel declared the reunited city as its capital. After 1967, highly disputed Jewish Israeli neighbourhoods were built in the Eastern part of Jerusalem. Today, the city is highly segregated between

Jewish and Arab neighbourhoods, and the authorities in many respsects – for example, building permits, public services, or infrastructure development (Wari 2011) – generally favour Jewish communities. Thus, the ethno-religious segregation lines of the city largely mirror socio-economic stratifications as well.

This fragmentation is also discernable in OSM. Browsing the OSM map view, Arab neighbourhoods – with many details missing, such as footpaths, house numbers, amenities, or road names – appear emptier than Jewish districts. Such map data inequalities between different districts can also be analysed more systematically: Figure 2.1 presents a map showing the statistical districts of Jerusalem, colour-coded with the predominate population group in blue (Jewish Israeli), and the minority group in green (Arab Palestinian) districts. The Arab Palestinian inhabitants of Jerusalem account for about one-third of the city's population (Jerusalem Institute for Policy Research 2015). The cartogram distortion by population numbers (Figure 2.2) gives an impression of the spatial distribution of both groups. In OSM data, however, Arab neighbourhoods are strongly underrepresented in terms of data density (Figure 2.3).[3] Why is that the case? Is it coincidence? A consequence of discrimination among mappers? Of unequal access barriers? Of disinterest among Palestinians? The mere map or the database behind it does not provide answers when it comes to these contextual questions. In 2013, one of the authors conducted a series of field interviews with sixteen of the 40 most active OSM mappers in the region of Israel and Palestine. These interviewees were all Jewish Israeli citizens, and there was no indication of significant mapping activity by Palestinians on OSM. As stated above, OSM prefers contributors to map the areas they are most familiar with. Since Palestinians are absent from OSM, their neighbourhoods often lack representatives who feel entitled or competent to contribute to the map. The actual reasons for the absence of Palestinians from OSM, however, remain somewhat unclear. There was no evidence for acts of exclusion or repulsion of Palestinian mappers from within the OSM community. To the contrary, Israeli OSM users repeatedly welcomed the idea of joint cross-border mapping efforts, and they assisted the bulk import of data for the Palestinian territories to OSM that was donated by the NGO JumpStart International in 2009. One delegate of this NGO, who was also a highly active OSM mapper, even made some effort to kick-start a local Palestinian community, but without much success.

3 The presented cartogram of OSM data in Jerusalem is based on data from 2015. Since then, many buildings have been added in the Arab parts of the city, which would relativize the differences between Arab and Jewish neighbourhoods to some extent.

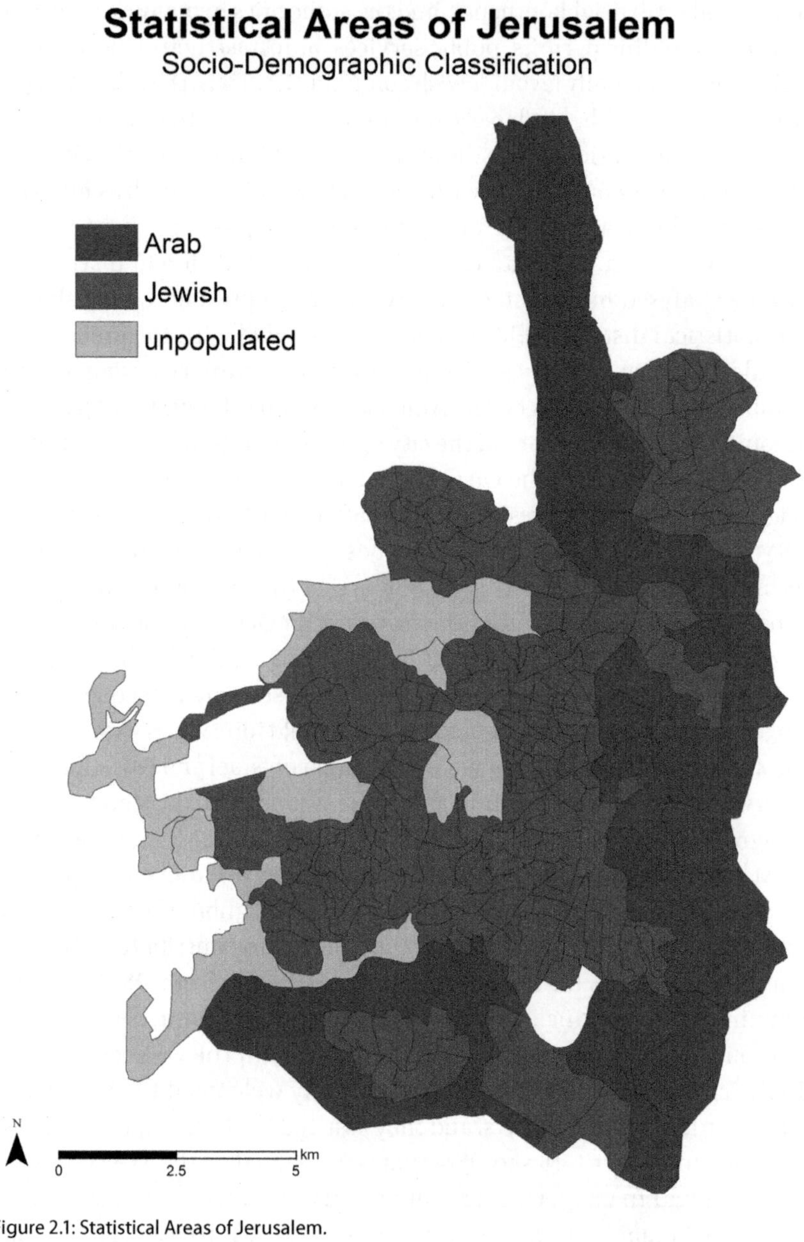

Figure 2.1: Statistical Areas of Jerusalem.

From a socio-economic perspective that is sensitive to the manifold manifestations of digital divides, it is tempting to interpret the finding of contributing Israelis and absent Palestinians as a result of unequal access barriers to the digital sphere. While we acknowledge the explanatory power

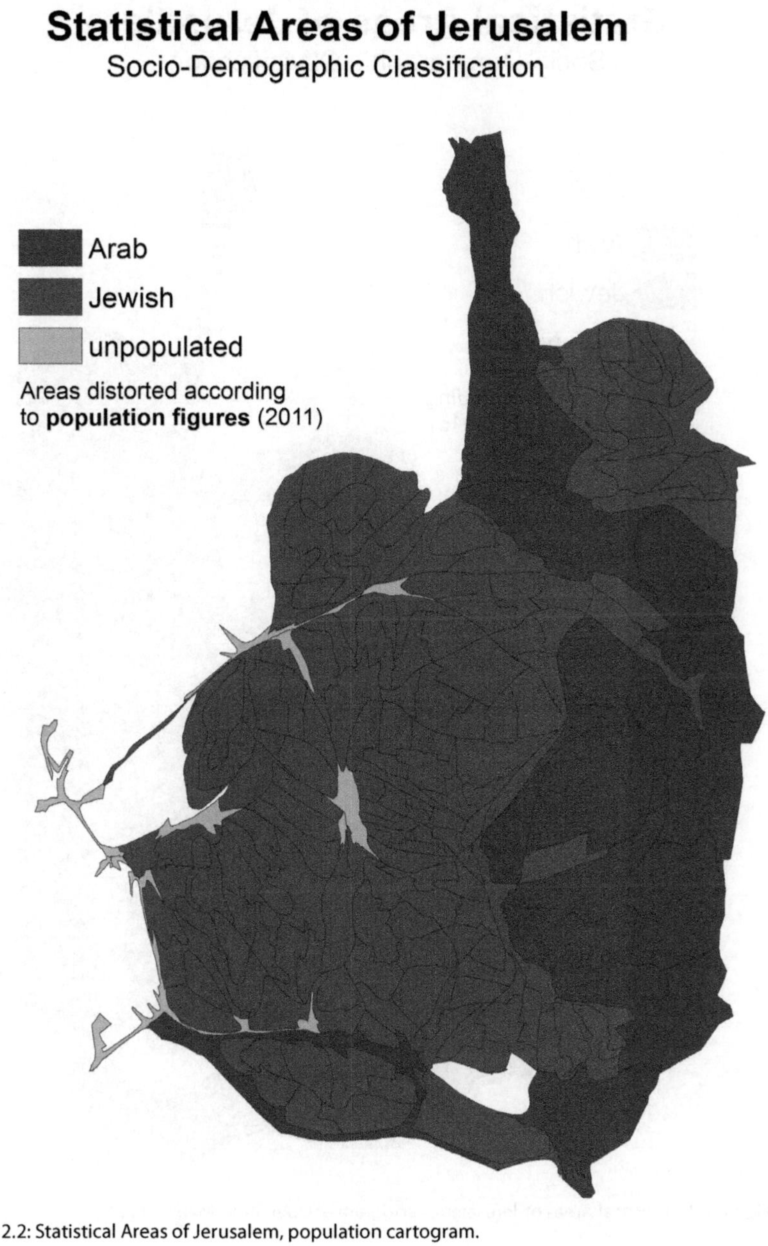

Figure 2.2: Statistical Areas of Jerusalem, population cartogram.

of this approach to some extent, socio-economic status is probably not a predetermining factor here. In addition, even despite the principally inviting atmosphere of OSM for Palestinians, there might be plausible reasons for them not to join the project. As one of the authors has argued

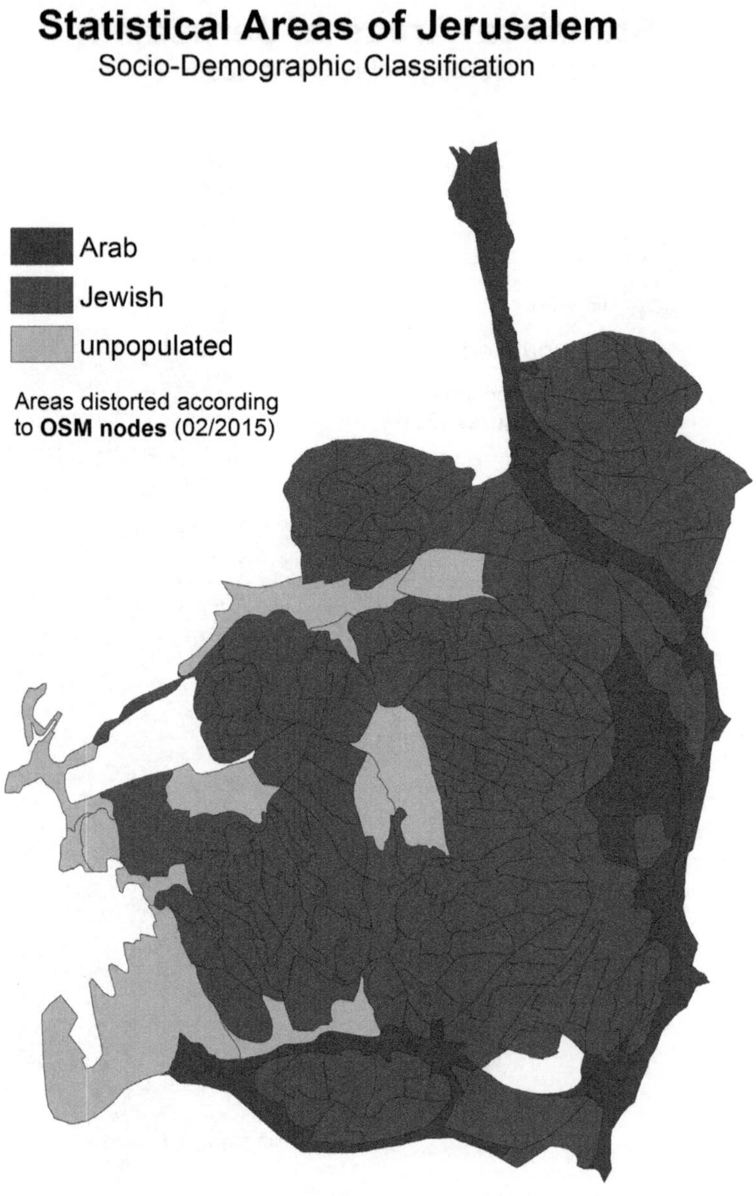

Figure 2.3: Statistical Areas of Jerusalem, cartogram of OpenStreetMap content.

elsewhere (Bittner 2017b), the conservative approach of OSM to stick to 'facts on the ground' produces exactly the kind of map that is hard to accept (and unlikely to be promoted) from a Palestinian point of view. The resulting map uncritically represents highly debated features such as refugee camps, the Israeli separation barrier, and Jewish settlements in the West Bank,

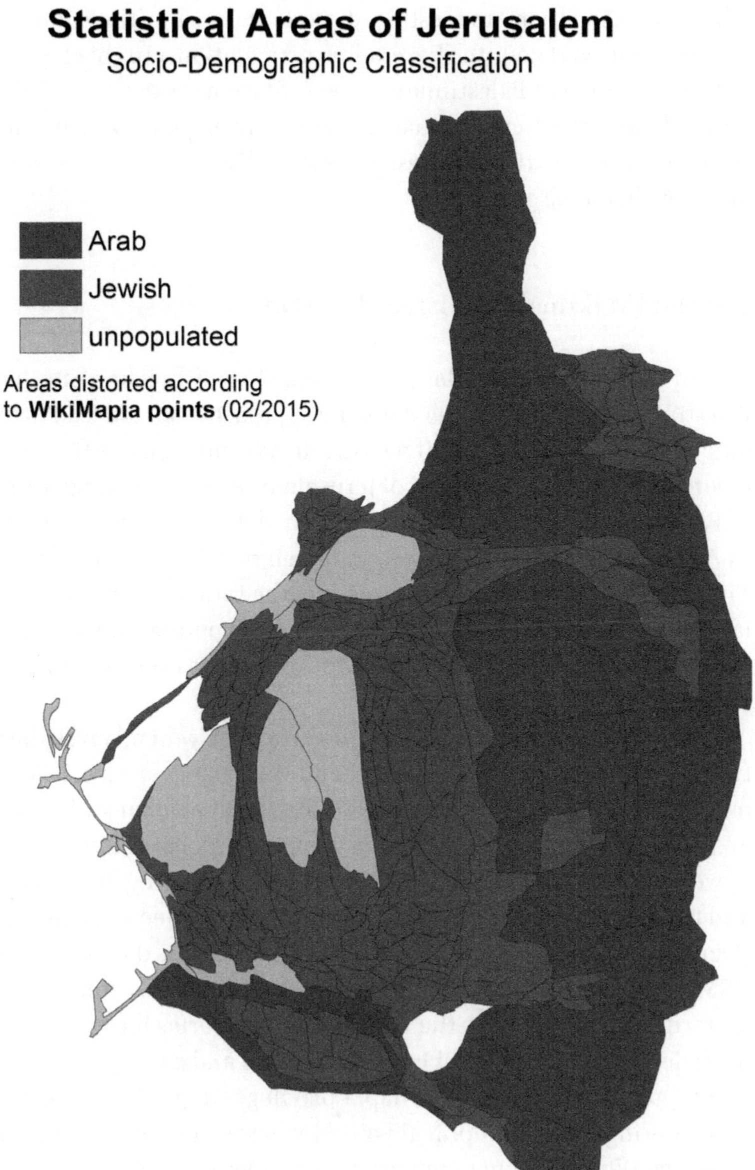

Figure 2.4: Statistical Areas of Jerusalem, cartogram of Wikimapia content.

while omitting Arab localities that were destroyed and abandoned in the course of the war of 1948–1949. Thus, we should keep in mind that the epistemological stance of a certain mapping project itself could predetermine who is likely to contribute. In the case of OSM, the epistemology of sticking to the facts on the ground reproduces spatial orders and leaves the power relations behind them unquestioned, while at the same time marginalizing contestations and counter-hegemonic articulations. In this light, it is not such a surprise that Palestinians are absent from OSM. From a Palestinian point of view, the factual situation itself may be perceived as wrong and so every map that sticks to these facts is itself a wrong map, or at least not one that should be supported.

OSM and Wikimapia in Israel/Palestine

If OSM's agenda works in fact as an exclusionary mechanism that keeps Palestinians from OSM, other mapping approaches should lead to different data distributions. Indeed, if we turn to Wikimapia, a different picture emerges. The Wikimapia map of Jerusalem seems to be largely an Arabic or Palestinian project. In contrast to OSM, data density is generally higher in Arab Palestinian than in Jewish Israeli neighbourhoods (Figure 2.4). This impression is also underpinned by the ratios of Hebrew and Arabic languages in the data: while OSM in Jerusalem contains much more textual information in Hebrew script than in Arabic, the opposite holds true for Wikimapia (Bittner 2017a).

Examining such spatial inequalities on a more general level, the finding that OSM is more of a Israeli project and Wikimapia more of a Palestinian one is reaffirmed by the data distributions on a regional scale (Figures 2.5 and 2.5a). On OSM, Israel is generally mapped in higher detail than Palestine (if we leave aside the sparsely populated desert covering the south of Israel and the special case of the Gaza Strip, which has been discussed above) (Figure 2a). Although Wikimapia contains much less data in general than OSM, the spatial distribution of these data obviously follows different patterns. On Wikimapia, the Palestinian territories have been edited in more detail than the area of Israel (Figure 2.6 and 2.6a).

As illustrated above, Wikimapia privileges a qualitative approach to representing the geographical world by describing places. If we browse along the Wikimapia map in Palestinian cities such as Nablus or Hebron, we notice a vast number of polygons, often with rather sloppy geometries, representing single houses whose owners are named in the descriptions. While this kind of personal information is very common in the Palestine map on Wikimapia, it can hardly be found in Israel. Furthermore, there are traces of edit wars on some disputed features on Wikimapia. Har Homa, for example, a Jewish Israeli neighbourhood in East Jerusalem that might be considered an illegal settlement from a Palestinian point of view, has

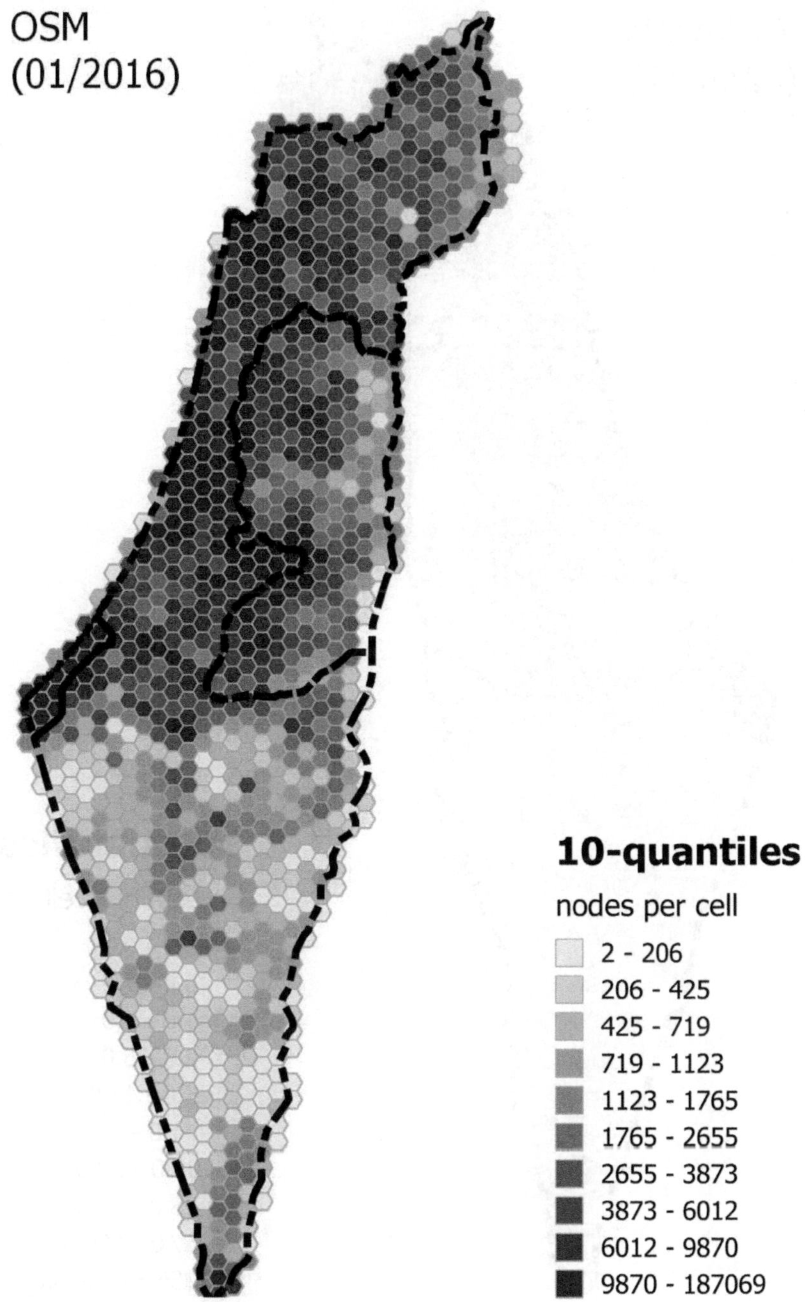

Figure 2.5: Spatial distribution of OSM content in Israel/Palestine.
Figure 2.5a: Legend.

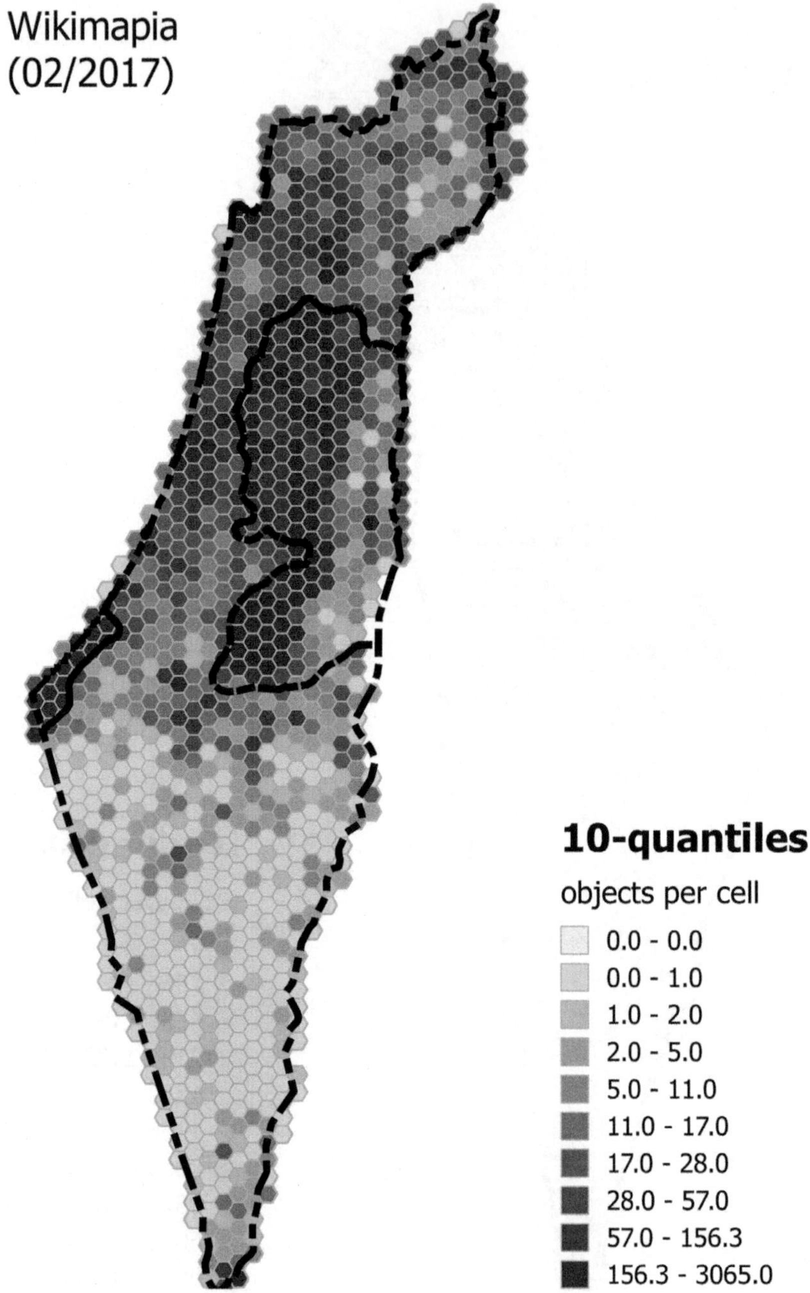

Figure 2.6: Spatial distribution of Wikimapia content in Israel/Palestine.
Figure 2.6a: Legend.

often been altered back and forth, bouncing between pro-Palestinian and pro-Israeli descriptions. In contrast to Wikimapia, OSM's data structure is not built for rich semantic descriptions of places, and the community generally rejects personal information such as names of residents on the map. This illustrates how Wikimapia and OSM enable very different forms of mapping practices, and each arguably meets the demands of different social groups.

Conclusion

Modern cartography has been a tool to exert and manifest political power, which is particularly virulent in a postcolonial regional context such as the Middle East. Recently, however, classic power structures in cartography have been shaken by the advent of collaborative Web 2.0 mapping platforms such as OSM and Wikimapia. Instead of governmental surveys or big commercial publishing houses, online communities of enthusiastic volunteers, people without formal cartographic expertise devoting their leisure time to contribute geographical data, produce these maps. In the era of Web 2.0, similar reconfigurations of actor constellations for other media and their societal implications have often been discussed, especially during the Arab Spring, when questionable terms such as 'Twitter Revolution' or 'Facebook Revolution' were frequent in public debates. Notwithstanding the actual impact of social media for those uprisings and the remaining question as to the Arab Spring's outcome, shifting access barriers to the production and dissemination of information seems to bear the potential to subvert established power (im)balances. In the case of cartography, the opening of cartographic tools and practices resonates with interventions of critical cartographers, who have questioned the elitism and conservatism of traditional cartography for a long time. The promise of the so-called democratization of cartography comprises a fostering of local empowerment and emancipation, giving a (cartographic) voice to marginalized and exploited groups and actors, and subverting the power imbalances between those who map and those who are mapped. In this chapter, we have reviewed some of these promises in the examples of the mapping platforms OSM and Wikimapia within the regional context of Israel and Palestine. Since the question of territory is a crucial point of contention between Zionism and Palestinian nationalism, maps in this context are highly politicized by nature and have often been employed as tools to rhetorically delegitimize, silence, or annex the other side. In this context, and in face of the current power balance between Israel and Palestine, we might expect the Web 2.0 mapping platforms to be a key vehicle for Palestinian spatial narratives.

Yet, as the case studies of OSM in the Gaza Strip and Jerusalem suggest, the mere possibility to take part in cartographic practices does not automatically cause Palestinians to do so. To the contrary, OSM in the region is largely driven by an Israeli mapping community, and the map of the Palestinian territories is mostly the result of external initiatives by the international development and relief sector. Thus, OSM in Palestine does not stand for local empowerment or the emancipation of marginalized groups, even though the Gaza Strip has become one of the most detailed sections of the Middle Eastern OSM map. Rather, we could think of it as a continuation of postcolonial mapping traditions, as spatial knowledge generated by Western actors and for a Western audience.

As we learned from OSM in the highly segregated city of Jerusalem, the absence of Palestinians from OSM can lead to strong distortions of map content along socio-spatial cleavage lines. Palestinian neighbourhoods of Jerusalem contain significantly less data than Jewish ones, which is why the map presents much more of an Israeli city of Jerusalem than a Palestinian one.

At first sight, these results correspond with general discussions of the digital divide, which marks the gap between those who can participate in digital media and those who remain behind manifold access barriers along socio-economic fault lines. Israelis, notwithstanding fierce domestic social inequalities, certainly enjoy in sum more socio-economic privileges than Palestinians do and it is surely no coincidence that the Israel OSM community consists predominantly of young, educated males of Jewish descent (Bittner 2017b).

However, as the counterexample of Wikimapia illustrates, socio-economic structures are by no means a determining factor for participation patterns in online mapping platforms. On Wikimapia, Palestinian areas are generally mapped in higher detail than Israeli ones, both at the local scale of Jerusalem as well as on the regional level of Israel and the Palestinian territories. This distribution not only subverts general socio-economic structures, it even reverses them.

Addressing the reasons for the differences between Wikimapia and OSM, we suggest considering the epistemological stances of both map platforms. While OSM pursues an agenda of creating a geometrically accurate representation of the world according to facts on the ground, Wikimapia rather seeks to collect semantic descriptions of places. We argue that OSM's approach inherently tends to reproduce spatial orders and leaves the power imbalances that underly these orders unquestioned, which results in a map that is hard to accept and probably unworthy of support from a Palestinian point of view. Wikimapia's agenda, however, is generally more open for confrontational, playful, or even artistic forms of mapping.

In conclusion, we may affirm that collaborative online mapping is certainly challenging established ideas, power (im)balances, and production chains in the domain of conventional cartography. Without question, this development bears great potential to overcome postcolonial cartographic traditions and to empower marginalized voices. Yet the digital sphere is neither a power vacuum bare of hierarchies and gatekeepers, nor is it detached from socio-economic inequalities of the world. Depending on the local context, these intervening variables can distort patterns of exclusion and inclusion significantly. Moreover, we have to acknowledge that mapping is not a uniform practice and accordingly that a map can take very different forms of spatial media. We have seen that OSM's focus on geometric accuracy is fundamentally different from Wikimapia's aim for semantic richness. Hence, we should also keep in mind that each form of cartographic practice would receive approval or consent from different individuals and social groups.

References

Maoz Azaryahu and Arnon Golan, '(Re)naming the Landscape: The Formation of the Hebrew Map of Israel 1949–1960', *Journal of Historical Geography*, 27(2) (2001), pp. 178–195.

Yoram Bar-Gal, 'From Berlin to Jerusalem. Professor David Amiran and the Atlas of Israel', *Erdkunde*, 58(1) (2004), pp. 31–41.

—, 'The Blue Box and JNF Propaganda Maps, 1930–1947', *Israel Studies*, 8(1) (2003), pp. 1–19.

Christian Bittner and Boris Michel, 'Kritische Kartographien der Stadt', in *Handbuch kritische Stadtgeographie*, ed. by Bernd Belina, Matthias Naumann, and Anke Strüver (Münster: Westfälisches Dampfboot, 2014), pp. 64–70.

—, 'Partizipative Methoden der Kartographie', in *Die Erforschung der sozialräumlichen Wirklichkeit. Qualitative Methoden in der Geographie und der sozialraumsensiblen Kulturforschung*, ed. by Jeannine Wintzer (Berlin and Heidelberg: Springer, 2018).

Christian Bittner, 'Diversity in Volunteered Geographic Information: Comparing OpenStreetMap and Wikimapia in Jerusalem', *GeoJournal*, 82(5) (2017a), pp. 887–906.

—, 'OpenStreetMap in Israel and Palestine – "Game Changer" or Reproducer of Contested Cartographies?', *Political Geography*, 57 (2017b), pp. 34–48.

—, 'Social Structures of Wikimapia... Towards "Wikimapia Studies"?', presentation at American Association of Geographers (AAG) – Annual Meeting (Boston: 5 April 2017c). Online Resource: https://www.youtube.com/watch?v=m4u3UwZFnps (accessed 18 November 2020).

Jeremy Black, 'Government, State, and Cartography: Mapping, Power, and Politics in Europe, 1650–1800', *Cartographica*, 43(2) (2008), pp. 95–105.

Jordan Branch, *The Cartographic State: Maps, Territory and the Origins of Sovereignty* (Cambridge: Cambridge University Press, 2014).

Nama Budhathoki and C. Haythornthwaite, 'Motivation for Open Collaboration: Crowd and Community Models and the Case of OpenStreetMap', *American Behavioral Scientist*, 20(10) (2012), pp. 1–28.

Nama Budhathoki, Bertram Bruce, and Zorica Nedovic-Budic, 'Reconceptualizing the Role of the User of Spatial Data Infrastructure', *GeoJournal*, 72(3–4) (2008), pp. 149–160.

Noga Collins-Kreiner, Yoel Mansfeld, and Nurit Kliot, 'The Reflection of a Political Conflict in Mapping: The Case of Israel's Borders and Frontiers', *Middle Eastern Studies*, 42(3) (2006), pp. 381–408.

Jeremy W. Crampton and John Krygier, 'An Introduction to Critical Cartography', *ACME: An International E-Journal for Critical Geographies*, 4(1) (2005), pp. 11–33.

Jeremy W. Crampton, 'Cartography: Maps 2.0', *Progress in Human Geography*, 33(1) (2009), pp. 91–100.

Jeremy W.Crampton et al., 'Beyond the Geotag: Situating "Big Data" and Leveraging the Potential of the Geoweb', *Cartography and Geographic Information Science*, 40(2) (2013), pp. 130–139.

Karen Culcasi, 'Constructing and Naturalizing the Middle East', *The Geographical Review*, 100(4) (2010), pp. 583–597.

—, 'Mapping the Middle East from Within: (Counter-)Cartographies of an Imperialist Construction', *Antipode*, 44(4) (2012), pp. 1099–1118.

Catherine Delano-Smith, 'Prehistoric Cartography in Asia', in *Cartography in the Traditional East and Southeast Asian Societies*, ed. by John Brian Harley and David Woodward (Chicago: University of Chicago Press, 1994), pp. 1–22.

Caitlin Dewey, *Google Maps Did Not "Delete" Palestine – But It Does Impact How You See It*, 2016. Online Resource: https://www.washingtonpost.com/news/the-intersect/wp/2016/08/09/google-maps-did-not-delete-palestine-but-it-does-impact-how-you-see-it/ (accessed 18 November 2020).

Martin Dodge and Chris Perkins, 'The "View from Nowhere"? Spatial Politics and Cultural Significance of High-Resolution Satellite Imagery', *Geoforum*, 40(4) (2009), pp. 497–501.

Martin Dodge and Robert M. Kitchin, 'Crowdsourced Cartography: Mapping Experience and Knowledge', *Environment & Planning A*, 45(1) (2013), pp. 19–36.

Georg Glasze and Chris Perkins, 'Social and Political Dimensions of the OpenStreetMap Project: Towards a Critical Geographical Research Agenda', in *OpenStreetMap in GIScience: Experiences, Research, and Applications*, ed. by Jamal J. Arsanjani et al. (Cham, Heidelberg, New Nork, Dordrecht, and London: Springer, 2015), pp. 143–166.

Georg Glasze, 'Kritische Kartographie', *Geographische Zeitschrift*, 97(4) (2009), pp. 181–191.

Mark Graham, Stefano de Sabbata, and Matthew A. Zook, 'Towards a Study of Information Geographies: (Im)mutable Augmentations and a Mapping of the Geographies of Information', *Geography and Environment*, 2(1) (2015), pp. 88–105.

Muki Haklay and Claire Ellul, 'Completeness in Volunteered Geographical Information: the Evolution of OpenStreetMap Coverage in England (2008–2009)', *Journal of Spatial Information Science*, Discussion Forum (2010).

Stuart Hall, 'The West and the Rest: Discourse and Power', in *Formations of Modernity*, ed. by Stuart Hall, reprinted (Cambridge: Polity Press, 2011), pp. 275–331.

John B. Harley, 'Cartography, Ethics and Social Theory', *Cartographica*, 27(2) (1990), pp. 1–23.

—, 'Deconstructing the Map', *Cartographica*, 26(2) (1989), pp. 1–20.

—, 'Maps, Knowledge, and Power', in *The Iconography of Landscape*, ed. by Dennis Cosgrove (Cambridge: Cambridge University Press, 1988a), pp. 277–312.

—, 'Silences and Secrecy: The Hidden Agenda of Cartography in Early Modern Europe', *Imago Mundi*, 40 (1988b), pp. 57–76.

Jerusalem Institute for Policy Research, *Statistical Yearbook of Jerusalem: Chapter III – Population*, 2015. Online Resource: http://en.jerusaleminstitute.org.il/?cmd=statistic.567 (accessed 18 November 2020).

Richard L. Kagan and Benjamin Schmidt, 'Maps and the Early Modern State: Official Cartography', in *Cartography in the European Renaissance: History of Cartography Volume Three (Part 1)*, ed. by David Woodward (Chicago: University of Chicago Press, 2007), pp. 661–679.

Andrew Keen, *The Cult of the Amateur: How Today's Internet is Killing our Culture* (New York: Doubleday/Currency, 2007).

Christine Leuenberger and Izhak Schnell, 'The Politics of Maps: Constructing National Territories in Israel', *Social Studies of Science*, 40(6) (2010), pp. 803–842.

Joanna C. Long, 'Rooting Diaspora, Reviving Nation: Zionist Landscapes of Palestine–Israel', *Transactions of the Institute of Britisch Geographers* (2008), pp. 61–77.

Joel L. Morrison, 'The Science of Cartography and its Essential Processes', *Cartographica*, 14(1) (1977), pp. 58–71.

Pascal Neis, *A Comparative Study Between Different OpenStreetMap Contributor Groups – Outline 2016: Neis One!*, 2016. Online Resource: http://neis-one.org/2016/11/osm-contributor-groups/ (accessed 18 November 2020).

Pascal Neis and Alexander Zipf, 'Analyzing the Contributor Activity of a Volunteered Geographic Information Project – The Case of OpenStreetMap', *IJGI*, 1(2) (2012), pp. 146–165.

Pascal Neis, Dennis W.G. Zielstra, and Alexander Zipf, 'Comparison of Volunteered Geographic Information Data Contributions and Community Development for Selected World Regions', *Future Internet*, 5(2) (2013), pp. 282–300.

OpenStreetMap. Online Resource: https://www.openstreetmap.org/ (accessed 18 November 2020).

OpenStreetMap. Online Resource: http://osmstats.neis-one.org/ (accessed 18 November 2020).

Tim O'Reilly, *What Is Web 2.0: Design Patterns and Business Models for the Next Generation of Software*, 2005. Online: http://www.oreillynet.com/pub/a/oreilly/tim/news/2005/09/30/what-is-web-20.htm

Giovanni Quattrone, Afra Mashhadi, and Licia Capra, 'Mind the Map', *Proc. of the ACM International Conference on Computer Supported Cooperative Work and Social Computing* (2014), pp. 934–944.

Haviv Rettig Gur, *Israeli-Palestinian Conflict Rages on Wikipedia – Israel – Jerusalem Post*, 2010. Online Resource: http://www.jpost.com/Israel/Israeli-Palestinian-conflict-rages-on-Wikipedia (accessed 18 November 2020).

George Ritzer and Nathan Jurgenson, 'Production, Consumption, Prosumption: The Nature of Capitalism in the Age of the Digital "Prosumer"', *Journal of Consumer Culture*, 10(1) (2010), pp. 13–36.

J.M. Roberts, *The Triumph of the West* (London: British Broadcasting Corporation, 1985).

Arthur H. Robinson and Barbara B. Petchenik, *The Nature of Maps: Essays Toward Understanding Maps and Mapping* (Chicago: University of Chicago Press, 1976).

Arthur H. Robinson, *Elements of Cartography* (New York: John Wiley & Sons, 1953).

—, *The Look of Maps: An Examination of Cartographic Design* (Madison: University of Wisconsin Press, 1952).

Robert A. Rundstrom, 'Counter-Mapping', in *International Encyclopedia of Human Geography*, ed. Robert M. Kitchin and Nigel Thrift (Oxford: Elsevier, 2009), pp. 314–318.

Renate Steinmann et al., 'Gender Dimensions in UGC and VGI: A Desk-Based Study', *GI_Forum 2013 – Creating the GISociety*, pp. 355–364.

Jeffrey Stone 'Imperialism, Colonialism and Cartography', *Trans Inst Br Geogr*, 13 (1988), pp. 57–64.

Winichakul Thongchai, *Siam Mapped: A History of the Geo-Body of a Nation* (Honolulu: University of Hawaii Press, 1994).

Alvin Toffler, *The Third Wave* (New York: Bantam Books, 1980).
Yair Wallach, 'Trapped in Mirror-Images: The Rhetoric of Maps in Israel/Palestine', *Political Geography*, 30(7) (2011), pp. 358–369.
Shahd Wari, 'Jerusalem: One Planning System, Two Urban Realities', *City*, 15(3–4) (2011), pp. 456–472.
Bridgette Wessels, 'The Reproduction and Reconfiguration of Inequality: Differentiation and Class, Status and Power in the Dynamics of Digital Divides', in *The Digital Divide: The Internet and Social Inequality in International Perspective*, ed. by Massimo Ragnedda and Glenn W. Muschert (Abingdon and New York: Routledge, 2013), pp. 17–28.
Wikimapia statistics, *Placed Added*, 26 January 2017a. Online Resource: http://wikimapia.org/stats/action_stats/?fstat=101&period=3&year=2009&mont_h=6 (accessed 18 November 2020).
Wikimapia history, online Resource: http://wikimapia.org/object/history/list/?object_type=&id=1945421 (accessed 18 November 2020).
—, *Users Count*, 14 November 2017b. Online Resource: http://wikimapia.org/stats/action_stats/?fstat=6&period=3&year=2009&month=6 (accessed 18 November 2020).
Denis Wood, *Rethinking the Power of Maps* (New York: Guilford, 2010).
—, 'The Map as a Kind of Talk: Brian Harley and the Confabulation of the Inner and Outer Voice', *Visual Communication*, 1(2) (2002), pp. 139–161.
Raoul Wootliff, *MSNBC Apologizes for 'Completely Wrong' Maps of Israel*, 2015. Online Resource: https://www.timesofisrael.com/msnbc-apologizes-for-completely-wrong-maps-of-israel/ (accessed 18 November 2020).

About the Authors

Christian Bittner is a research fellow at the University and State Library Bonn. He finished his PhD at the Institute of Geography, FA-Universität Erlangen-Nürnberg. His research dealt with the social implications of digital geographies and newer cartographic practices. He investigated exclusions in volunteered geographic information in Israel and Palestine. He is interested in the research fields of technology and society, critical geographic information systems, cartography and the geospatial web, and, more generally, in relations between technology and society as well as political geographies of the internet.

Georg Glasze is a full professor of cultural geography at FA-Universität Erlangen-Nürnberg. He is interested in the 'making of geographies'; in other words, how certain spaces and places are (re)produced, distinguished, valuated, institutionalized, and constituted through socio-technical processes. His research aims to uncover to what extent the production of spaces is part of the (re)production of specific social structures and the production of spaces therefore has to be scrutinized as elements of power relations. His working group in Erlangen focuses on cultural, political and digital geographies.

3. Taking the Battle to Cyberspace: Delineating Borders and Mapping Identities in Western Sahara

Frederik von Reumont

Abstract
Examining the case of Western Sahara, this chapter takes a closer look at the connection between identity, territory, and borders, as well as their construction in mapping processes. Analysing different web maps, I will show their rootedness in the dominant political world views and contrast them with the perspectives of the parties of conflict (Sahrawi Arab Democratic Republic and Morocco). The results show significant differences between proprietary and crowdsourced web maps as well as the cartographic representations authored by the factions. The delineation of borders is directly linked to territorial identity and nation-building. This demonstrates the important role of web maps and other digitally disseminated maps in the definition of identity and (self)-governance.

Keywords: Cartography, Western Sahara, Google, Wikimapia

Introduction

From a historical perspective, I consider maps as tools for institutions to sustain and demonstrate power over a territory (Harley 1989). Culcasi states that 'elite classes (whether they were ministers, monarchs or the state) have traditionally controlled the production and distribution of maps' (Culcasi 2012, p. 1104). The arrival of new technologies, however, made the map-making process and, at the same time, the distribution of maps, available to a much wider audience. This made possible a certain culture of counter-mapping, taking advantage of these new technologies

Strohmaier, A. and A. Krewani (eds.), *Media and Mapping Practices in the Middle East and North Africa: Producing Space*. Taylor & Francis Group, 2021
DOI 10.5117/9789462989092_CH03

and making the mapping processes available to the subaltern or oppressed groups and individuals. Following Culcasi, any form of mapping that goes against the norm of an established mapped perspective can be considered counter-mapping in a 'broader or more subtle' sense (Culcasi 2012, p. 1105). One of the key technologies in counter-mapping has been the use of the internet. However, the internet has also been one of the key technologies to spread the view of the powerful. Wood called Google Maps 'a faithful servant of the state, if ever there was one' (Wood 2010, p. 38). The battle about name places such as the Persian versus Arabian Gulf has led to popular internet hoaxes (Wikipedia 2019). Although there is a comical component involved, the goal is obviously to maintain sovereignty of interpretation. The internet offers a wide variety of maps and thus constructs of the world. This is especially visible in disputed areas of conflicts in the context of the nation state.

In the following, I argue that this battle is not only about the maps themselves, but that the maps are used to shape reality by using the strong connection between the map and what it represents: the territory. In the case of Western Sahara, I will show how maps mirror current differing world views and how they are used as a means to construe and thus construct reality.

The Role of Maps in Forming National Identities

The vast majority of map users seem to assume that maps are truthful representations of the Earth's surface. This results in the perception of a strong link between the map and the reality it represents. Anderson shows how mappings change the imagination of the territory. Their influence can be far-reaching to the point that they shape the language used to talk about territory. Maps can shape the linguistic expression of these new imaginations by coining a different vocabulary. As a powerful incarnation of the 'map as a logo' (Anderson 1983, p. 175), maps can also shape the imagination of the nation state. The map as a logo is ubiquitously disseminated and can 'form a powerful emblem of [...] nationalism' (Anderson 1983, p. 175). Anderson directly links the case of a brutal territorial struggle (in this case, Indonesia) to the depiction of the corresponding logo map defining the imagination of the territory of the nation state (Anderson 1983, footnote). The often quoted 'power of maps' (e.g. Wood 1992) then resides in the ability to impose non-material concepts – such as borders delineating an independent nation state – to a tangible thing belonging to a material reality, namely terrains or territories. The power of maps results from their strong connection to the real world. In the map, we can equip a piece of land with an idea. Conversely, we can apply an idea or idealist world view directly to a representation of the real. The power of maps is to convince their users that this is there, and thus maps 'affirm the existence of the things on them' (Wood 2010, p. 34). Despite the efforts of critical and post-representational cartography to deconstruct the map and question its relation to reality, this does not seem to have great impact outside academia. Maps are still largely seen as representations of reality. In this sense, map readers take borders that are inscribed in the map's visuals as realities, even when they are not visible in the field (in reality) or are weakly enforced, by a lack of fences or guards, for example. Then the map is all the more important, since it is the only proof or documentation of a status of reality that should only ideally be attained but does not exist (yet). The map is, at the same time, documentation and expression of an unreal ideal. In this sense, it is a document of utopia.

The role of borders in a map is to define a group of persons by the means of terrain. Maps group individuals by the territory they occupy, more or less arbitrarily, and those groups' occupation of the territory is naturalized and de-historicized. Ultimately, the nation state embodies its belonging to the territory. The role that maps oftentimes assume in this situation is to stabilize this unstable idea, 'to give the elusive idea of the state concrete form' (Wood 2010, p. 31). Maps give states the appearance of 'facts of nature, as real enduring things, things like mountains' (Wood 2010, p. 33). The coherence of the power exerted by states over a certain territory would be 'very hard to imagine without the intercession of the map' (Wood 2010, p. 34). Using these mechanisms, the map 'turned the fiction [...] into a fact' (Wood 2010, p. 34). This capacity of maps, along with the suppression of alternative perspectives, 'substantively underwrites the reproduction of the social relations of power' (Wood 2010, p. 35). Referring to Winichakul and Anderson, Wood argues that, early on, state leaders saw the merit of the map as something that turns the *idea* of the state into a *geo-body*, a mappable and thus real thing that can – e.g. as a logo – be used to pin identity to territory. This was the case in early American portraits; maps of America in the background of these portraits identify the sitter as American. By 'assuming [...] the prior existence' (Wood 2010, p. 32) of the relationship between people and territory (e.g. the geo-body), the state obscures its origins in history and thus naturalizes itself as something ahistorical and almost mythical. Once established, however, the connection of the people with the land they live on is not only folkloric in the sense of a mythical connection between person and land. It has real effects on the livelihoods of individuals and peoples as soon as a governing law is executed by a ruling power. In the era of the nation state, law is tied to territory. Thus, just a step over the border, whether imagined or not, can, in extreme cases, be a matter of life and death. This seems to be felt throughout communities that consider themselves deprived of the right to self-determination or self-governance.

Self-governance in the UN sense means the formation of or the voluntary integration into an independent nation state with a sovereign government. The United Nations has considered Western Sahara a non-self-governing territory since 1963. No administering power has been named for this territory since Spain ended its presence there (United Nations 2019). The United Nations maintains a peacekeeping mission (United Nations Mission for the Referendum in Western Sahara, MINURSO) with the goal of monitoring the ceasefire between the conflicting parties and carrying out a referendum concerning self-governance. One main issue on which the parties disagree is the identification of eligible voters. Persons from three groups have been identified, including persons living outside Western Sahara as refugees (MINURSO 2019). For the factions themselves the situation is much clearer.

Morocco apparently considers Western Sahara territory as the 'southern provinces of the Kingdom of Morocco' (Corell 2015, p. 1). Although Morocco participates in UN peace talks, it has repeatedly displayed difficulties agreeing to certain proposals. Although there is no official statement on the website of the Moroccan government, a featured map speaks an uncompromising language as to the ownership of the territory. The Moroccan government established a Royal Advisory Council for Saharan Affairs (CORCAS), and its websites leave no doubt that Morocco claims the territory of Western Sahara as part of the kingdom.

The proclamation of the government of the Saharawi Arab Democratic Republic (SADR), by contrast, claims 'the land of Saguiat el-Hamra and Rio de Oro' for the people of Western Sahara, already relating to the topography of the territory, namely the major rivers. It states that the Polisario Front, responsible for 'raising the flag of liberation', is carrying out the will of the Saharawi people to 'self-determination' (Lahlou 1976), while Morocco is accused of 'illegally and forcefully' occupying Western Sahara (Bashir 2017). Morocco is further accused of trying to outnumber the Saharawi population on its own territory by the means of 'over 120,000 settlers who are the eyes and ears of the oppressor' (Bachir 2017). The means of conveying the faction's respective messages are often maps.

Analysing Internet Mappings

In the following, I will analyse mappings presented on the websites of the three main stakeholders in the conflict of Western Sahara: the United Nations, Morocco, and the Polisario Front/the government of SADR. I will also consider mappings of the most popular internet platforms globally – Google Maps, Yandex, as well as the OpenStreetMap and Wikipedia – as representatives of open, crowdsourced platforms.

United Nations

The United Nations has a single map version it refers to when discussing Western Sahara (Figure 3.1). This map is placed on the website of the peacekeeping mission (MINURSO). The map accentuates the territory of Western Sahara in a light yellow colour, while the rest of the land masses are coloured in a darker brownish tone. An outline marking its borders with neighbouring states in dots and dashes further defines the territory. Morocco is depicted in the same brownish colour as the rest of the world outside Western Sahara, and the boundary between the two is the same as that between independent states. The territory is labelled 'Western Sahara' with capital and lower-case letters, while the names of the neighbouring states are written out in all capital letters. Besides roads (paved and unpaved), cities, rivers, *wadis*, and the locations of MINURSO activity, the map also displays the berm. The berm is a heavily mined strip in Western Sahara stretching over the entire length of the territory from south to north, into Morocco. It separates the Moroccan-controlled part of the territory from the scarcely inhabited part that is controlled by the Polisario Front. The map also indicates five locations where the berm is permeable. According to this map, another five roads cross the berm where it is impenetrable. The symbol representing the berm is a thick, solid orange line that is only interrupted where the passing points are located.

The rather muted colours of the map and its reduced artistic design suggest a matter-of-fact attitude towards the disputed political situation. The colour scheme, use of symbols, and lettering are all typical of the corporate design of UN maps. However, the demarcation of the territory

Figure 3.1: The MINURSO map representing the UN perspective on the conflict.

of Western Sahara makes clear that it is not considered part of Morocco. The lettering of the territory's label using lower case letters indicates the lack of clarity in the current situation. In addition, the distinct demarcation stressing the existence of the berm may be interpreted as a controversial political statement. A statement added to the map relativizes these political assertions: 'The boundaries and names shown and the designations used on this map do not imply official endorsement or acceptance by the United

Nations'. (Figure 3.1). The statement is italicized but uses the smallest font size on the map.

Morocco

The official website of the Moroccan government provides a kind of blueprint map for matters concerning Western Sahara (Figure 3.2). This is a map displaying the Moroccan territory, including Western Sahara in a brownish-beige colour with a solid white border, making it stand out against neighbouring states. The map shows no inland border. It features no additional information as to the topography or administrational units. The only label we can find within the territory designates the capital Rabat with a label and a point symbol using the colours and the star of the national flag.

This map suggests Morocco, including Western Sahara, is a homogenous monolithic whole without any possible fissures. All the maps on the different websites of the Royal Advisory Council for Saharan Affairs (CORCAS) use this blueprint and fill it with different information on Western Sahara such

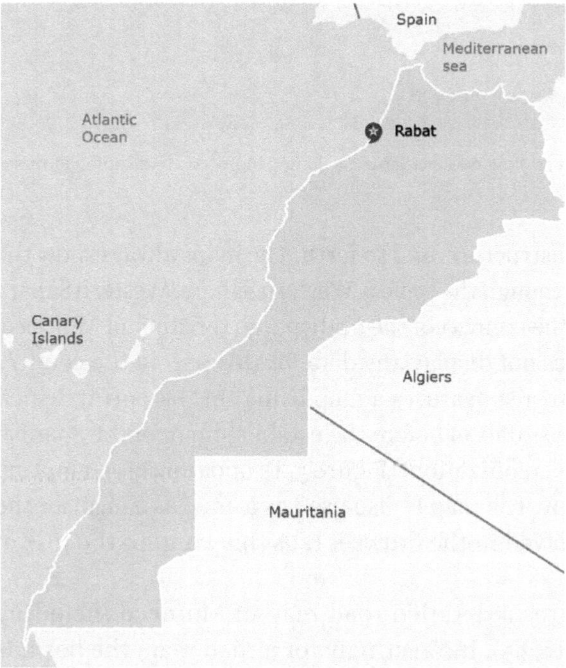

Figure 3.2: The main map of the official site of the Moroccan government. Source: Map of Morocco.

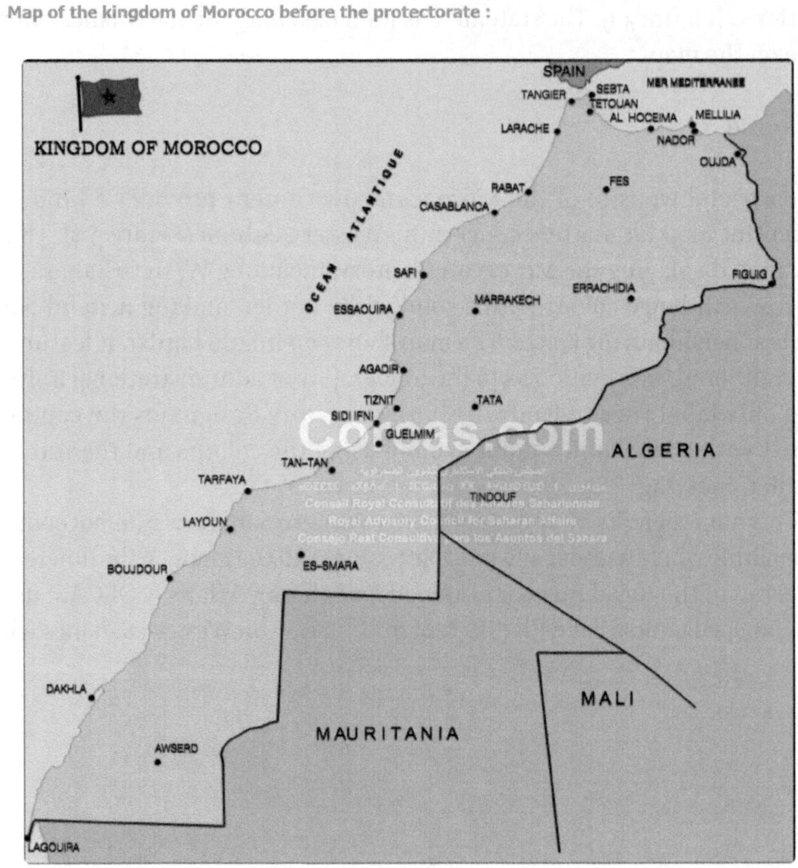

Figure 3.3: Moroccan maps always show Western Sahara including Morocco's undisputed territory. Source: Sahara Portal.

as geology, climate, infrastructure, and so forth. The maps always show the established whole, even though the topic is Western Sahara. Western Sahara is always visualized in the context of the undisputed territory of Morocco, even when the map does not display any data for this part of the country (Figure 3.3). The website also features a map using this blueprint, which is supposed to show the situation before the establishment of the Spanish 'protectorate' or 'before colonization' (Figure 3.4), apparently attempting to establish the geo-body. The map is also used as a 'map as a logo' for the website's favicon, displayed in the browser tabs that contain the title of the website.

CORCAS also features a detailed road map of Morocco including Western Sahara (Figure 3.5). Interestingly for a road map, the berm is not shown. The road layout differs considerably in some parts from the

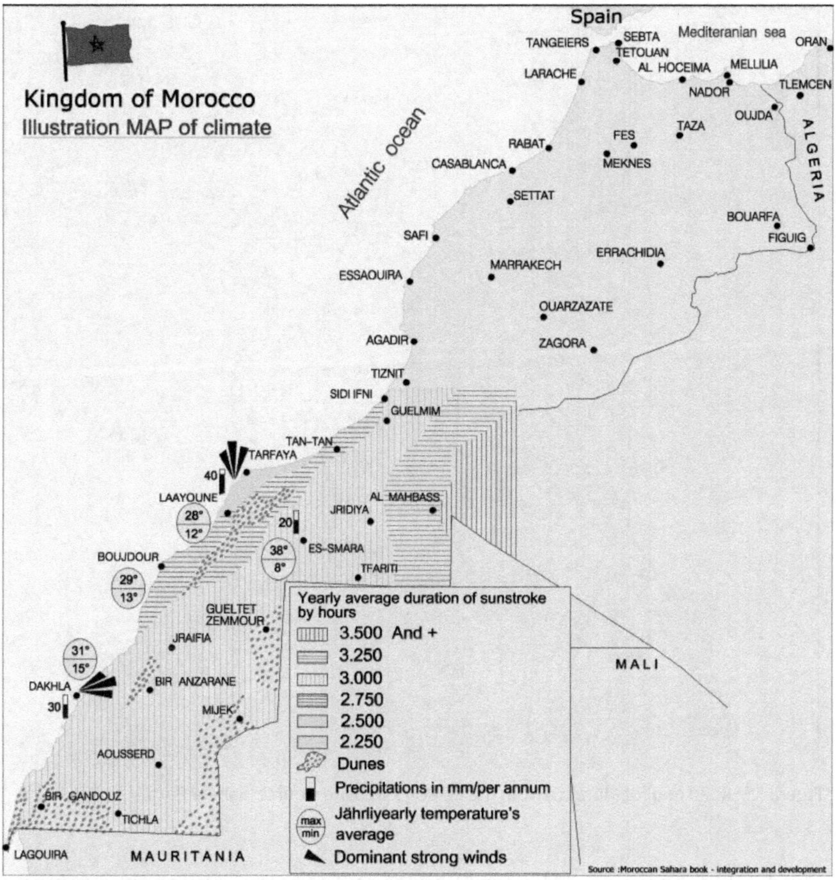

Figure 3.4: Even if the map's data focusses on Western Sahara only, Morocco's undisputed territory is always part of the map. Source: Sahara Portal.

situation shown in the UN map. Some roads that cross the berm in areas that the UN map marks as impenetrable are left out (Agwanit–Zbayra and Agwanit–Agailas), while others are left in (Zug–Tichla, Agwanit–Guelta Zemmur, Mahbas–Tindouf).

The undisrupted road network impressively shows the coherence between the territories, merging them into one. This tight network, tying together the two regions, ends abruptly at the national borders to Mauritania, whose existing roads are coloured in yellow, which is not in use for the Moroccan roads, rendering them almost invisible against the brownish background marking Mauritania. The lack of connections between Morocco and surrounding countries, and the use of a very different colour scheme for the roads within and outside of the Moroccan claimed territory, visually

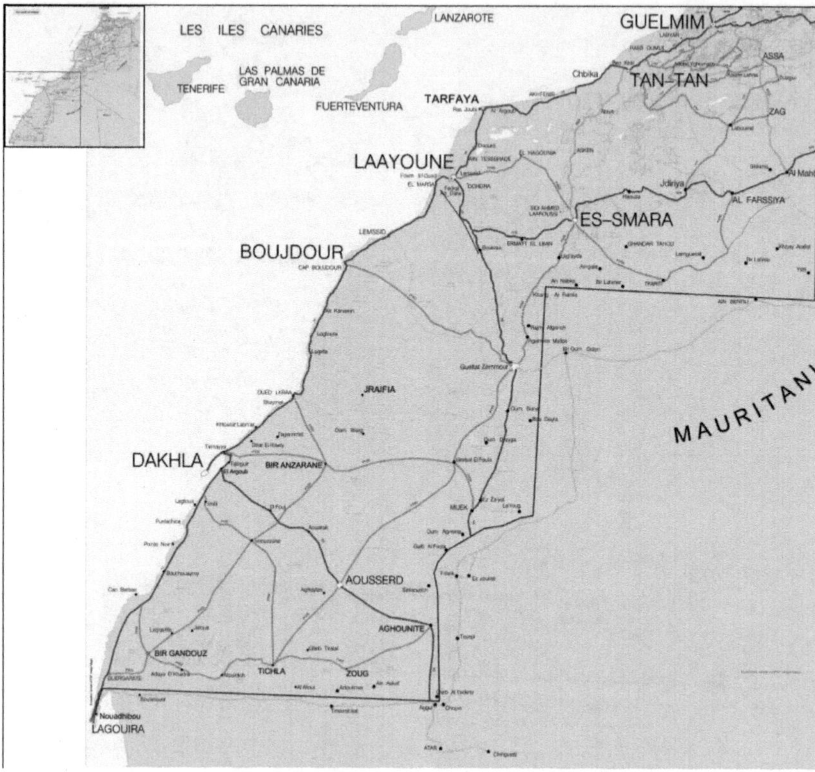

Figure 3.5: Road map of Morocco from the CORCAS website. Source: Sahara Portal.

emphasize the unity of Western Sahara and Morocco and their communal border to the neighbouring states.

Another interesting map on the CORCAS website is an interactive web map using Google Earth satellite imagery; this map displays important cities in Morocco and in Western Sahara using Morocco's national flag as icons (Figure 3.6). Also on display in this map are the Western Saharan refugee camps and the government-in-exile site in Tindouf (Algeria) marked with illegible icons and the Algerian flag, respectively. One of the photos shows an armed, dark-skinned fighter in a desert landscape, and another shows light-skinned men behind a fence, apparently imprisoned. In the map, all the borders of Morocco and Western Sahara are omitted completely; only the ones between Mauritania and Algeria are designated.

The photos of warriors and prisoners in the Tindouf camps are the only photos in a map whose topic is supposedly the cities of Western Sahara. This is especially striking since the cities themselves are not endowed with any

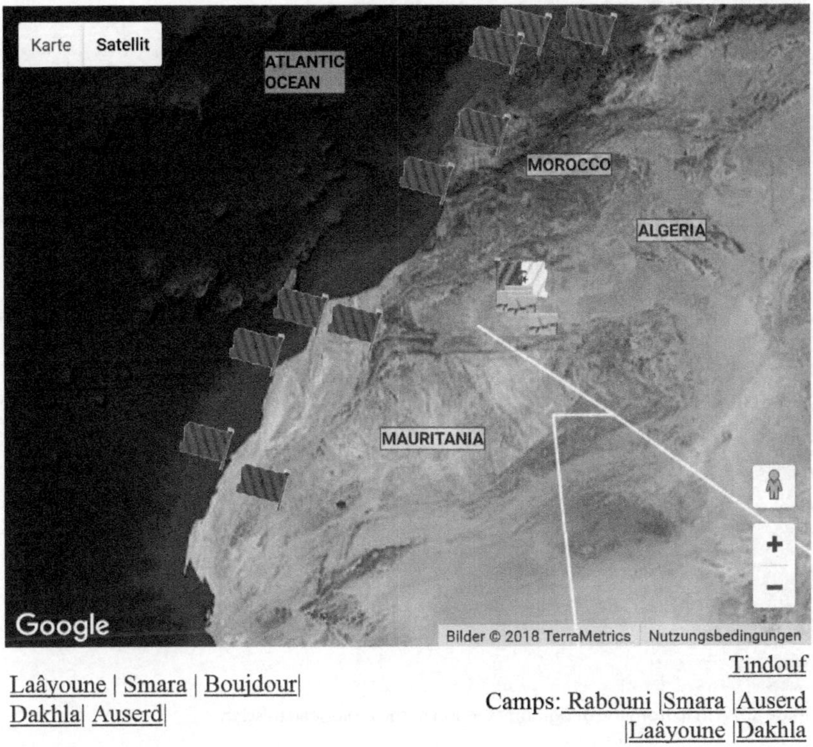

Figure 3.6: The map from the opening page of the CORCAS. Source: Sahara Portal.

illustrations at all. Therefore, besides stressing the union of the territories of Western Sahara and Morocco, this map also seems to contain an accusation against Algeria, an ally of the Polisario Front. The pictures of warriors and prisoners can be interpreted as an allusion to Moroccan prisoners of war in the Tindouf camps (Usher 2003). The close proximity of the pictures to the Algerian flag in the map places the responsibility for these irregularities on Algeria in a quite literal sense.

The Polisario Front

The maps featured on websites related to the Polisario Front or SADR usually also speak unequivocally.

The official site of the Ethiopian mission to the Western Saharan government, in its section about the geography of the territory, uses the UN map described above. It is an exact copy of the map, though in a comparatively low resolution. This makes some parts, especially the smaller font sizes,

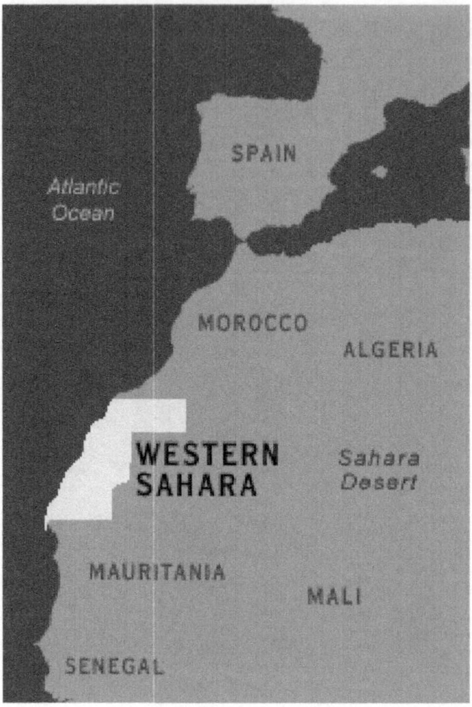

Figure 3.7: A map from the geography section of the Ethiopean mission.

illegible. The use of the UN map underlines the objectivity of the statement the map makes, the source of the map justifying its internationally accepted correctness. The small size conveniently renders the relativizing statement illegible. A second map on the same site makes the statement of an independent Western Sahara even clearer (Figure 3.7). In it, the territory appears in a bright yellow, the only distinguishable shape in the map besides the continental outlines. All other countries are marked only by their names as labels; no further borders are displayed.

The maps on the website also tend to show the hostile role of Morocco, not only towards Western Sahara. In the history section of the website, a map can be found that shows the member states of the African Union (AU) in green (Figure 3.8). Non-members are presented in grey, which is only Morocco, isolating it from the rest of Africa, conveniently ignoring the fact that Morocco joined the AU in January 2017. The colonial ambitions of the 'oppressor' Morocco are documented with the help of a map showing 'Great Morocco' (Lakhal 2016) as encompassing not only Western Sahara but also Mauritania as well as parts of Mali and Algeria (Figure 3.9). In this map, Morocco is presented in a light colour, making it stand out against the other African countries,

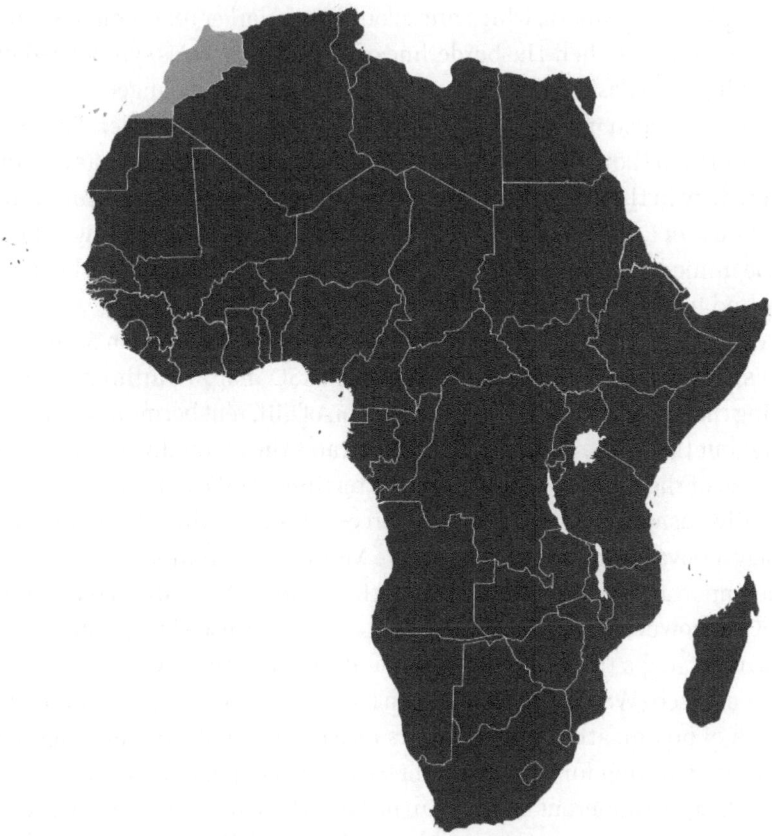

Figure 3.8: Morocco as an outsider on the African continent. Source: Saharawi Mission.

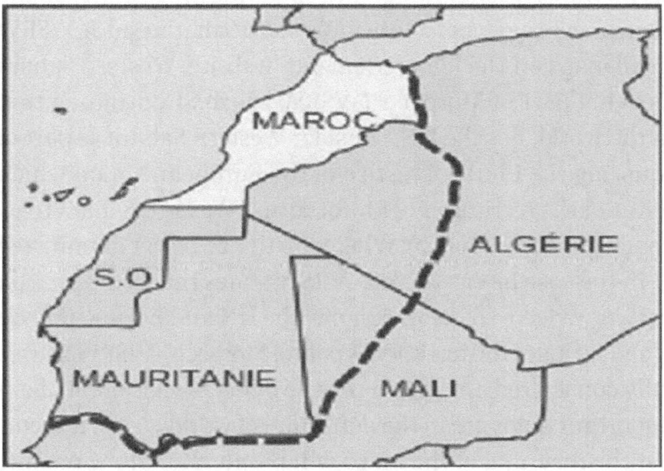

Figure 3.9: Morocco as an outright threat to other African nations claiming territory of several countries. Source: Saharawi Mission.

including Western Sahara, which are coloured in a darker tone, emphasizing their belonging together. The borderline of 'Great Morocco' is symbolized in a heavy, light red dashed line, interpretable as a symbol of danger.

The website also makes use of the 'map as a logo' in its header. This is a flash animation showing a small-scale map with Western Sahara at its centre. The territory in the map radiates an animated green ray of light similar to a lighthouse or a ship's radar screen. The animation and contrasting colour scheme immediately draw the eye to Western Sahara, courageously holding its ground in the face of danger.

The website of the Saharan Delegation in Spain (Delegación Saharaui para España) features, along with the MINURSO map, an animated map showing the history of Morocco's construction of different berms in six stages throughout the 1980s. The animation illustrates the gradually diminishing influence of the Polisario Front over the territory of Western Sahara, as it is literally pushed aside by Moroccan forces. This gives the impression of a very aggressive strategy enforced by the Moroccan military.

The importance of the role of maps in the context of defining a territorial identity is, however, underlined not necessarily in the use of maps but in the attacking of maps that show the opponent's perspective. Western Sahara Resource Watch (WSRW) is, according to its own definition, 'an international network of organizations and activists researching and campaigning the companies working for Moroccan interests in occupied Western Sahara' (WSRW 2007a). Numerous articles can be found here that criticize the use of maps displaying Western Sahara as part of Morocco. Two recent examples, published in 2018 on the WSRW website, are the cases of HeidelbergCement and Vigeo Eiris, two transnational companies operating in Western Sahara. HeidelbergCement owns a cement factory in Western Sahara, and as WSRW points out, '[o]n all maps on HeidelbergCement's website, Western Sahara is illustrated as being part of Morocco' (WSRW 2007b). Continuing that '[n]o country in the world, nor the UN, consider Western Sahara as part of Morocco' and quoting the Court of Justice of the European Union which 'clarified the status of the territories of Morocco and Western Sahara to be "distinct and separate"' (WSRW 2007b). Where can this be better illuminated than on a map? In the case of Vigeo Eiris, WRSW states that '[t]he map on the front page of www.vigeo-eiris.com ignores the UN and EU position on Western Sahara and includes the territory as part of Morocco' (WSRW 2007c).

These carefully considered attacks on the use of maps on the web demonstrate how important maps are in the definition of world views. It seems to be feared that the meaning of the maps will proliferate into economic and exploitative practices of international companies.

Non-Faction Media

An important question seems to be whether these mapped-out discourses have any influence on the mappings created by media that are not officially part of the conflict. In the following, some highly frequented providers of information are analysed; namely Wikipedia, Google Maps, Yandex, and the OpenStreetMap.

Wikipedia

In this analysis, I concentrate on the English-language articles in Wikipedia concerning Western Sahara and Morocco because they are the longest, contain the most maps, and are probably the most frequented of all the different language articles available. The two topmost maps in the English Wikipedia article on Morocco are probably also the most visible ones. Their purpose is to provide a general orientation to the reader. The first map shows mainly Africa and Europe in shades of grey. Morocco is coloured in dark green while Western Sahara is coloured in light green. An explanatory text below the map states that Western Sahara is 'a territory claimed and mostly controlled by Morocco as its Southern Provinces' (Wikimedia 2017). Below that map, the reader finds a map showing Morocco as a white space with some cities inscribed in it. Here, Western Sahara is distinctly separate from Moroccan territory. This map is attributed to the UN Office for the Coordination of Humanitarian Affairs. Additional maps of Moroccan climate[1] and ethnolinguistic groups[2] are constrained to this 'internationally recognised' territory of Morocco. A map titled 'Western Sahara' is based on the map in use by the UN peacekeepers,[3] MINURSO, using more or less the same rhetoric: Western Sahara is a clearly distinct territory separated from Morocco by colour and borderline. Immediately below this map, however, there is another one showing the 'Regions of Morocco' (Figure 3.10). This rather colourful map is subtitled 'the 12 official administrative Regions of Morocco'. It shows Morocco and its boundaries, including Western Sahara. No separation of the two territories is indicated.

1 Wikimedia Commons, Online Resource: https://commons.wikimedia.org/wiki/File:Morocco_K%C3%B6ppen.svg.

2 Wikimedia Commons, Online Resource: https://commons.wikimedia.org/wiki/File:Morocco_ethno_1973_all_v2.svg.

3 Wikimedia Commons, Online Resource: https://commons.wikimedia.org/wiki/File:Westernsaharamap.png.

Figure 3.10: Wikipedia also features this map of Morocco contradicting most other maps on the website where Moroccan territory is concerned. Source: Wikimedia Commons.

The article on Western Sahara features most prominently the map based on the MINURSO map that is also used in the article on Morocco.[4] The description below the map lists as 'country' both Morocco ('as its "southern provinces"') and the SADR ('in the free zone'). Further maps put forward in the article include a British historical map of the Western Sahara region from the nineteenth century that shows no borders.[5] There is also a map showing the situation in north-western Africa in the year 1912, in which French colonies and 'French Morocco' are displayed in shades of green,

4 Wikimedia Commons, Online Resource: https://commons.wikimedia.org/wiki/File:Westernsaharamap.png.

5 Wikimedia Commons, Online Resource: https://commons.wikimedia.org/wiki/File:Western_Sahara_1876.png.

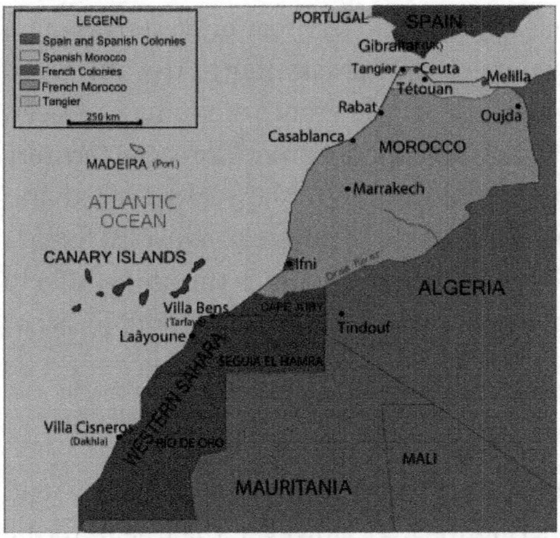

Figure 3.11: These maps emphasize Western Sahara as separated from Morocco. Source Wikimedia Commons.

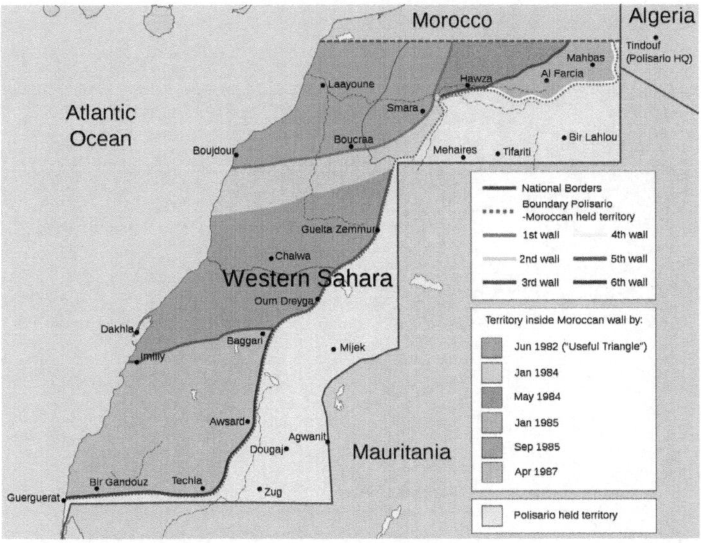

Figure 3.12: These maps emphasize Western Sahara as separated from Morocco. Source: Wikimedia Commons.

whereas Spain and its colonies, including Western Sahara, are coloured in a contrasting red tone, clearly separating Western Sahara from Morocco (Figure 3.11). The colouration ties Western Sahara and even the southern part of Morocco called 'Cape Juby' in this map to Spain, whereas Morocco is associated more closely with the green-coloured French colonies. Another map shows the progression of Moroccan military forces into the territory in the 1980s and the 'system of the Moroccan Walls', including the current berm

(Figure 3.12). This map divides a clearly separate territory of Western Sahara into a colourful patchwork of occupation zones similar to the animated map described in the section about the Polisario Front's mapping practices.

These maps stressing Western Sahara's distinction from Moroccan territory are followed by a series of four maps, titled '4 ways to show Western Sahara'. The maps show Western Sahara and Morocco 'separately', Western Sahara as 'disputed territory', Western Sahara as separate only in the parts de facto controlled by the Polisario Front, and Western Sahara 'as part of Morocco'.

Web Maps

Google Maps shows Western Sahara as a separate territory with mainly solid lines to indicate national borders (Figure 3.13). However, the line marking the border to Morocco is dashed. This seems to suggest that the territory of Western Sahara or the border between the two territories is under dispute.

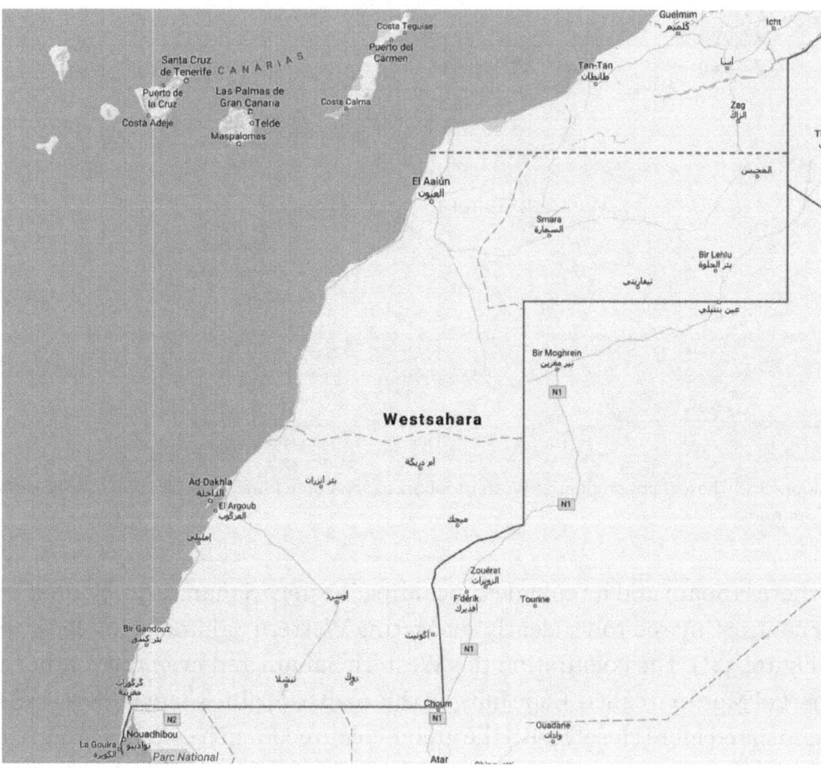

Figure 3.13: Google's version of the conflict.

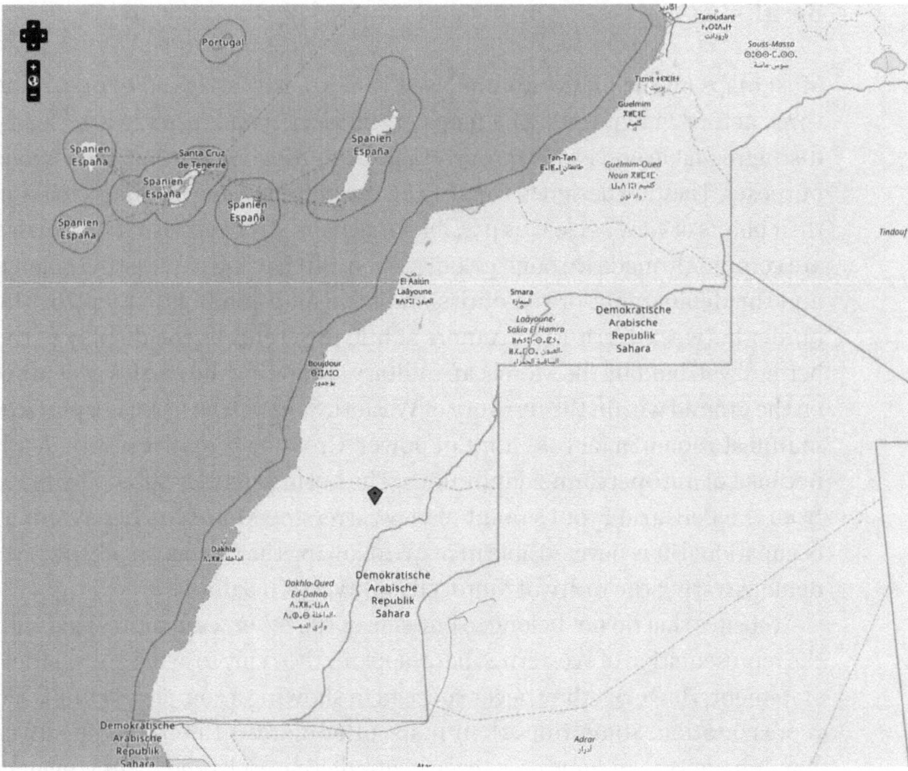

Figure 3.14: OpenStreetMap's version of the conflict.

The community-based and mainly crowdsourced OpenStreetMap shows the berm as the border between Morocco and the SADR, using the same symbolization as for any other national border (Figure 3.14). This establishes SADR as an independent state, but only on parts of the claimed territory. A dispute is not indicated.

The Russian web map provider Yandex changed policies during the process of writing this paper. Up to mid 2017, the Yandex map did not feature Western Sahara at all. Its database yielded no results for Western Sahara. The territory was not marked on the maps and was completely integrated with Morocco. This has now changed, and Western Sahara's border with Morocco is indicated with a line style identical to other national borders; the city of Laayoun is – judging from the font size and weight – designated as capital of the territory. In addition, the lettering of the place label 'Western Sahara' is not distinguishable from other independent nation states' labels. No dispute is indicated.

Results

Most maps on the platforms analysed can be easily assigned to one of the three actors' viewpoints. The maps of the factions themselves are clearly distinguishable and easily interpretable according to their mostly obvious purposes. They are designed to speak an uncompromising language, making their points of view as clear as possible. The maps' strongest visual statements are commonly made through colourization, often using contrasting colours and the deployment or the omission of the map symbol delineating the border between territories. A more subtle difference is the display of the berm, a wall built by the Moroccan military separating the parties of conflict on the ground within the territory of Western Sahara. The berm is a physical manifestation of a factual limit of power. Crossing it may result in death because of anti-personnel landmines. The berm is mostly shown by maps from the Polisario Front's point of view, stressing Morocco's behaviour as occupational. It is never visible in maps following the Moroccan perspective, demonstrating the unity of Morocco and Western Sahara.

Websites that do not belong to one side or the other seem to struggle with the representation of Western Sahara, apparently trying to avoid any political statement. However, they never succeed in showing a completely unbiased representation. Sometimes their maps offer many different perspectives simultaneously – in some cases even contradicting each other. This is evident in the case of Wikipedia. The articles on Western Sahara and Morocco are each biased towards the respective perspective. Although Wikipedia mostly features maps that attempt to represent the UN point of view (which might be considered a bias in its own right), at least one map clearly follows the Moroccan perspective, displaying Western Sahara as the 'Southern Provinces'. The article on Western Sahara makes use of a historical map that alludes to the status pre-dating the established nation states, attacking Moroccan views of a naturalized unity. This is admittedly a very subtle bias, and the strategy in this article seems rather to avoid a political stance in this matter. The maps attempt this not by endorsing a seemingly objective point of view, but by showing different options for mapping Western Sahara. However, by rendering each option equally valid, no party of the conflict will be able to agree with this compromise. A concordant point of view is not achieved.

At other times, the struggle shows in policy changes over time. Yandex, for example, applies opposing concepts at different times. Another solution in the contest for neutrality seems to be to present the area as ambiguously as possible, such as in the case of Google Maps. The dashed line between

Morocco and Western Sahara leaves the question unanswered as to whether Western Sahara belongs to Morocco. Yet another strategy is to attempt a representation not of the political ideas defining a territory but the factual realms of power, such as in the OpenStreetMap. Here the berm represents a factual border that fully mirrors neither the Moroccan nor the Western Saharan claims, but a factual division of the control over the territory. However, this statement is so unpolitical that it suggests a resolved situation for a highly disputed war zone. This is, in effect, not unpolitical at all.

Discussion

In the case of Western Sahara, where there has been a long-lasting ceasefire, the battle over the territory continues on the internet – and maps play an important role in these disputes. The factions use maps to try to produce certainty about the shapes of the nation state. They are supposed to establish valid perspectives and to refute the opposition. In the age of the internet, this battle might have intensified. An abundance of maps can be produced and distributed, and their visualized messages can be made quickly accessible in a cognitive sense. Nevertheless, the possibility of proliferation of maps is always also an opportunity for the opponent and has to be closely watched.

Anderson's theories about the role of maps in the imagination of the nation state still hold ground, as outlined at the beginning of this chapter. However, in recent times, researchers have considered maps processes rather than unchanging immutable entities (e.g. Kitchin and Dodge 2007). This relativizes the claim to the absolute truth of a map and reveals its inherent perspectives. In the struggle to establish enduring truths, the processual characteristics of web mapping emerge. A mapping is the expression of ideas that are considered valid at a certain time and by certain people. Internet mappings, in their abundance and topicality, make that point clear to the astute user of the different channels of the internet. Caquard describes the tension of the map in the era of the internet between being an instrument to resist power and to reaffirm power. Both are possible at the same time using the same platforms depending on the political position of the map authors (Caquard 2011). This might entail a generally more critical or even sceptical reading of maps, which might deprive the map of its striking power. In any case, internet mappings have replaced the state as sole reference point for defining territories and nations with the arrival of Web 2.0 (Caquard 2014). This makes it even more important to deliver clear messages. As Culcasi shows, counter-mappings may reflect a difference between the self-chosen definitions of the territory and the externally imposed ones (Culcasi 2012). Analysing mappings of the internet, as in this case, yields interesting insights into the perspectives of 'the other', which might not necessarily be explicitly expressed in words. These wordless perspectives in turn reveal ideals and utopias of non-state actors that could not have been expressed and delivered before the era of the internet. The sovereignty of the nation state has become vulnerable in the area of dissemination of territorial interpretations.

References

Benedict Anderson, *Imagined Communities*, revised edition (London: Verso, 1983).

Radhi Bachir, 'Western Sahara: Self-Determination Delayed', *Pambazuka News*, Online Resource: https://www.pambazuka.org/democracy-governance/western-sahara-self-determination-delayed (accessed 11 November 2020).

Karen Culcasi, 'Cartographies of Supranationalism: Creating and Silencing Territories in the "Arab Homeland"', *Political Geography*, 30 (2011), pp. 417–428.

Sébastien Caquard, 'Cartography I: Mapping Narrative Cartography', *Progress in Human Geography*, 37(1) (2011), pp. 135–144.

Hans Corell, *The Responsibility of the UN Security Council in the Case of Western Sahara*, Online Resource: http://www.embajadasaharauimexico.org/wp-content/uploads/2016/01/hans-Correll-international-judicial-monitor.pdf (accessed 11 November 2020).

—, 'Cartography II: Collective Cartographies in the Social Media Era', *Progress in Human Geography*, 38(1) (2014), pp. 141–150.

John Brian Harley, 'Deconstructing the Map', *Cartographica*, 26(2) (1989), pp. 1–20.

Rob Kitchin and Martin Dodge, 'Rethinking Maps', *Progress in Human Geography*, 31(3) (2007), pp. 331–344.

Bir Lahlou, *Proclamation of the First Government of the Saharawi Arab Democratic Republic*, Online Resource: http://www.arso.org/03-1.htm (accessed 11 November 2020).

Malainin Lakhal, *Morocco's Colonialist Agenda Must not Succeed*, Online Resource: http://saharawi.net/en/?p=694 (accessed 11 November 2020).

Maroc, *Map of Morocco*, Online Resource: http://www.maroc.ma/en/content/map-morocco (accessed 11 November 2020).

MINURSO, *United Nations Mission for the Referendum in Western Sahara: Background*, Online Resource: https://minurso.unmissions.org/background (accessed 11 November 2020).

Sahara Cities, *Western Sahara Cities*, Online Resource: http://www.sahara-villes.com/en/WesternSaharacities/tabid/228/Default.aspx (accessed 11 November 2020).

Sahara Portal, *Before Colonisation*, Online Resource: http://www.sahara-online.net/Default.aspx?tabid=710 (accessed 11 November 2020).
Saharawi Mission. Ethiopia-African Union, *Geography*, Online Resource: http://www.sadr-emb-au.net/geography/ (accessed 11 November 2020).
The United Nations and Decolonization, *Non-Self-Governing Terrorities*, 2019, Online Resource: https://www.un.org/dppa/decolonization/en/nsgt (accessed 11 November 2020).
Sebastian Usher, *Polisario Releases Moroccan POWs*, Online Resource: http://news.bbc.co.uk/2/hi/africa/3200039.stm (accessed 11 November 2020).
Western Sahara Resource Watch, *About Western Sahara Resource Watch* (WSRW), 2007a, Online Resource: https://wsrw.org/en/about-us (accessed 11 November 2020).
Western Sahara Resource Watch, *What is Heidelberg Cement Doing in Occupied Western Sahara?*, 2007b, Online Resource: https://wsrw.org/en/archive/4124 (accessed 11 November 2020).
Western Sahara Resource Watch, *Vigeo Eiris Goes Back On False Claim*, 2007c, Online Resource: https://wsrw.org/en/archive/4095 (accessed 11 November 2020).
Wikipedia, *Persian Gulf Naming Dispute*, 2019, Online Resource: https://en.wikipedia.org/wiki/Persian_Gulf_naming_dispute (accessed 11 November 2020).
Denis Wood, *The Power of Maps* (London: Routledge, 1992).
—, *Rethinking the Power of Maps* (New York: Guilford Press, 2010).

About the Author

Frederik von Reumont is a research fellow and PhD candidate at the Institute of Geography Education at Cologne University. Trained in geography as well as cartography, he is interested in the potential of digital and other maps in critical geo-education. In his research, he currently focuses on the use of comics to convey geographical meaning. He experiments with the combination of maps and comics to try to exploit the cognitive advantages of both modalities.

4. Wargaming the Middle East: The Evolution of Simulated Battlefields from Chequerboards to Virtual Worlds and Instrumented Artificial Cities

Janina Schupp

Abstract

Shortly after the end of a tank combat during the Gulf War, a team of US Army historians, scientists, and engineers flew to Iraq to gather detailed data of the battle. The collected information was used to create an exact virtual simulation of the combat for training. The mapping capability – offered by the resulting simulation game 73 Easting – to visualize the battlefield from any position and point in time revolutionized military exercises. With ongoing conflicts in the Middle East, these digital training cartographies are now linked to real bodies and vehicles through digital and mobile technologies during live training in artificially constructed villages. This chapter analyses this evolution and critically investigates the growing 'gamification' ensuing in these representations of Middle Eastern battlefields.

Keywords: Wargames, Middle East, interactive battlespace, live simulations

> In war the experienced soldier reacts in the same way as the human eye does in the dark: the pupil expands to admit what little light there is, discerning objects by degrees, and finally seeing them indistinctly. By contrast, the novice is plunged into the deepest night. [...] It is immensely important that no soldier, whatever his rank, should wait for war to expose him to those aspects of active service that amaze and confuse him when he first comes across them. (Clausewitz 1989, p. 122)

Strohmaier, A. and A. Krewani (eds.), *Media and Mapping Practices in the Middle East and North Africa: Producing Space*. Taylor & Francis Group, 2021
DOI 10.5117/9789462989092_CH04

The act of playing at war is deeply engrained in human history and has persisted to the present day in both professional and hobby culture. Wargaming allows the participant to become an active part of a warfare situation and thus 'provides a unique learning experience that leads to a deeper and more personal understanding and appreciation of warfare than can be attained by any other method short of actual participation on the field of battle' (Perla 2011, p. 19). Wargaming is a unique tool for grasping the complexity of war, understanding the effects of modern weapons and strategies, and preparing soldiers for deployment on the battlefield. Thus, war games improve logical thinking under intense circumstances and enhance personal comprehension of combat. The war game, however, provides this new knowledge and skills at the price of compressing the complexity of reality:

> The war game gives us a chance to play at one of the most terrible creations of the hand of man, a creation capable of destroying all we have ever built and, at the last, the very world on which we live. It also gives us a chance to reduce the terrifying dimensions of this monster of our own making to a more human scale. [...] Its weaknesses lie in its inability to re-create the actual physical conditions under which such decisions must be made or the actual consequences, of those decisions. Then again, perhaps those too are strengths. (Perla 2011, p. 277)

Environments and humans have to be reduced and compressed to their most crucial features to be accommodated as components of a game or model of reality. These compressions widen the gap between the experience of a real, original space and the perceptions generated by its artificial doppelgänger. What is left out and what is considered relevant enough to be included in the artificial reproduction are both equally important factors to consider and have an impact on the distortion of reality. As Roger Dean Smith describes,

> real systems are extremely complex and a determination must be made about the details that will be captured in the model. Some details must be omitted and their effects lost or aggregated into other variables that are included in the model. In both cases, an inaccuracy has been introduced. (Smith 2009, p. 11)

Violence, pain, death, and guilt are the most evident realities of conflict that fall victim to this process of simulation, as their reproduction is often hindered by technological limitations or judged less relevant for the training of strategy and tactics. In addition, the association between war and game also reveals a fundamental enjoyment in the competition of such a compressed and so-called sanitized war – a feeling that has been purposely reinforced by militaries up to the present day. As H. G. Wells's miniature wargaming manual *Little Wars* expresses:

> Here is the premeditation, the thrill, the strain of accumulating victory or disaster – and [...] no shattered fine buildings nor devastated country sides, no petty cruelties, none of that awful universal boredom and embitterment, that tiresome delay or stoppage or embarrassment of every gracious, bold, sweet, and charming thing, that we who are old enough to remember a real modern war know to be the reality of belligerence. (Wells 1913, p. 97)

However, war games reduce not only the destruction and emotional extent of real war but also the representation of actual urban spaces and populations that will be encountered in future conflicts. The artificial constructs developed for training purposes are based on – and in turn produce – imaginary cartographies, which are often several degrees removed from their real geographical counterparts.

Preconceived notions that Western militaries have about the terrain and population of the Middle East can influence the artificial training cartographies produced to grasp foreign countries. According to Edward W. Said, such an imagined or 'imaginative geography'

> legitimates a vocabulary, a universe of representative discourse peculiar to the discussion and understanding of Islam and of the Orient. [...] We need not look for correspondence between the language used to depict the Orient and the Orient itself, not so much because the language is inaccurate but because it is not even trying to be accurate. What it is trying to do [...] is at one and the same time to characterise the Orient as alien and to incorporate it schematically on a theatrical stage whose audience, manager, and actors are [the Western world]. (Said 1978, p. 71)

The application of such an imagined Middle Eastern geography and Orientalist language in military simulations thus functions as 'an accepted grid for filtering through the Orient into Western consciousness' (Said 1978, p. 6). The resulting military simulations in turn transfer these imaginative constructs of the Middle East to the soldiers during training. The cartographies encountered by soldiers during the training hence shape their expectation and imagination of these foreign countries even before deployment.

Throughout history, technological progress has determined the ways in which these imagined geographies of enemy territories were produced by Western militaries. This chapter will trace the technologies used for spatial simulation throughout history in order to reveal the evolution of these artificial and imagined cartographies built by Western militaries and their effect on the perception of the Middle East.

A Historical Introduction: Early Terrain Representation from Abstraction to Map Use

Tracing back the origins of mapping geographical spaces and humans in wargaming is an almost impossible task, as warfare has influenced games – and vice versa – long before this phenomenon was documented (Allen 1987; Perla 2011; Lewin 2012; Creveld 2013). The first known ancestor of today's wargaming practices is the game Wei Hai, c. 500 B.C.,[1] frequently credited to the Chinese general Sun Tzu (Perla 2011, p. 29), which later developed into the Japanese game Go. While Wei Hai relied on coloured stones to represent armies, a more elaborate four-sided board game with miniature soldiers, elephants, and cavalry appeared in India around the sixth century under the name Chaturanga. Chaturanga is often claimed to be the ancestor of modern chess, which in turn marked the beginning of Western military wargaming. Chess represented its contemporary state of warfare, with kings, towers, and knights mirroring the components of conflicts at the time.

Chess and its wargaming ancestors and descendants have thus always reflected contemporary warfare practices, and war games have increasingly aimed for accuracy in relation to real warfare and geography. However, the terrain of the battlefield was still visually reduced to a grid of squares with colour coding, and no spatial or geographical features were attributed to the simulated combat space. The main difficulty of early wargaming was thus the accurate mapping of terrain specificities and of the spatial positions of military elements, such as soldiers, machinery, and provisions. Only important military figures were personified, while the general population was not captured by the simulation. These strong reductions, resulting from the finite technological means available, communicated a limited impression of foreign lives and environments.

1 Wei Hai is variously dated in publications as originating in 3000 B.C., 1000 B.C., and 500 B.C. Since Sun Tzu lived around 500 B.C., this approximate date is used here.

In the mid eighteenth to mid nineteenth century, wargaming was primarily expanded in Germany. Early progress in terrain representation was made by Johann C.L. Hellwig in 1780, followed by Georg Venturini in 1797, who both extended the board (to 1666 and then 3600 squares, respectively), introduced different colours to differentiate terrain conditions, and included pieces to represent bridges, wagons, or fortifications, amongst other features. In Prussia in 1812 the most prominent early war game, the Kriegsspiel, was invented by civilian war counsellor Baron George Leopold von Reiswitz. The Kriegsspiel exchanged the rigid chessboard for an open table with several topographical terrain pieces that could be rearranged at will to create different landscapes. The terrain squares were made of wood and painted plaster and represented streets, villages, rivers, or forests. The simulated terrain thus became not only more detailed but also three-dimensional to reflect the heights of mountains. The concept of spatial representation developed by Reiswitz was a typology of spatial features that could, in their combination, create an imagined landscape or approximately reproduce the real geography of a deployment terrain. Reiswitz himself called the selection of shapes 'types', 'since, like the letters of printing works, which serve to reproduce writing, these types and their combination can reproduce a spatial area' (Reiswitz 1812, p. 11).[2] The Kriegsspiel also created a variety of playing pieces detailing the particularities of different military units; for example, artillery or cavalry, and the comprehensive effects of weaponry and movements. While his game had little success at first, his son Lieutenant George Heinrich Rudolph Johann von Reiswitz developed a revised version in 1824, which rose to fame. The main adaptation was again the terrain representation, which, in the revised version, was finally able to use accurate topographic maps instead of fabricated typologies, and employed rulers and dividers to calculate distances of shots and troop movement. The Kriegsspiel pioneered many design characteristics and war game processes that have become core components of war games to the present day.

The tracing of early wargaming history shows that the mapping of enemy terrain in war games has always depended on and adapted to the technologies of representation available during each period. Thus, the later rise of computers deeply influenced the accurate mapping of terrains and simulation of combat activities in war preparation.

2 'Typen [...] weil eben so wie die Buchstaben in der Druckerey, dazu dienen eine Schrift nach zu bilden, durch diese Typen und ihre Zusammenstellung eine Gegend nachgebildet wird' (translation by the author).

The Computer Age: Battlefield Cartographies Go Virtual and Interactive

Earlier military war games allowed for the testing of strategies, but as soon became apparent, military combat also requires a coordinated interaction and cooperation of the individual soldiers in one unit. Modern war fighting is 'rooted in each individual's performance with his weapon system', but 'its scope is far greater', including 'the coordination of that individual's activities with others in the crew, and that crew's interaction with other crews' and so on, and to 'be successful at warfighting combatants must master these interactions at all levels' (Thorpe 1987, p. 1). As war games increasingly changed focus from the omniscient viewpoint of the commander to the limited viewpoint of the individual soldiers on the battlefield, simulation developers looked for ways to create shared simulated environments to train skills and teamwork at these lower levels. With the rise of computer technology, commercial video games began not only to provide cheap, high-performing technology but also created new elements that could be conveniently transposed for such training – especially from first-person shooter games, which emerged in 1974 with the Maze Game (Greg Thompson et al.) and Spasim (Jim Bowery). These games created a feeling of being an active agent in a battlefield space and enabled the player/soldier to acquire an embodied combat experience. The US Army hence discovered computer games as a 'new and better means for training', especially since 'people in those arcades are volunteers', even 'paying for the use of the machines, and two-thirds of these games are military in nature' (TRADOC Commanding General Starry, cited in Ludvigsen 1981, p. 16). The new aim of the military thus became to tap into the spatial-mapping capabilities and the natural appeal of video games from the beginning of the 1980s.

When Atari released the arcade game Battlezone in 1980, the US Army immediately requested a prototype military version (Schachter 1983, p. xii). The game, designed by Ed Rotberg, uses vector graphics to create a 3D wireframe view of a battlefield with mountains, which is navigated from a first-person perspective located inside a tank. The game hence 'provided an environment where the player had complete freedom of movement over the ground in any direction' (Poole 2000, p. 125). Early versions even had a mock periscope attached to the cabinet frame to be used as a tank 'viewfinder' and thus generated the 'feel of a genuine battlezone: limited vision, nasty surprises, panicky adjustments while the enemy tank wheels slowly round to get you in its sights' (Halter 2006, p. 123). The military version, named the Bradley Trainer, was conceived as a training resource for soldiers operating the Bradley Fighting Vehicle – since real training was highly costly and video games offered great potential to develop 'incredible hand, eye, and brain coordination' (Reagan, cited in Halter 2006, pp. 117–118). The military version of Battlezone in the end only existed in two copies and was never used for training. However, it highlighted the turn of the military's interest towards entertainment technologies. The increasingly sophisticated commercial simulation games introduced useful elements for military training, especially terrains navigable from a first-person perspective, training in the use of virtual weaponry, restricted battlefield intelligence, and competition between players through networking technology, all while creating the pleasure and excitement of using these features over and over again.

The Simulator Network (SIMNET) project by Jack Thorpe subsequently created the first military network for simulators in 1983, when the networking of more than one simulator or game terminal became possible through the early forms of the Advanced Research Projects Agency Network (ARPANET), which later became the internet (Miller and Thorpe 1995).

With the SIMNET project, simulator design shifted from a tradition of fidelity to an emphasis on interaction between players and use of entertainment effects for a new sense of immersion in a collective virtual battlefield. Instead of building an accurate replica, the project explored 'how much reality was needed to fool people into believing they were actually firing on targets' and incorporated 'some of the attractiveness of video game technology into the drudgery of military drills' (Voss 1993, p. 10). For SIMNET, Robert Jacobs invented the concept of 'selective fidelity' – also referred to as a '60 percent solution' (Voss 1993, p. 11). Instead of replicating the tank interior accurately – like previous flight simulators, which had reproduced all cockpit details – only elements fundamental to the execution of tasks were recreated. Other elements, such as switches, were simply shown as pictures or painted on. Instead, a focus was placed on employing entertainment technologies to simulate a believable battlefield environment and atmosphere that triggers soldiers to think they are in an actual battlefield. For example, in an early prototype, a graphic video filmed in the US Grafenwöhr training centre was employed to simulate the shooting and crash of the tanks, and screens became defunct to simulate damage. As McDonough notes: 'We basically turned the tank gunnery into a video game' (cited in Voss 1993, p. 11). The crews looked out onto the virtual battlefield through television screens that were mounted in the position of the usual tank lookout windows. The magazine *Wired*, as a result, named SIMNET the first creation of cyberspace, stating that 'William Gibson didn't invent cyberspace, Air Force Captain Jack Thorpe did' (Hapgood 1997, p. 116).

To describe how engaging this new shared mapping of a battlefield with enemies and foes felt, developer Jim McDonough describes a situation that occurred during tests: 'The hatch in the SIMNET simulator doesn't open. [...] One commander got so excited while looking through his commander observation point, [...] he bashed his head against the top of the simulator, forgetting that he was in a simulator' (cited in Voss 1993, pp. 12–13).

The SIMNET project was one of the early military simulations that deliberately integrated video gaming techniques for virtual, interactive battlefield mapping. During the conception of SIMNET, a prototype computer display was also developed to allow an impartial observer to fly over the battlefield and look at the position of units. The observer could even switch to the perspective of a specific tank, thus embodying its view. This display technology allowed the first virtual and interactive mapping of real Gulf War combat a few years later.

73 Easting:[3] Gulf War History Mapped as a Virtual, Interactive Battlespace

Shortly after the end of a successful tank combat during the Gulf War in early 1990, a team of US Army historians, scientists, and engineers flew to Iraq to gather detailed data on the battle through interviews, terrain inspection, aerial photos, communication recordings, bunker locations, tank shots, and destroyed targets. The collected information was then used to create an exact virtual simulation of the combat for review and training, including geographical terrain, soldier and vehicle movements, communications, and use of weaponry. The 73 Easting battle simulation generated a 'flying carpet' together with a timeline function, which allowed commanders to examine the virtual version of the successful combat and understand the terrain and battle movements that had occurred. The reconstruction of the battle was played like a live SIMNET simulation and recorded. As Walter N. Lang describes:

> It was designed to help commanders prepare for future battles and provide the 'wherewithal' for 'time travel' – that is to see the battle from any perspective at any time, many times, anywhere, in real-time or even faster than real time. [...] Commanders can cruise a computer-generated battlefield that shows the deployment and operations of both allied and

3 The Battle of 73 Easting was named after its location on a north-south grid line of military area maps.

enemy orders of battle in two-dimensional and three-dimensional views. The simulated battlefield can be visually displayed from any viewpoint – air or ground – or the overall situation can be seen on a digitized map. (Lang 1991, p. 12)

Sounds of explosions and original radio transmissions of crews were also integrated, thus creating a multidimensional simulation of the battle, which allowed for fast-forwarding or rewinding the battle and inspecting specific events, movements, or use of weapons (Orlansky and Thorpe 1992, p. i.). Alternative scenarios could be played to test out different courses of action, and various details could be filtered out for analysis. The capability to capture and visualize the battlefield and the historical as well as practice missions from any position and point in time was revolutionary and introduced new 2D and 3D mapping into virtual training. As Jesse Orlansky and Jack Thorpe describe the achievement in their report:

[T]he product, a 'moving picture' permits observers to review what actually happened at any moment from any position or vehicle on any side of the battle. No previous battle has been recorded in greater detail or can be re-created dynamically for study, review and 'what-if' analyses. (Orlansky and Thorpe 1992, p. i)

However, the reception of the new technology was split. While the military applauded the new technological advancements of 73 Easting's video game-style virtual reality, the wider public saw the first risks of a virtualization and gamification of real combat with real death tolls. This was reflected in the technology magazine *Wired*:

The flesh of real young men was there inside those flaming tank-shaped polygons, and that flesh was burning. That is what one knows – but it's not what one sees. What one really sees in 73 Easting is something new and very strange: a complete and utter triumph of chilling, analytic, cybernetic rationality over chaotic, real-life, human desperation. [...] This is virtual reality as a new way of knowledge: a new and terrible kind of transcendent military power. (Sterling 1993, p. 48)

While the move from tabletop to computer games brought the technologies for a more detailed spatial and temporal simulation of the Middle East, these elaborate and scientifically precise cartographies did not provide a more humane representation of the foreign terrain and population. The data gathered as the basis for the simulation significantly only focused on the terrain specificities and Western combat movements that were deemed relevant for the battle. Meanwhile, the civilian population, socio-economic context, and political motivation of the enemy did not inform the simulation design. Led by analytical motivations, the design of the virtual simulation of Iraq focused on military manoeuvres and treated the foreign population as a calculable opposing force. The displacement of civilian lives and the misery resulting from the violence of war were omitted in the clean-cut virtual cartography. The view generated was thus a detached division between the victorious Western force and anonymous, dehumanized opposing players.

The Introduction of Commercial Video Game Spaces in Military Training

As 3D environments emerged in the entertainment sector in the 1990s, the US military, constrained by budget cuts, began to acquire and adapt an increasing number of computer games. The first 3D game to be altered for military use was Doom II (id Software 1994) in 1995. The game, originally based on four extraterrestrial marines combatting demons in outer space, came with a level editor that enabled the acquisition and easy modification of the game for marine use. The Marines Scott Barnett and Dan Snyder subsequently adapted the game to represent an urban environment with an opposing military force and a wealth of Marine logos. Player avatars were given the skin of a scanned 'G.I. Joe' action figure, and battlefield sounds, including military communication channels, were added (Richard 1999, p. 160). The original game producers 'gave their blessing though they thought the idea of someone using their game to train soldiers was a joke' (Kushner 2004, p. 194). The rise of more detailed game designs and modification tools (such as level editors, scenario builders, and avatar skins) thus enabled a growing transferability of entertainment games into military use and fuelled the exchange between both industries:

> [I]t is apparent that the entertainment industry has emerged as an innovative source of modeling and simulation technology. [...] Though the two communities differ widely in their structures, incentives, and motivations, opportunities may exist for the entertainment industry and the defense modeling and simulation community to work together to advance the state of the art in modeling and simulation technology. (National Research Council et al. 1997, p. 13)

One of the earliest products resulting from the newly invigorated collaboration between commercial and military sectors was Full Spectrum Command (Institute for Creative Technologies 2003), a game simultaneously developed as a training simulation and as the spin-off game Full Spectrum Warrior (Institute for Creative Technologies 2004) for the Xbox. The game marked the 'first military training application published for a commercial game console' (Institute for Creative Technologies 2004). The narrative setting is an invented Middle Eastern country – 'Zekistan' in the commercial adaptation and 'Kazar' in the army version. The game was developed in close collaboration with the Army Infantry School at Fort Benning, and motion captures of real soldiers were used for the avatars. The army's participation was a big incentive for the commercial audience and contributed to large sales numbers: 'It's just real cool that you know this is made by the military' (Sam Vongillen [a player], cited in Ross and Rackmill 2005).

At the same time, the US military also conceived its first game destined for the entertainment market: America's Army (2002). To create a game on par with entertainment standards, a team of designers, programmers, and artists was recruited from video game companies, including Electronic Arts and Sony, and the game was built using the most advanced 3D engine at the time, the Unreal Engine 2 (developed by Epic Games for commercial FPS games). The game is programmed to restrict all players to the perspective of US Army soldiers, while they appear to others as the terrorist enemy. The game thus projects the army as an enlightened, heroic institution, while the enemy is here again often framed as the 'other' from the Middle East.

Although the game was initially meant for entertainment audiences to boost recruitment, it also offered the potential to extend the gaming pleasure to soldier training. The main advantage of entertainment games was thus perceived by the army as their capability to entice the soldier subconsciously into training through the immersion provided by games. This process was replicated in America's Army, which was altered into an army training simulation and was also provided free of charge to US allies:

> We want our systems as immersive as games. We want them entertaining, so that work is play and people don't leave. In short, we want our training systems so immersive, soldiers forget to eat. (Zyda et al. 2004, p. 2)

Technology provided the tools for a more realistic cartography of Middle Eastern geographies and the simulation of populations. However, in contrast, the new generation of war games steered towards vagueness and additional abstractions. Since the Middle East was not concretely labelled as the enemy in either Full Spectrum or America's Army, the fabricated vague terrorist enemies and fictional Middle Eastern-ized geographical settings generated blurred stereotypes. By avoiding a direct use of real geographical maps and concrete enemy imagery, the games thus extended the imagined geography of the Middle Eastern 'other' through a host of invented characteristics. Crucially, the emergence of these commercial military wargaming products further naturalized the idea that Western nations will face the inhabitants of the Middle East in future combats (Graham 2006, p. 128). Through the extended cooperation with the entertainment industry and the commercialization of these war games, the novel imagined cartographies created in the two games not only filtered into the soldiers' imaginations of the Middle East but also into the imaginations of Western and even global civilian audiences.

Mapping at a Distance: Increased Terrain Size and Geographical Accuracy through Extended Virtual Worlds and Remote Sensing Technologies

With rising casualties from roadside Improvised Explosive Devices in the Middle East, training simulations had to be created specifically for 'convoy training, with some important urban elements' – a feature not yet available in the limited terrains of America's Army, in which 'a soldier driving a virtual truck would drive to the edge of the virtual world in less than a minute' (Curry et al. 2016, p. 14). Larger virtual interactive battlefields of the virtual Middle East with diverse terrains were thus required, and avatars and missions needed to be modifiable, since multiple nations now shared the same software for training. Scenarios also had to be constantly customized to keep trainees on their toes. Solutions to these needs arrived with the commercial game Operation Flashpoint (2001) by Bohemia Interactive, which brought the first extensive terrains (25 km x 25 km terrain areas) and in-built model, terrain, and mission editors to the market. Bohemia Interactive adopted the entertainment game into the military version Virtual Battle Space (VBS) in 2004–2005. A second version, Virtual Battle Space 2 (2007–2008) was conceived with 200 km x 200 km terrains and became a worldwide success. The game was initially delivered to the US Marine Corps, but quickly also sold to the US Army, the British Ministry of Defence (MoD), and the Canadian Army (Bohemia Interactive Australia 2012). Many NATO countries, including France, soon followed, and the VBS series has since become the most widely used training simulation software across the world.

While the terrain size in war games was continuously expanded, the variety of atmospheres and natural, built, and human environments was also refined – including in the newest instalment Virtual Battle Space 3 in 2013 and Virtual Battle Space 4 in 2020. The software now accurately maps a variety of climates, weather and lighting conditions, environments, and ambient lives, and calculates their influence on the virtual soldiers and vehicles. Notably, real geography can now be used as the basis for these virtual environments. Terrains can be imported from a variety of terrain generation software, such as TerraTools by TerraSim, making it possible to simulate and dissect any area of the world from satellite imagery. As Bruce Sterling already critically noted a few years prior,

> what does this mean? It means that soon there will be no such thing as 'unknown territory' for the United States military. In the future – soon, very soon – the United States military will know the entire planet just like the back of its hand. It will know other countries better than those countries know themselves. (Sterling 1993, p. 49)

A detailed knowledge of geographical specificities can now be acquired by any soldier without ever having to set foot in the Middle East or make contact with the real population. Rather than research on the ground, technologies of remote sensing now serve as the basis for the simulated, virtual battle spaces. These resulting imagined geographies of the Middle East, although now anchored in reality, are notably forged from a detached perspective at a distance. As a result, the accuracy in terms of geography is heightened, while the accurate depiction and cultural understanding of the population fade into the distance. Other war games, such as the Bilateral Negotiation Trainer (BiLAT) by the Institute for Creative Technologies (2004–2008) aimed to bring in these cultural elements, but the majority of simulations prioritize geographical accuracy and environmental omniscience over human terrain mapping and cultural understanding.

The Integration of Movement and Kinaesthesia

Due to the hardware-bound video game simulators of previous decades, the soldier was limited to navigating these virtual worlds while actually sitting behind a desk and staring at a screen. Consequently, new developments have pushed in the direction of wearable gaming technologies for individual soldiers, specifically through head-mounted displays and wearable sensors so soldiers can physically enter and move in the replicated terrains. In early 2011, the US Army invested $57 million into the creation of the Dismounted Soldier Training System (DSTS), the 'first-ever, fully immersive virtual simulation training program for soldiers' (Intelligent Decisions 2011). For the project, ex-game developers were hired 'to leverage the capability of a multibillion-dollar gaming industry, basically cherry-picking the best available technology that is out there' (Colonel Francisco Espaillat, cited in Burnett 2011). The first versions of the DSTS used the entertainment graphics engine CryENGINE 3, previously employed for the FPS game Crysis 2 (Electronic Arts 2011). In the simulation system, soldiers wear high-performance computers on their backs and helmet-mounted displays and stereo headsets to be audiovisually immersed in the virtual world. Meanwhile, they are also equipped with a range of motion sensors and a dummy weapon, allowing them to move within a designated ten-foot matted training area as their movements are projected into the virtual world. Due to the limited spatial radius, a joystick is used to recreate walking and running. Although freed from the computer desk, the physical movement of the player is thus still significantly reduced to each soldier's ten-square-foot area in the gaming experience.

One company attempting to alleviate this problem is Motion Reality, with its military product VIRTSIM. Although their general approach – a stereo head-mounted display with 360-degree view – is similar, VIRTSIM innovatively integrates Motion Reality's motion capture with 3D technologies that were previously used for James Cameron's film *Avatar* (2009). During the military exercise, reflective markers are placed on the bodies of up to thirteen soldiers to track their movements and turn them into their own avatars. VIRTSIM thus enables soldiers to actually jump, run, or crawl through a basketball court-sized training hall while in the virtual world (Motion Reality 2015). A further addition is the reproduction of physical pain when struck by bullets or explosions through shock-delivering electrodes. It hence 'immerses soldiers from head-to-toe, creating the closest possible experience to an actual mission' (Miller, cited in Raytheon 2012).

The integration of the whole body allows the simulated cartographies of the Middle East to be more freely explored and hence perceived as more realistic spaces on a sensory level. The discrepancies between these authored versions and the real Middle East thus become increasingly cloaked for the military user, as the sensual immersion grows in authenticity.

However, although these virtual simulations have now reached an extremely high level of accuracy, they are yet unable to integrate and reproduce real physical movement in the simulated battlefield to a larger extent – and within touchable natural and built environments. These aspects are only currently available in live simulations in which soldiers now fully become their own avatars.

Live Simulations: The Transfer of the Visual and Virtual Mapping Heritage into Real Space

Over the years, and with ongoing conflicts in the Middle East, digital cartographies have matured and have been linked to real bodies, architectures, and vehicles through digital and mobile technologies during live simulations. These live simulations produce exercises during which soldiers execute their combat tasks inside purpose-built artificial cities. Thus, live simulations make every effort to engage the soldier in an environment that is as close to reality as possible, including the use of real equipment, recreated architecture, and role-played populations. One example of these physically reproduced Middle Eastern spaces can be found at the United States National Training Center (NTC) in the Mojave Desert in California. Here, the US Army built a range of Iraqi towns after the turn of the millennium and later adjusted these mock environments to also reflect Afghan spaces during the War in Afghanistan (2001–2014). Over thirteen mock towns and villages were constructed at the NTC over the years, and each of them was and is variably used as Iraq, Afghanistan, or other fictive Middle Eastern settings. Role-players, both from Middle Eastern and Western backgrounds, generate a simulated Middle Eastern population that interacts with the soldiers during the training scenarios.

While previous war games separated the game terrain and avatars through technology from the real environment and soldiers, the new highly instrumented live simulations, conversely, turn players into their own avatars as they manoeuvre through the space of the game. In these modern live simulations, every movement and weapon use of the soldier is now mapped, documented, and remediated through a network of technological devices during exercises. Akin to Virilio's 'logistics of perception', the military motivation behind this evolution of training mapping is the will 'to see all, to know all, at every moment, everywhere' and 'forever rule out the surprise, the accident, the irruption of the unforeseen' (Virilio 1994, p. 70). Modern military live simulation centres, one might argue, generate a twofold *visual* and *virtual* mapping of their combat geographies to monitor and optimize tactical procedures and individual combat performance. On one level, a *visual mapping* of the battlefield is produced through a range of camera types, including IP cameras and infrared cameras. The recorded footage is edited in the operational centre to produce short films for after action review. On a second level, a *virtual mapping* takes place through a digital reproduction of the training terrain, which is generated in 2D and 3D in the central operating system. The system obtains information on the battlefield action through a range of sensors distributed all over the real architecture. Players are located outdoors via GPS and indoors through radio waves and are equipped with a sensor harness that links to the laser tactical engagement simulation system of the employed weapons and vehicles. The soldiers' movements are constantly tracked in real time, while the path of their shots is registered by the harnesses and sensors mounted on buildings and vehicles. The central operating system then calculates building destructions, injuries, and deaths, which are signalled back to the soldier through a status report device – a sort of military player interface. All effects and movements are reproduced in a virtual version of the combat terrain with avatars paralleling the soldiers' actions. However, this dual visual and virtual mapping system only serves to monitor the exercises, while the constructed actual space embodies the imagined geography of the Middle East.

The modern military live simulations have thus transferred the heritage of war game technologies from board games, maps, and cyberspace into one all-encompassing real space. The imagined cartography of the Middle East is now transferred into a real space that the soldier can occupy. At the NTC, soldiers, for example, spend around eighteen days immersed in these artificial spaces during their training rotations (Tan 2015). As a result, these live simulations add realistic emotions and physical sensations to the training, such as cold temperatures, physical movements, hunger, stress, tiredness, and confusion. Thus, in a sense, they provide a more realistic emotional environment than virtual simulations and enable the transferral of wargaming into real three-dimensional, sensory-immersive mock cities, including 'real' populations. The 'play space' (Nitsche 2009, p. 16) is no longer situated in front of a computer screen and detached from the simulated Middle East, but instead, the soliders physically inhabit the fictional cartography of the Middle East, and real space and imagined spaces are merged into one.

Technological innovations have always led to new perceptions of conflicts and a distancing from the physical and optical reality of the battlefield. Communication evolutions such as the telephone allowed orders from a distance; remote weapons, from guns to bombs, enabled impersonal attacks; surveillance became effectuated remotely through the invention of planes, drones, and camera lenses; and now the Middle East can be experienced in 'laser tag'-style war games in Western countries. Terrain refinement and soldier mapping has evolved, as this chapter has demonstrated, from the 'flat worlds' (Connor) of tabletop and virtual war games to more recent live simulations, which have brought back real human bodies into the simulation of the Middle East in the form of role-players, a continuation of the war game tradition of reducing real human lives and geographical space into a simulatable, manoeuvrable imagined geography of the Middle East.

Two opposed tendencies can be found in the evolution of these imagined geographies: a significant distance from the actual reality of the Middle East persists, while the soldiers themselves lose their critical detachment from the simulation. The sensory and emotional engulfment in the imagined geographies that simulations consistently strive for obliterates the necessary distance to identify the abstractions, omissions, and compressions of the real Middle East that are performed by the simulations. The imagined geographies generated by the simulated Middle East in turn shape the expectations and imaginations of the soldiers that are then applied in future conflicts. These military mapping practices enlisted in the duty of war preparation hence demonstrate 'the extraordinary power and performative force of colonial modernity' (Gregory 2004, p. 4):

> Its constructions of other cultures – not only the way these are understood in an immediate, improvisational sense, but also the way in which more or less enduring codifications of them are produced – shape its own dispositions and deployments. (Gregory 2004, p. 4)

The simulated Middle Eastern spaces run the risk of being internalized by soldiers, especially since they often have no prior personal experience of the foreign land and culture. These engrained experiences then have the potential to turn into self-fulfilling prophecies during deployment, when soldiers encounter the real population and Middle Eastern spaces and perceive them through the experiential reading grid and imagined geography acquired through the training simulations. As Lippmann already warned, the 'picture inside' – in this case the picture created through the wargaming of the Middle East – 'so often misleads men in their dealings with the world outside' (1998, p. 30).

References

Thomas B. Allen, *War Games: Inside the Secret World of the Men Who Play at World War III* (Munich: Heinemann, 1987).

Atari, Battlezone, 1980.

—, Bradley Trainer, 1980.

Bohemia Interactive, Operation Flashpoint, 2001.

—, Virtual Battle Space, 2004–2005.

—, Virtual Battle Space 2, 2007–2008.

—, Virtual Battle Space 3, 2013.

—, Virtual Battle Space 4, 2020.

Bohemia Interactive Australia, *White Paper: VBS2, Release Version 2.0,* 2012, Online Resource: http://www.readbag.com/distribution-vbs2-docs-vbs2-whitepaper (accessed 13 November 2020).

Jim Bowery, Spasim, 1974.

Richard Burnett, 'Army Developing First Virtual-Reality Training System for Infantry in Orlando', *Orlando Sentinel,* Online Resource: http://articles.orlandosentinel.com/2011-06-09/business/os-video-training-mission-20110609_1_virtual-reality-training-gaming-technology-simulator (accessed 13 November 2020).

Roger Caillois, *Man, Play and Games* (Chicago: University of Illinois Press, 2001).

James Cameron, *Avatar,* 20th Century Fox, 2009.

Carl von Clausewitz, *On War* (Princeton: Princeton University Press, 1989).

Steven Connor, 'Flat Life', *Steven Connor,* no date given, Online Resource: http://stevenconnor.com/flat.html (accessed 13 November 2020).

Martin van Creveld, *Wargames: From Gladiators to Gigabytes* (Cambridge: Cambridge University Press, 2013).

John Curry et al., 'Commercial-Off-the-Shelf-Technology in UK Military Training', *Simulation & Gaming,* 47(1) (2016), pp. 7–30.

Electronic Arts, Crysis 2, 2011.

Stephen Graham, 'Cities and the "War on Terror"', *International Journal of Urban and Regional Research,* 30(2) (2006), pp. 255–276.

Derek Gregory, *The Colonial Present: Afghanistan, Palestine, and Iraq* (Malden, Oxford, and Carlton: Blackwell, 2004).

Ed Halter, *From Sun Tzu to XBox: War and Video Games* (New York: Thunder's Mouth Press, 2006).

Fred Hapgood, 'SIMNET', *Wired Magazine,* 5(4) (1997), p. 116.

Johan Huizinga, *Homo Ludens: A Study of the Play-Element in Culture* (London: Routledge, 2003).

id Software, Doom II, 1994.

Institute for Creative Technologies, Bilateral Negotiation Trainer (BiLAT), 2004-2008.

Institute for Creative Technologies, '*Full Spectrum Video Games*', Online Resource: https://ict.usc.edu/prototypes/full-spectrum/ (accessed 13 November 2020).

Intelligent Decisions. 'U.S. Army to Use First-Ever, Fully Immersive Virtual Simulation Training for American Soldiers', Online Resource: http://www.intelligent.net/news/press-releases/us-army-use-first-ever-fully-immersive-virtual-simulation-training-american (accessed 13 November 2020).

David Kushner, *Masters of Doom: How Two Guys Created an Empire and Transformed Pop Culture* (New York: Random House, 2004).

Walter N. Lang, 'Simulation Makes Possible... A Mobile Electronic Battlefield', *National Defense*, 11 (1991), pp. 10–13.

Christopher George Lewin, *War Games and Their History* (Stoud: Fonthill Media, 2012).

Walter Lippmann, *Public Opinion* (New Brunswick: Transaction, 1998).

Eric C. Ludvigsen, 'Combat in a Box', *Army Magazine*, 31(8) (1981), pp. 14–21.

Marine Doom and Marine Corps Modeling and Simulation Management Office (MCMSMO), 1995.

Duncan C. Miller and Jack Thorpe, 'SIMNET: The Advent of Simulator Networking', *Proceedings of the IEEE*, 83(8) (1995), pp. 1114–1123.

Motion Reality, VIRTSIM, 2010.

National Research Council, Committee on Modeling and Simulation: Opportunities for Collaboration Between the Defense and Entertainment Research Communities, Computer Science and Telecommunications Board, Comission on Physical Sciences, Mathematics, and Applications, *Modeling and Simulation – Linking Entertainment and Defense* (Washington DC: National Academy Press, 1997).

Michael Nitsche, *Video Game Spaces: Image, Play, and Structure in 3D Worlds* (Cambridge: MIT Press, 2009).

Jesse Orlansky and Jack Thorpe, *73 Easting: Lessons Learned from Desert Storm via Advanced Distributed Simulation* Technology (Alexandria: Institute for Defense Analyses, 1992).

Peter Perla, 'The Art of Wargaming – A Guide for Professionals and Hobbyists', *Peter Perla's The Art of Wargaming – A Guide for Professionals and Hobbyists*, ed. by John Curry and Peter Perla (Bristol: History of Wargaming Project, 2011).

Steven Poole, *Trigger Happy: Videogames and the Entertainment Revolution* (New York: Arcade, 2000).

Raytheon Company. 'Raytheon, Motion Reality Demonstrate Fully Immersive, Small Unit Tactical Training System at US Army's Bold Quest', Online Resource: http://investor.raytheon.com/phoenix.zhtml?c=84193&p=irol-newsArticle&ID=1749291 (accessed 13 November 2020).

George Leopold von Reiswitz, *Taktisches Kriegs-Spiel oder Anleitung zu einer-mechanischen Vorrichtung um taktische Manoeuvres sinnlich darzustellen* (Berlin: Gadicke, 1812).

—, Kriegsspiel, 1812.
Birgit Richard, 'Norn Attacks and Marine Doom', *Ars Electronica: Facing the Future: A Survey of Two Decades*, ed. by Timothy Druckrey (Cambridge: MIT Press, 1999), pp. 153–163.
Brian Ross and Jill Rackmill, 'Full Spectrum Failure? Army Video Game Draws Fire', *ABC News*, Online Resource: http://abcnews.go.com/WNT/Investigation/story?id=787575&page=1 (accessed 13 November 2020).
Edward W. Said, *Orientalism* (New York and London: Penguin, 1978).
Bruce J. Schachter, *Computer Image Generation* (Hoboken: Wiley, 1983).
Roger Dean Smith, *Military Simulation & Serious Games: Where we Came from and Where we Are Going* (Orlando: Modelbenders Press, 2009).
Bruce Sterling, 'War Is Virtual Hell', *Wired Magazine*, 1(1) (1993), pp. 46–51.
Michelle Tan, 'Combat Training Rotations Will Increase to 18 Days', *Army Times*, Online Resource: http://www.armytimes.com/news/your-army/2015/02/06/combat-training-rotations-will-increase-to-18-days (accessed 13 November 2020).
TerraSim, TerraTools, c.1996–now.
Greg Thompson et al., Maze Game, 1974.
Jack A. Thorpe, *The New Technology of Large Scale Simulator Networking: Implications for Mastering the Art of Warfighting*, Defense Advanced Research Projects Agency, 1987, Online Resource: https://apps.dtic.mil/dtic/tr/fulltext/u2/a278600.pdf (accessed 13 November 2020).
Sun Tzu, Wei Hai, c. 500 B.C.
Unknown, Chaturanga, c. 500.
Unknown, Chess, c. 500.
Unknown, Go, c. 500 B.C.
US Army, *America's Army*, MOVES Institute, 2002.
US Army and Intelligent Decisions, Inc., *Dismounted Soldier Training System (DSTS)*, 2011.
USC Institute for Creative Technologies and Quicksilver Software, *Full Spectrum Command*, 2003.
USC Institute for Creative Technologies and Pandemic Studios, *Full Spectrum Warrior*, 2004.
Paul Virilio, *The Vision Machine* (London: British Film Institute, 1994).
Linda D. Voss, *A Revolution in Simulation: Distributed Interaction in the '90s and Beyond* (Arlingtion: Pasha Publications, 1993).
Herbert George Wells, *Little Wars – A Game for Boys from Twelve Years of Age to One Hundred and Fifty and for That More Intelligent Sort of Girls Who Like Boys' Games and Books* (London: Palmer, 1913).
Michael Zyda et al., 'From Viz-Sim to VR to Games: How We Built a Hit Game-Based Simulation', *Organizational Simulation: From Modeling & Simulation to Games*

& Entertainment, ed. by William B. Rouse and Kenneth R. Boff (Hoboken: Wiley, 2005), pp. 553–590.

About the Author

Janina Schupp is an affiliated lecturer in the Department of Architecture and the Centre for Film and Screen at the University of Cambridge. She holds a PhD in Architecture (Cantab) and works as a documentary film producer for ZDF German Television. Her research focuses on representations of urban spaces and the home in screen media. She held fellowships at the Library of Congress in Washington, DC in 2015-2016 and the Camargo Foundation in France in 2017. In 2019, she was a visiting scholar at Nanjing University, China. Her recent publications include the edited volume *The Everyday in Visual Culture* (2021).

Part II

Movements

5. Iranian Internet Cinema, a Cinema of Embodied Protest: Imperfect, Amateur, Small, Unauthorized, Global

Hamid Naficy

Abstract
This chapter examines the emergence of what I call 'internet cinema' by focusing on Iran in the late 2000s, particularly since the 2009 disputed re-election of Mahmoud Ahmadinejad as president, which unleashed the opposition Green Movement undergirded by this cinema. Iran provides an example of the use of the new internet-driven digital global media in support of democratic ideals that, soon after the Green Movement, rocked many authoritarian regimes of the Middle East and North Africa.

Keywords: Internet cinema, Iran, Green Movement, amateur

The Emergence of a Citizen's Medium

With the success of the Iranian Revolution in 1979, the Islamic Republic took over the broadcast media and placed it under direct control of the 'supreme leader'. Having gotten rid of one authoritarian regime with a monopolistic state-run broadcast media under the Shah, Iranians did not take this lying down. Throughout the 1980s and 1990s, they engaged in a creative cat-and-mouse game with the state to create thriving black markets for alternative media, particularly for video – by means of a mushrooming video club scene – and satellite television – by means of clandestine receivers at home and Persian extraterritorial satellite stations (Naficy 2012a, pp. 339–348). However, they did not stop there. In the 2000s, Iranians developed a third front in the struggle to create an alternative citizen-driven public medium.

Strohmaier, A. and A. Krewani (eds.), *Media and Mapping Practices in the Middle East and North Africa: Producing Space*. Taylor & Francis Group, 2021
DOI 10.5117/9789462989092_CH05

This was emblematized by the emergence of the internet cinema with which Iranians worldwide challenged the Islamic regime's broadcasting monopoly and mono-vocalism during the presidency of Mahmoud Ahmadinejad, particularly after his disputed re-election (Naficy 2012b, pp. 348–362).

This was a new, legitimate, artistic, and expressive form, regardless of its political uses. Iranian cosmopolitanism, the financial wealth of the country, and the widespread penetration of the internet and its various modalities of connectivity and interactivity drove the emergence of the internet cinema. In 2005, CIA officers at the Open Source Center trolling the internet discovered that Persian/Farsi was among the top five languages in the blogosphere, offering valuable textual and audiovisual information and insight about Iranian people and their sentiments (Shane 2005). In 2008, a Harvard University study further showed that the Persian blogosphere is 'a large discussion space of approximately 60,000 routinely updated blogs featuring a rich and varied mix of bloggers' that ranges from President Mahmoud Ahmadinejad to students (Kelly and Etling 2008, p. 2). Still later studies showed that Iran has 'the most internet users in the Middle East, approaching 30 million' and that Persian is reported as the 'fourth most popular Blogging language on the Internet' (Shahab and Mousoli 2010).

However, as Sreberny and Khiabany rightly note, these figures are ambiguous, as it is not clear how the number of bloggers is calculated and what constitutes an Iranian blog. If the physical location of bloggers is the criteria (counting only those inside Iran), the calculation would ignore the large number of bloggers in the diaspora; if the language of blogging is considered (Persian/Farsi), then all those Iranians, dual national Iranians, and non-Iranians who write in other languages are ignored (2010, pp. 35–36). However they are counted, Iranian bloggers formed a formidable presence in the blogosphere in particular and in cyberspace in general. Because of this presence and because of the strangling of other forms of journalism, the internet became a vital source of information and activism in the 2000s for all sides and a highly contested public diplomacy sphere. The 25-minute film *Iran: The Cyber-Dissidents* (2006), produced by Vivien Altman and reported by Mark Corcoran for Australian Broadcasting Corporation, deals with this burgeoning phenomenon.

What I call 'internet cinema'[1] consists of the totality of the digital media connected by the internet – blogs, films, social networks, and the sharing and streaming of videos – which have turned the internet's virtual space into a vast online public sphere that, in turn, encourages discursive formations and

1 This chapter is a modified version of a section from Nafcy 2012a, pp. 348–362.

political activism in society, such as the Green Movement, which demand government accountability and democracy. The internet thus became social, not only in its virtuality but also in its actuality. The streets, in turn, became virtual, both in their powerful representations on the internet and in their power to represent. This movement posed the biggest challenge in the lifespan of the Islamic Republic of Iran, not only to its legitimacy but also temporarily to its existence.

Mainstream Western media and exiled Iranian media lauded the power of this new virtual and discursive space ad infinitum, which prematurely and erroneously dubbed the new protests a 'Twitter Revolution', ignoring the fact that technology and media by themselves do not make a revolution (Acuff 2010). In fact, Twitter played a relatively small role at the time inside Iran, partly because of government censorship, but it proved more effective outside the country in publicizing the events (Esfandiari 2010). While Twitter and Facebook were used admirably to exchange information, file news reports and images, and organize protests domestically and internationally (Hashem and Najjar 2010), the social space and the physical place remained high on the agenda of the protesters, emblematized by a street placard carrying the following slogan, addressed to the regime: 'We'll give you back the web sites and mobile phones, but we won't give you the country' (*'Seit O' mobeil arzunitun, keshvar nemidim behetun'*), referring, on the one hand, to the power of the government to shut down the internet and mobile phones and, on the other hand, to the power of citizens to withhold support and to fight back.

The negative consequences of the shift in US and Iranian public diplomacies, from gathering secret information from enemy countries through espionage to collecting information from public sources in those countries, such as through the internet, for mobilizing soft power, was that it politicized, even militarized, all public spheres, including the streets and the internet. Both the people and the government in Iran became camera-shy in public places in major cities. If someone took out a camera to take a picture in the streets, both government agents and passers-by harassed the person – one fearing that the image would be uploaded to social networking sites or to oppositional sites, feeding the gathering anti-regime dissent; the other fearing government surveillance and future arrests. As a result, some of the early protest videos were filmed almost clandestinely, from private places like balconies and from inside apartments and offices, where the filmers were not exposed. Nevertheless, the government attempted to remove, impede, or block these public sources of information and to surveil, track down, imprison, and severely punish the internet operators and users it considered to be aiding the dissidents. Ironically, the US sanctions against the Islamic Republic of Iran, too, initially contributed to the suppression of the digital uprising of Iranians. Although the sanctions law, dating from the Clinton presidency, prohibited Americans from exporting goods and services to Iran, it allowed certain exceptions, among them 'information and informational materials' (Parsi, Elliot, and Disney 2010, p. 164). The problem was that internet technology, or any technology developed in the 1990s, was apparently not covered under this exception, with the result that companies such as Microsoft and Google denied their instant messaging to Iranians (MSN Messenger and Google Talk) because these depended on user downloads, which were interpreted as constituting not information but prohibited service. Even Twitter's legal status came under question until the Obama administration removed the doubt, when in the aftermath of the 2009 elections, it asked Twitter to 'forego routine maintenance in order to continue providing uninterrupted service to Iranians' (Parsi, Elliot, and Disney 2010, p. 165). Six months later, the Iranian Digital Empowerment Act, spearheaded by Representative Jim Moran, which 'authorized downloads of free mass market software by companies such as Microsoft and Google to Iran necessary for exchange of personal communications and/or sharing of information over the internet such as instant messaging, chat and email, and social networking' (Parsi, Elliot, and Disney 2010, p. 166), was passed by the US Congress.

The regime's impediments and censorship efforts did not deter the opposition, either; instead, it encouraged a creative but sometimes deadly and violent game of cat and mouse – both on the internet and in the streets. Protesters found ever more ingenuous alternatives to stay in touch, reconnect, and coordinate actions, and they sought new strategies, internal and international, to create an alternative non-governmental mediascape to publicize their own activities and grievances, organize their protests, and pressure and punish the regime. Many resorted to proxies, which redirected them to banned sites, or used anonymizers, which concealed the identities of senders and recipients. Foremost among these anti-censorship software tools were Tor, Psiphon, Mixminion, Incognito, Freegate, and the purportedly more ingenuous programme Haystack, designed especially for Iran by American hacktivist Austin Heap (Dobson 2010). Many, too, took to the streets to agitate and to record events. One placard in a street demonstration around the time of the disputed 2009 election carried the pithy slogan, 'My [mobile] phone, my medium' ('*Telefon-e man, resaneh-ye man*'), which summarized the defiance of the protesters against the state and its centralized 'big' media, the broadcast media. Amplifying that message, another said 'Every Iranian, a historiographer' ('*Har Irani tarikhnegar*'). A cartoon echoed these sentiments by depicting an Iranian woman wearing a crown of light and posing like the US Statue of Liberty, but instead of holding a lit torch, she holds up a mobile phone emanating waves.

This was the moment for the efflorescence of a new 'little medium', the internet cinema, with its simple, ubiquitous equipment – a mobile phone or a consumer model digital camera or recorder – to replace the formerly powerful little medium of analogue audio cassette, which Ayatollah Khomeini had employed so effectively to energize the Islamic Revolution in the late 1970s (Naficy 2011). This was a new era, necessitating a new medium, for a new cause. Indeed, each protester became a digital medium and a historiographer by taking to the streets, digital camera, mobile phone, or recorder in hand, defying the government's threats of force and terror and its monopoly on the big media. Oppositional film-making against the Islamic Republic of Iran took varied forms: documentaries, fiction films, agitprop films, animated films, internet films, and music videos. Here, I discuss the latter two.

Production, Distribution, Exhibition, Reception

Emboldened by their initial successes in publicizing the protests, internet cinema filmers began to record their videos from the turmoil in the streets, at ground level, instead of from the safety of the high-rise buildings and from behind windows and curtains as before. With this shift from the private to the public space, and from a distant view to a close-up view of the subjects, came other fundamental political and aesthetic changes in what was recorded. Social barriers and divisions, such as those separating genders, crumbled. Women therefore became such a strong and defiant presence both in the streets and on internet cinema's community videos. Hamed Yusefi's 25-minute film, *The Aesthetics of Political Protest in Iran* (*Zibashenasi-ye E'terazat-e Siasi dar Iran*), aired by BBC Persian on 22 July 2010, provides insightful analysis of these points.

In addition, instead of recording from the point of view of an outsider observing news events, community videographers began to record from the point of view of insiders, engaged in action. Their recording mode thus shifted from direct cinema's fly-on-the-wall observational filming of outside events to cinéma-vérité's provocation-cum-recording of pro-filmic events (Nichols 2001). As such, theirs was not a simple recording of protests that someone else had organized, but an embodied form of protest in whose organization they themselves participated, with all the subjectivity, ambiguity, and hapticity – roughness, shakiness, and out of focus and chaotic aiming and muffled sounds – that these entailed. The result was unedited raw footage of affect, not edited films of a polished presentation. These proved to be very powerful when uploaded onto the video sharing and social networking sites, bypassing government control and censorship.

Unlike traditional films, which are highly planned affairs and involve industrial procedures, high-tech equipment, and numerous skilled personnel both in front and behind the camera, led by a hierarchical command structure, internet films are often a result of individual amateur filmers from around the world sharing their spontaneously shot scenes on video sharing and social media sites. These unplanned scenes become films not by the action of these filmers, but by the actions of others who may utilize the scenes filmed by disparate and unrelated filmers in constructing other longer films of their own or by the actions of the viewers as they surf the net, visiting this or that social media platform. If the first kind of film is a material, 'produced' film consisting of scenes contributed by multiple filmers, the second kind of film is an immaterial 'received' film created by each viewer in the act of receiving or spectating multiple unrelated film scenes. The result is not a single, authoritative film in either case, as each film is the result of the actions of one person, which, at least in the case of the received film, will be entirely individual, personal, immaterial, and immanent to the spectating subject.

However, the affective power of these internet films is often diluted when Western television channels broadcast them, for they often suppress the original Persian language soundtrack, replacing it with their own reporters' voice-over narrations or commentaries. To be sure, the video of Neda Agha-Soltan's on-camera death as broadcast by Western news media was very powerful, but this power came from its tragic content, timing, brief length (40 seconds), and the favourable semiotics of Agha-Soltan's youth, beauty, fair skin, and Westernized veil and attire, which made her a subject with whom Westerners could easily identify (Sabety 2010; Afshar 2010). Its power also came from the video's powerful visuality, the way it clearly witnessed the last eye contact of the dying young woman with the camera, as life departed from her body. However, if viewers saw the original raw footage of her on the internet, without the foreign reporters' voice-over intervention, they would hear the voices of the bystanders – all male – gathered around her trying to console her, communicate with each other, and save her, as their voices rise into desperate cries, making the video much more compelling in its emotional and visceral impact. Here is my transcription of the bystanders' utterances, some of which are not sufficiently audible:

– 'Let's get someone to carry her away...' [to a hospital?]
– 'Neda! Watch her eyes, watch her eyes.'
– 'Her eyes are turning in.'
– 'Neda, don't be afraid, don't be afraid.'
– 'Oh, oh, oh, OH!' [voices rise to desperate shouts]
– 'Press on her, press on her.' [hands press on her chest, presumably the site of her gunshot wound]
– 'Neda, stay, stay, stay, STAY!' [man's voice rises to a frantic and horrifying shout]
– 'Open her mouth, open her mouth.'[2]

This sort of affective internet cinema augured not only entirely new relations with its social subjects, such as Neda as described above, but also new relations of production, distribution, exhibition, and reception with its viewers. While the site of production of the internet cinema in its initial phases was primarily public places – either in the streets or underground – the site of reception was chiefly private places – homes and other private or semi-private locations. Production – filming – was ad hoc, spontaneous, amateurish, and without official permission, bypassing the strict government approval and censorship apparatuses for public filming. As such, internet cinema is inherently an unauthorized, underground cinema. Its technology of recording is no longer optical and chemical, involving celuloid film or videotape, but is entirely electronic and digital. This technological shift facilitates the globalized production, distribution, and reception of the internet cinema videos.

The production of this cinema occurred in two distinct phases. In the first phase, people with digital cameras recorded the events without editing and uploaded them to internet sites, hence my designation of these practitioners as filmers or videographers, not film-makers. In the case of Agha-Soltan's video, an anonymous person recorded the incident on a mobile phone. In a growing practice, a second phase evolved, during which others, who were not

2 To see the video, go to: *Neda Agha-Soltan Shot in Iran*, http://www.dailymotion.com/video/x9oltq_neda-agha-soltan-shot-in-iran_news.

necessarily involved in the first phase, compiled, aggregated, repurposed, edited, and uploaded these internet videos to create other fictional or nonfictional videos. This secondary phase turns multiple individual scenes recorded by disparate filmers into a single film.

Distribution of internet cinema videos is simultaneously both individualized and globalized. The digitized recording of events and their uploading onto social networking, video sharing, and video streaming sites as well as their publication by global news media outlets facilitates this globalized individuality or individualized globalization. In this way, ordinary videographers become not only extraordinary producers but also extraordinary distributors, who bypass the traditional commercial or governmental distribution systems with their complicated monopolistic techno-political economies. In the case of Agha-Soltan's video, the anonymous filmer forwarded the video to the British newspaper *The Guardian*, the network Voice of America, and five other individuals. Significantly, according to *The Guardian*, one of these latter individuals – not the filmer – uploaded the video onto Facebook, showing the two-step production of internet films. From that one posting, copies spread to YouTube and were broadcast within hours by CNN. This video was then used in, or served to inspire, numerous types of internet films – fiction, documentary, music video, and animated films. It is thus that this single video went viral and became simultaneously an icon, index, and symbol of the post-election protest and of the democratic aspirations of Iranians – the Green Movement – fulfilling all three definitions of the sign in Charles Sanders Peirce's semiology.

The exhibition venue of the internet cinema vastly differed from other forms of cinema. It shifted from the stationary brick-and-mortar cinemas where people watch films collectively according to a schedule, to small and mobile handheld devices, on which individuals with internet access can watch films at any time, anywhere in the world. This meant that internet cinema's exhibition was freed from both physical location and physical structure, becoming global and virtual. In short, in all its phases of production, distribution, exhibition, and reception, internet cinema moved from professional and formal to amateur and personal forms, marking the ultimate triumph of modernity's individuality in Iranian cinematic expression.

Because of their highly individualized production; their spontaneous, embodied, and amateur aesthetics; their small-scale, short length, and unedited raw footage; and their mobility, first-phase internet videos cannot satisfactorily be labelled as cinema. However, I am applying that term to them because they are a new form of audiovisual expression on their own, and they spawn, in their second phase, other longer, more robust, and professional works, which are exhibited both on the internet and in the cinemas. The internet cinema thus added another type to the typology of cinemas that emerged in the Islamic Republic of Iran. Whatever the most accurate terminology, the fact remains that instead of being dismissed by the world on the grounds of their amateur, improvised aesthetics and brief bursts of raw footage of affect, internet videos gained an added value precisely because of those characteristics of imperfection, which were reminiscent of the Third Cinema's aesthetics of gritty imperfection that in the 1960s and 1970s, had countered both the reigning oppressive political regimes in the world and mainstream cinema's oppressive aesthetic regime of polish and perfection. In fact, the aesthetics of imperfection, smallness, and embodiment of these videos authenticated them fully, as intimate and defiant documents of their filmers' presence on the ground, documenting, provoking, and protesting against a seemingly hegemonic and intractable state and its media.

International recognition of Iranian internet cinema of protest was swift, not only in the reiteration and transmedial dissemination of its videos but also in the awards received from film festivals and media and journalism associations – another reason to call them internet cinema. For example, in 2009 Long Island University bestowed its prestigious journalism award, the George Polk Award for Videography, to the video of Neda Agha-Soltan in recognition of 'the efforts of the people responsible for recording the death of 26-year-old Neda Agha-Soltan at a June protest in Tehran, Iran, and uploading the video to the Internet'.[3]

Internet films, such as Agha-Soltan's video, soon became the contents of other protest films and videos made in different forms – fiction, documentary, animated, and music videos – and some were used, reused, remade, remixed, homaged, signified upon, and repurposed so many times and so transmedially and rhizomatically in a meandering global chain reaction that they became viral and infectious, even though virality has usually been associated with comedy (Gurney 2011).[4] One of these is *The Green Wave* (2010), which premiered internationally at the International Documentary Film Festival Amsterdam in 2011. Directed by Iranian-born German filmmaker Ali Samadi Ahadi, the film, according to its press book, is a 'touching documentary-collage' that illustrates the dramatic 2009 presidential election

3 See: https://liu.edu/polk-awards/past-winners#2009

4 *Neda*, an animated version, offers a historical background of the incident: http://www.youtube.com/watch?v=wXN_yCSbUYk&feature=related.

protests and expresses the feelings of the people involved in the Green Movement. Emblematic of internet cinema,

> Facebook reports, Twitter messages and videos posted in the internet are included in the film composition, and hundreds of real blog entries serve as reference for the experiences and thoughts of two young students, whose story is running through the film as the main thread. The film describes their initial hope and curiosity, their desperate fear, and the courage to yet continue to fight.[5]

These fictional storylines were animated using motion comic technique, and augmented by interviews with prominent human rights campaigners and exiled Iranians, such as Shirin Ebadi, Mohsen Kadivar, Payam Akhavan, Mehdi Mohseni, and Mitra Khalatbari.

These extended internet cinema films, too, received international praise and prizes. For example, in November 2010 the International Documentary Film Festival Amsterdam recognized the importance of these internet films when it gave its Unlimited Award to *The Silent Majority Speaks* (2010) by The Silent Collective, a 93-minute film anthology consisting of fourteen films made on mobile phones and similar devices by anonymous Iranians about the post-election protests and the state violence against protestors. The 5000-euro cash award given to the film was an initiative of Hivos Cultural Fund, one of the NGOs that the Iranian intelligence ministry had identified as supporting a velvet revolution against the regime. The award was given to an Iranian women's rights activist, Mahboubeh Abbasgholizade, condemned in absentia in Iran to two-and-a-half years of prison and 30 lashes, who dedicated her film to the anonymous citizen filmers.[6] In addition, in November 2010, the Foreign Press Association named Iranian journalist Saeed Kamali Dehghan, who regularly writes for *The Guardian*, 'journalist of the year' for his coverage of the disputed presidential election and gave the top award for best TV Feature/Documentary film to the HBO film *For Neda*, which Anthony Thomas had co-directed with Kamali Dehghan (Batty 2010). All these are examples of the internet cinema in its various modalities.

5 *The Green Wave* press book: http://www.thegreenwave-film.com.

6 'Jayezeh-ye Sinemai Bara-ye "Filmsazan-e Nashenas-e" Iran', http://zamaaneh.com/news/2010/11/post_14948.html.

Hailing/Counterhailing

Music videos became one of the most powerful forms of the internet cinema, driven by the global popularity of the music video form, the emergence of dynamic Iranian rock music groups inside Iran and in the diaspora, and the continuing disenchantment of Iranians with their regime. These music videos became effective means of counterhailing the state and its repressive apparatuses. Blurred Vision, a Toronto-based rock band formed in 2007 by brothers Sepp and Sohl, who fled Iran with their family in 1986 (and do not reveal their last name to safeguard family members' security in Iran), produced one of the more powerful music videos supporting the protesters of Ahmadinejad's re-election. This video became very popular if not viral. On 30 January 2010, the band remade the famous Pink Floyd song 'Another Brick in the Wall', popular in Iran, into a clear-eyed new protest anthem, 'Another Brick in the Wall (Hey, Ayatollah, Leave Those Kids Alone!)'. In it, they intercut footage of their purported clandestine performance of the song inside Iran, filmed by a fan on a mobile phone, with those of street clashes filmed by anonymous filmers, and an overbearing Ayatollah character who orders the security police to shut down the performance. Within a short time, the music video had been played an impressive 383,394 times on YouTube; however, the brothers were more impressed by the responses from inside Iran.[7] Sepp explained:

> We get a lot of emails, especially from the younger guys, and I remember we were in London for a film festival where we were there to receive an award for best video and Sohl was translating an email into English. And as he was translating he started crying. The email said, 'It is you guys out there that can keep this going for us, that can keep our voice alive. We're here sort of isolated from the rest of the world, we've been shut down and shut off from the rest of the world and all we can say is just keep our voice alive, keep going to allow us to reach this point of freedom. (quoted in Macedo 2010)

That YouTube was often shut down by the Islamic government made such a response all the more significant, since users had to go through extra steps to get around censorship (without censorship, the hits the video received would have been much higher). Not all the comments were positive,

7 Blurred Vision's *Another Brick in the Wall*'s music video is at: http://www.dailymotion.com/video/xcp5ua_blurred-vision-another-brick-in-the_music.

however, as some people accused the band of being involved in American soft-power public diplomacy against Iran, of being 'backed by the CIA and the Pentagon and making a fortune off the U.S. government', a charge the band denied, stating that it donates most of the song's download proceeds to Amnesty International. 'It's an ethos of the band that awareness can change the world and music is our tool and platform to do that,' said Sepp, 'and in my opinion it's truly working because the dialogue has begun.' (quoted in Macedo 2010)

'Hey Executioner, Get Lost!' ('*Hey Jallad, Gom Kon Gureto!*') (2011) is another music video on the protest movement that tapped into Pink Floyd's 'Another Brick in the Wall' song and quoted a line from Blurred Vision's music video version of it, except 'Hey Executioner' is more incendiary than its predecessor. It identifies President Ahmadinejad and Supreme Leader Khamenei as the executioners in its title, urging them to leave a country that is no longer theirs. The video clips of the street demonstrations that accompany the lyrics include images of orderly protests, which are disrupted by police violence, causing the angry demonstrators to turn on the uniformed and plain-clothes police officers, whom they beat up mercilessly and whose van they turn over. The video's message is no longer protest but revolt, summarized in the lyrics, 'We are full of anger, like the typhoon/ We are engaged in a relentless battle/ Listen to me my enemy, there is no place for you in my homeland/ Hey executioner, get lost'.[8]

Yet another effective counterhailing song, which spawned music videos on YouTube, was 'Thorn and Riffraff' ('*Khas O' Khashak*'), which Iranians produced to counter what Ahmadinejad had once labelled the thousands of demonstrators to his re-election, as 'a few thorns and riffraff'. The song turns Ahmadinejad's own words against him and his regime, addressing them in its refrain: You are the thorn and riffraff/ You are lower than dirt/ I am the aching lover/ Ablaze, bright, and full of fervor.[9]

Clearly, these music videos and many others like them, which incorporate internet cinema footage, are examples of citizens withdrawing their support from the regime, and they constitute one of the most effective peaceful means of combatting intolerant and undemocratic regimes. However, these videos took one step further from those that did not feature music video footage. All of them reversed the interpellation that Althusser describes in

8 *Hey Executioner, Get Lost* music video is at: http://www.youtube.com/watch?v=wR77TN1_d8k&feature=player_embedded#at=60.

9 For *Khas o Khashak*, go to: https://www.youtube.com/watch?v=xtLp2VMYF50.

his famous illustration of a police officer calling a citizen: 'Hey! You, there' (Althusser and Balibar 1971). By turning to respond to the police officer call in that example, the passer-by is hailed and becomes a subject of the state. Instead of being hailed by the state and its agents ('Hey! You, there'), in these videos it is the protesters and their global sympathizers and collaborators who counterhail the state, not only in the video's titles but also in their frequent shouted refrains addressed to the regime: 'Hey, Ayatollah, Leave Those Kids Alone!', 'Hey Executioner, Get Lost!', and 'You are the thorn and riffraff / You are lower than dirt'. It is in this power not only to withdraw consent but also to counterhail the state, of speaking truth to power, that internet cinema finds its political fulfilment.

The Iranian government moved to stifle this new form of cinema, both in its production and dissemination. It began arresting virtual and social activists and videographers, and it arrested an increasing number of bloggers, whose labour complemented the internet cinema, sentencing them to jail terms ranging from a few months to many years. In 2004 it arrested some 20 bloggers and internet journalists (Tehrani 2008), and between 2000 and 2006 Iran became the top censor of the internet in the Middle East (El Gody 2007, p. 223), thwarting the democratic potential of the internet as well as of Iranians. Within a few years, Reporters Without Borders was able to label the Islamic Republic of Iran, along with thirteen other countries, as 'enemies of the Internet' (Shahab and Mousoli 2010). In September 2008, prominent Iranian Canadian blogger, Hosain Derakhshan (nicknamed blogfather, aka Hoder.com), was arrested inside Iran on a variety of charges, among them, collaborating with hostile foreign powers, including Israel. Derakhshan had been the editor of the Internet and Film section of *Doniya-ye Tasvir* (*World of Pictures*) magazine in Tehran and after emigrating to Canada, had introduced 'a simple but groundbreaking' way to show Persian letters and characters on the internet, which contributed to the prominence of Iranians in the blogosphere. This charge of spying was apparently due to a well-publicized trip of his in 2006 to Israel (on his Canadian passport) whose aims were to show 'his 20,000 daily Iranian readers what Israel really looks like and how people live there', and to 'humanize' Iranians for Israelis (Theodoulou 2008). A controversial figure among the blogging community, he had toned down his anti-regime blogging and began supporting it, and he returned to Iran from his self-imposed exile, hoping to work inside the country. He was jailed for nearly two years before finally being sentenced to an incredible nineteen and a half years of imprisonment on charges of 'conspiring with hostile governments, spreading propaganda against the Islamic system, spreading propaganda in favour of counterrevolutionary groups, blasphemy, and creating and managing obscene Web sites' (Mackey 2010).

'Like journalists, bloggers have been treated for months as if they are enemies of the regime', Reporters Without Borders said. 'But the authorities have now started to impose much harsher sentences on them. Bloggers involved in censorship circumvention are being particularly targeted as they help their fellow citizens to gain access to banned information.'[10] Internet censorship took many forms, not only imprisonment but also more subtle and structural forms, such as censorship laws and regulations, content filtering, tapping and surveillance, infrastructural control, telecom control, and self-censorship (El Gody 2007, p. 231). Imprisonment, however, remained a common form of censorship. The Committee to Protect Journalists reported that by February 2010, there were 52 journalists in Iranian jails, a record high that accounted for a third of all the journalists imprisoned in the world.[11] A month later, Reporters Without Borders noted that

> with some sixty journalists and bloggers behind bars and another 50 forced to seek asylum elsewhere, the Islamic Republic of Iran has become the largest prison in the Middle East – and one of the world's largest prisons – for journalists and netizens.[12]

Undeterred, some Iranian activists came up with another tactic: they took legal action against the Islamic Republic's interests from outside Iran. In August 2010, jailed and tortured prominent journalist Isa Saharkhiz and his son, Mehdi Saharkhiz, filed a lawsuit in the US Federal Court in Alexandria, Virginia, against Nokia Siemens Networks and its parent companies, Siemens AG and Nokia Inc., alleging their complicity in the Iranian government's human rights violations through the spying centres that the companies had supplied. The journalist claimed that such centres had conducted surveillance, eavesdropping, and tracking of his mobile phone and other communications after the 2009 elections, resulting in his incarceration and torture. The plaintiffs demanded that Nokia Siemens Networks cease its support of the Iranian intercepting centres and use its connections within the Iranian regime to secure Saharkhiz's freedom. They also called on the United States's judicial system to hold accountable

10 https://rsf.org/en/news/one-blogger-freed-bail-another-blogger-denied-parole.

11 http://cpj.org/2010/03/with-52-journalists-in-jail-iran-hits-new-shamefu.php.

12 http://en.rsf.org/spip.php?page=article&id_article=36684.

business practices such as the ones engaged in by Nokia Siemens Networks in Iran.[13]

In another tactic, International Campaign for Human Rights in Iran called for the removal of Ezzatollah Zarghami, Director General of Voice and Vision of the Islamic Republic (VVIR, aka IRIB, Islamic Republic of Iran Broadcasting), the state's national broadcast monopoly, on the charge that in addition to its close cooperation with intelligence agents and interrogators, the VVIR, under Zarghami's direction, had 'systematically produced and broadcast programs aimed to target well-known personalities through attributing undue, libellous, and untrue matters to them' (Zarghami 2010). This was a reference to the latest iteration of *Identity* and *Armageddon* shows, a new series called *Sedition Documents* that purported to offer documentaries about reformists such as Ataollah Mohajerani, Abdolkarim Sorush, Akbar Ganji, Shirin Ebadi, Fatemeh Haqiqatju, and Mohsen Sazegara, many in diaspora, aired on main VVIR channels during prime time. Having no access to them to interrogate and force them into incriminating show trials, these state-sanctioned documentaries smeared these individuals by digging up dirt in their private lives and by falsely presenting them as immoral and hostile to Islam and to religion (Siamand 2010). The campaign further demanded that the Iranian parliament and judiciary launch an independent inquiry into VVIR's violations of the constitution and citizens' rights on behalf of the defamed individuals who could not defend themselves in the court of public opinion because of the government monopoly of all broadcast media.[14] Assessing the success or failure of these extraterritorial legal interventions in impeding Iranian government's illegal attacks on its own citizens requires further research.

Public Diplomacy and Soft-Power Politics

The Iranian regime took a page from the US government's soft-power public diplomacy rule book, went global, and began countering the US government's funding of anti-Islamic Republic and pro-democracy NGOs inside and outside Iran. This included supporting and funding NGOs in the West that

13 See: http://onlymehdi.posterous.com/tortured-prominent-iranian-journalist-sues-no.

14 'Iranian State TV Acts as an Arm of the Intelligence Apparatus', http://www.iranhumanrights.org/2010/08/iranian-state-tv-acts-as-an-arm-of-the-intelligence-apparatus/print/.

engaged in cultural programming that favoured the Islamic Republic.[15] It also began a wide-ranging 'campaign of harassing and intimidating members of its diaspora worldwide – not just prominent dissidents – who criticize the regime' (Bandow 2009). This unprecedented action consisted not only of slowing down the internet speed, blocking social networking sites (Facebook, Twitter), video-sharing sites (YouTube), internet telephone services (Skype), and cutting email service inside Iran, but also of tracking the activities of Iranians on networking sites worldwide, creating fake sites for the protesters to rope in more victims, monitoring Iranian protesters in diaspora (900 were tracked in Germany), videotaping their public demonstrations in order to harass them and their families at home, and sending them anonymous threatening emails to cease and desist from 'spreading lies and insults' (Fassihi 2009). All these measures may have been facilitated and enhanced by the Islamic Revolutionary Guard Corps' takeover of the telecom infrastructure.

It must be noted that the situation governing the internet in Iran is very complicated and conflictual not only within the opposing factions – pro- and anti-government – but also within each faction. For example, within the ruling faction, one faction bans Twitter while others turn to it as their favourite medium of communication with their cadres and to the nation. On 15 August 2018, Mohammad Jafar Montazeri, the attorney general, condemned the actions of some of the lawmakers to lift the ban on Twitter, calling it a 'highway to corruption' whose control is in the hands of 'the enemy'. His deputy accused Twitter of having had a major hand in the 'seditious' Green Movement, which led to its banning soon thereafter. Despite that ban, not only ordinary citizens and social activists but also high-ranking officials such as Abbas Ali Kadkhodai, the spokesperson for the Guardian Council, and even the supreme leader himself, Ali Khamenei, maintain their Twitter accounts and actively communicate with them.[16]

15 One NGO that purportedly received such help was the Center for Iranian Studies in Toronto; see: 'Is Toronto Cultural Centre Funded by Iran's Mullah Regime?', http://www.onlydemocracy4iran.com/2010/04/15/is-toronto-cultural-centre-funded-by-irans-mullah-regime/. National Iranian American Council (NIAC) has also been accused of promoting Islamic Republic causes by some exiles, even though it has received funding from the US government and other American foundations, which would militate against such cross-funding. NIAC denies any affiliations with the Iranian regime. See: 'EDITORIAL: Voice of the mullahs: Public diplomacy takes a pro-Islamist tilt', http://www.washingtontimes.com/news/2010/apr/14/voice-of-the-mullahs/; http://english.iranianlobby.com/page1.php?id=22&bakhsh=JOURNAL; http://www.niacouncil.org/site/PageServer?pagename=About_myths_facts.

16 See: 'Dadsetan-e Koll-e Iran: Telegram va Instagram Bala Ijad Kardehand', https://m.dw.com/fa-ir/iran/a-45090089?maca=fa-IR-Whatsapp-sharing&.

Kadkhodai went as far as to state that he does not consider using Twitter to be illegal.[17] The cat-and-mouse game around social media continues at all levels and in all sectors of society.

In 2010, Ahmadinejad's government attempted another tactic: it put on another charm offensive for the Iranians in diaspora who had previously been denigrated by the Islamic Republic leaders. It established the High Council of Iranians Abroad, which sought to attract thousands of professional Iranians and potential investors to visit, invest in, and return to Iran. It also sought to facilitate their travels by establishing a 24-hour hotline, establishing branch offices of the council in Iranian provinces, supporting the creation of 'Iran House' branches in foreign countries, and organizing in Tehran a massive Grand Conference of Iranians Living Abroad, offering to pay for participants' travel and accommodations.[18] However, instead of 'polishing Iran's image', the grand conference, 'ended up showcasing many of the country's bitter internal divisions' (Yong and Worth 2010), and the opposition was quick to post the participants' names and pictures on the internet in an effort to 'expose the double crossers'.[19]

Such politicization of digital media and the internet by the public diplomacy quartet of partners – Western powers, the Iranian regime, Iranian dissidents at home, and Iranian dissidents in diaspora – had the unfortunate result of deeply and structurally politicizing the Iranian domestic and diaspora media and internet cinema, whether their contents were political or not, degrading their professional impartiality and journalistic fairness, and undermining the media and civil society formations inside and outside the country, regardless of whether they had accepted American or Iranian governments funding or not, feeding Iranians' penchant for conspiracy thinking. It also made Iranian or binational Iranian intellectuals, academics, artists, bloggers, filmers, film-makers, and community activists, whether they were beneficiaries of public diplomacy funding or not, suspect to the Iranian and Western intelligence services or targets of their surveillance and recruitment, with serious consequences for their democratic aspirations and their lives.

17 See: 'Sokhangu-ye Showra-ye Negahban Estefadeh az Twitter ra Gheir-e Qanuni Nemidanad', http://www.bbc.com/persian/iran-45202575?ocid=wspersian.chat-apps.in-app-msg.whatsapp.trial.link1_.auin.

18 See: http://congress.iranianshouse.ir/; http://congress.iranianshouse.ir/.

19 'Barkhi Sherkat Konandegan-e dar Hamayesh-e Bozorg-e Iranian-e Moqime-e Kharej', http://greenprotests.blogspot.com/2010/08/blog-post_05.html.

References

Jonathan M. Acuff, 'Social Networking Media and the Revolution That Wasn't', in *Media, Power, and Politics in the Digital Age: The 2009 Presidential Election Uprising in Iran*, ed. by Yahya Kamalipour (New York: Rowman & Littlefield Publishers, 2010), pp. 221–234.

Sareh Afshar, 'Are We Neda? The Iranian Women, the Election, and International Media', in *Media, Power, and Politics in the Digital Age: The 2009 Presidential Election Uprising in Iran*, ed. by Yahya Kamalipour (New York: Rowman & Littlefield Publishers, 2010), pp. 235–249.

Louis Althusser and E. Balibar, 'Ideology and Ideological State Apparatuses (notes toward an Investigation)', in *Lenin and Philosophy and Other Essays*, trans. by Ben Brewster (New York: Monthly Review Press, 1971), pp. 127–189.

Doug Bandow, 'Iranian Thugs Take Crackdown Worldwide', *Cato At Liberty*, Online Resource: https://www.cato.org/blog/iranian-thugs-take-crackdown-worldwide (accessed 20 November 2020).

David Batty, 'Guardian Journalist Wins Award for Iranian Protest Coverage', *The Guardian*, Online Resource: http://www.guardian.co.uk/culture/2010/nov/24/guardian-journalist-award-iranian-protests (accessed 18 November 2020).

William J. Dobson, 'Needles in Haystack', *Newsweek*, Online Resource: http://www.newsweek.com/2010/08/06/needles-in-a-haystack.html (accessed 18 November 2020).

Ahmed El Gody, 'New Media, New Audience, New Topics, and New Forms of Censorship in the Middle East', in *New Media and the Middle East*, ed. by Philip Seib (New York: Palgrave, 2007), pp. 213–234.

Golnaz Esfandiari, 'The Twitter Revolution', *Foreign Policy*, Online Resource: http://www.foreignpolicy.com/articles/2010/06/07/the_twitter_revolution_that_wasnt (accessed 18 November 2020).

Farnaz Fassihi, 'Revolutionary Guards Extends Reach to Iran's Media', *The Wall Street Journal*, Online Resource: http://online.wsj.com/article/SB125730352972127145.html?KEYWORDS=farnaz+fassihi (accessed 18 November 2020).

Gurney, David, *Infectious Culture: Virality, Comedy, and Transmediality in the Digital Age* (PhD Dissertation, School of Communication. Evanston: Northwestern University, 2011).

Mahboub Hashem and Abeer Najjar, 'The Role and Impact of New Information Technology (NIT) Applications in Disseminating News about the Recent Iran Presidential Election and Uprisings', in *Media, Power, and Politics in the Digital Age: The 2009 Presidential Election Uprising in Iran*, ed. by Yahya Kamalipour (New York: Rowman & Littlefield Publishers, 2010), pp. 125–142.

John Kelly and Bruce Etling, *Mapping Iran's Online Public: Politics and Culture in the Persian Blogosphere*, Berkman Center for Internet and Society, Harvard Law School, 2008, Online Resource: http://cyber.law.harvard.edu/publications/2008/Mapping_Irans_Online_Public (accessed 18 November 2020).

Diane Macedo, 'Iranian Rockers Tear Down "The Wall"', *Fox News*, Online Resource: http://www.foxnews.com/world/2010/08/11/iranian-rockers-break-wall/ (accessed 18 November 2020).

Robert Mackey, 'Long Jail Term for Iran's "Blogfather"', *The New York Times*, Online Resource: http://thelede.blogs.nytimes.com/2010/09/28/long-jail-term-for-iranian-blogger/?ref=middleeast (accessed 18 November 2020).

Hamid Naficy, *A Social History of Iranian Cinema*, Volume 3*: The Islamicate Period, 1978–1984* (Durham: Duke University Press, 2012).

—, *A Social History of Iranian Cinema*, Volume 4*: The Globalizing Era, 1984–2010* (Durham: Duke University Press, 2012).

—, *A Social History of Iranian Cinema*, Volume 2*: The Industrializing Years, 1941–1978* (Durham: Duke University Press, 2011).

Bill Nichols, *Introduction to Documentary* (Bloomington: Indiana University Press, 2001).

Trita Parsi, David Elliot, and Patrick Disney, 'Silencing Iran's Twitterati: How U.S. Sanctions Muzzle Iran's Online Opposition', in *Media, Power, and Politics in the Digital Age: The 2009 Presidential Election Uprising in Iran*, ed. by Yahya Kamalipour (New York: Rowman & Littlefield Publishers, 2010), pp. 161–169.

Setareh Sabety, 'Graphic Content: The Semiotics of a YouTube Uprising', in *Media, Power, and Politics in the Digital Age: The 2009 Presidential Election Uprising in Iran*, ed. by Yahya Kamalipour (New York: Rowman & Littlefield Publishers, 2010), pp. 111–125.

Cameron J. Shahab and Reza Mousoli, 'Cat and Mouse in Cyberspace: A Case Study of China vs. Iran', *Iranian.com*, Online Resource: http://www.iranian.com/main/print/120645 (accessed 18 November 2020).

Scott Shane, 'A T-Shirt-and-Dagger Operation', *The New York Times*, 13 November 2015, Online resource: https://www.nytimes.com/2005/11/13/weekinreview/a-tshirtand-dagger-operation.html (accessed 18 November 2020).

Siamand. 'Patak-e Narm-e Televizioni', *Roozeonline*, Online Resource: http://www.roozonline.com/persian/opinion/opinion-article/article/2010/june/14//-5399ef79c5.html (accessed 18 November 2020).

Annabelle Sreberny and Gholam Khiabany, *Blogistan: The Internet and Politics in Iran* (London: I.B. Tauris, 2010).

Hamid Tehrani, 'Iran: A Long and Painful Story of Jailed Bloggers', *Global Voices*, Online Resource: http://globalvoicesonline.org/2008/12/18/iran-a-long-and-painful-story-of-jailed-bloggers/ (accessed 18 November 2020).

Michael Theodoulou, 'Iranian "Blogfather" Hossein Derakhshan is arrested on charge of spying for Israel', *The Sunday Times*, Online Resource: http://www.timesonline.co.uk/tol/news/world/middle_east/article5190462.ece (accessed 18 November 2020).

William Yong and Robert F. Worth, 'Iran Expatriates Get Chilly Reception', *The New York Times*, Online Resource: https://www.nytimes.com/2010/08/08/world/middleeast/08iran.html (accessed 18 November 2020).

Hamed Yusefi, *Zibashenasi-ye E'terazat-e Siasi dar Iran* [*The Aesthetics of Political Protest in Iran*], http://www.bbc.co.uk/persian/tv/2010/07/100721_green_art.shtml (accessed 18 November 2020).

Ezatollah Zarghami, 'Iranian State TV Acts as an Arm of the Intelligence Apparatus', *Iran Human Rights*, Online Resource: https://www.iranhumanrights.org/wp-content/uploads/Iranian-State-TV-Acts-as-an-Arm-of-the-Intelligence-Apparatus.pdf (accessed 20 November 2020).

About the Author

Hamid Naficy is Hamad bin Khalifa al-Thani Professor Emeritus of Radio/Television/Film at Northwestern University, Chicago and a leading authority in cultural studies of diaspora, exile, and postcolonial cinemas and media, and of Iranian and Middle Eastern cinemas. Naficy has published and lectured extensively, nationally, and internationally, on these and related topics. His English-language books are *An Accented Cinema: Exilic and Diasporic Filmmaking* (2001); *Home, Exile, Homeland: Film, Media, and the Politics of Place* (1999); *The Making of Exile Cultures: Iranian Television in Los Angeles* (1993); *Otherness and the Media: the Ethnography of the Imagined and the Imaged* (co-edited 1993); and *Iran Media Index* (1984). His latest work is the award-winning four-volume book *A Social History of Iranian Cinema* (2011–2012). He has also published extensively in Persian/Farsi, including a two-volume book on documentary cinema theory and history, *Film-e Mostanad* (1978-79). In addition, he has produced many educational, and instructional radio programs, films, and television programs for the training of physicians, teaching assistants, tutors, and for the education of the public.

6. From Amateur Video to New Documentary Formats: Citizen Journalism and a Reconfiguring of Historical Knowledge

Katarzyna Ruchel-Stockmans

Abstract

The Arab Spring of 2011 appeared on the news in countless video testimonials shot by participants and bystanders. Yet the sheer amount of such visual data, combined with the untranslatability and poor quality of the footage, delivers an often opaque and ambiguous material. Peter Snowdon's film *The Uprising* (2013) appropriates and reassembles this amateur footage. Compared to the 1992 film *Videograms of a Revolution* by Harun Farocki and Andrei Ujică, *The Uprising* raises the question of a possible knowledge shift produced and performed by the amateur video maker. In an analysis of these films, this chapter investigates the possibilities of documenting historical events through mobile video footage.

Keywords: Documentary, found footage film, Syria, amateur

The act of self-immolation by the fruit vendor Mohamed Bouazizi in December 2010 triggered a wave of demonstrations throughout Tunisia, mobilizing masses of people to oppose the failing government. The spirit of protest quickly spread to other countries in the Middle East and North Africa. Spontaneous demonstrations emerged and expanded not only in the streets but also through an unprecedented use of social media. As a result, virtually anyone in the world could instantly see the events by means of images and mostly mobile phone camera videos, recorded and uploaded to the internet on a massive scale from various epicentres of the protests. Many citizens turned into amateur journalists in protesting against

Strohmaier, A. and A. Krewani (eds.), *Media and Mapping Practices in the Middle East and North Africa: Producing Space*. Taylor & Francis Group, 2021
DOI 10.5117/9789462989092_CH06

authoritarianism, human rights violations, and dire living conditions using methods of civil disobedience combined with internet activism. There has been an intense debate about the role of the new media and electronic social platforms in sparking, maintaining, and spreading the revolutions of 2011 (Boëx 2013a; 2013b; Khatib 2013). After initial enthusiasm about the 'YouTube', 'Facebook', or 'Twitter revolutions', various commentators expressed their doubts about the actual impact of new media formats and social connectivity on the political events (Downey 2014; Harkin 2013; Hawkins 2014; Werbner et al. 2014). What is clearly novel and groundbreaking in this period of upheavals is the broad and instant availability of numerous audiovisual documents on the internet, making the participant-generated visuals the main source of knowledge about the revolutions and ensuing conflicts for the public outside of the region (Robertson 2015).

As Henry Jenkins, Sam Ford, and Joshua Green note in their book *Spreadable Media*, there is a direct journalistic value to the spontaneously created online image archive (2013, pp. 41–44). Video activists using the internet bypass the normal news channels, creating a democratic peer-to-peer communication (Jenkins, Ford, and Green 2013). Yet this vast repertory of clips, photographs, and sounds defies the more established forms of knowledge production, such as film, documentary or journalistic report (Zabunyan 2016, pp. 7–36). This chapter aims to investigate the role of online video sharing from the post-2011 MENA (Middle East and North Africa) region, especially Syria, in building new models of historical knowledge production. It encompasses an analysis of three specific case studies: Peter Snowdon's film *The Uprising* (2013); the short film series by the Abounaddara collective; and the Syrian Archive launched by Hadi Al Khatib. These recent phenomena can be seen as novel forms of historical representations. Instead of a conventional film or documentary narrative, they stem from and react to the amateur video culture of sharing, tagging, commenting on, and appropriating visual material. Throughout this analysis, I trace a possible shift in knowledge production precipitated by civic resistance and alternative modes of information exchange. Mobile phone images as instruments in the hands of participants and eyewitnesses yield a form of knowledge that holds an emancipatory promise and at the same time is bound with new risks and challenges. The three case studies considered here allow to demonstrate the paradigmatic changes in visualizing dissent and conflict, while they avoid any authoritative elucidation of the events in the Middle East and North Africa. Approaching these new documentary formats through the perspective of art and media theories will allow to interrogate the ways in which these practices respond to crises (whether seen as political upheaval or humanitarian emergency) while at the same time mobilizing and critically revising the newly emerging participatory cultures.[1]

Vernacular Videos by Citizen Journalists

Many commentators agree that the widespread protests of 2011 were marked by an unprecedented role of media activism, yet the debate about what this role precisely is still continues. As virtually everyone has a mobile phone at hand, it has become common to record an event and upload the videos or photographs to the internet (Haskins 2007; Maier-Rothe et al. 2014). During the upheavals in the MENA region, citizen journalism emerged as an attractive alternative to professional media. Accessible and uncontrollable, it was believed to offer a democratizing and emancipatory tool in the hands of the people. Soon enough, citizen journalism was also questioned for its partisan character. As a matter of fact, citizen journalists are mostly activists who are involved in the situations they portray, unlike professional journalists who are expected to keep an emotional and ideological distance from the subjects they cover. Moreover, citizen journalists are not professionals, and therefore the material they deliver is often fragmentary and inadequately contextualized for the public. At the same time, many clips found on the internet have to remain anonymous to protect the identity of their makers and therefore cannot be fact-checked. Videos appear and disappear in the mass of images uploaded and re-uploaded every minute on different social platforms. The sheer amount of audiovisual information makes it impossible to keep an overview of what is happening in different parts of the globe. There have been many different ways in which artists, film-makers, and researchers have responded to this evolution. This chapter is centered on three projects stemming from the online sharing of citizen-journalist videos. The first one, Peter Snowdon's *The Uprising* (2013) was made shortly after the eruption of protests as a found footage film consisting of various videos published online. The second, known as Abounaddara, is an anonymous video art collective operating in Syria and broadcasting online since 2011. The third, the Syrian Archive, is a researcher-led project of collecting and verifying videos of human rights violations in Syria since 2014. Each of these documentary formats activate different forms of engagement with the video material as they reassemble the unruly online visuals in distinct configurations.

1 The research presented in this chapter is part of my long term investigation of participative media and knowledge production in and as artistic practice (Ruchel-Stockmans 2012; 2015; 2017).

The Revolution Reimagined[2]

The British film-maker Peter Snowdon was one of those who followed the 2010–2011 protests in the MENA region from the outside. Working in his studio in Belgium, he compiled a selection of videos he found on the internet during the events in the Middle East and North Africa. The result is a full-length film entitled *The Uprising* (2013), which is based exclusively on found footage, devoid of any authorial commentary.

The Uprising (2013) consists of about one hundred clips of various origins found by Peter Snowdon on YouTube and other internet sources. Almost all videos relate to protests and upheavals in six Arab countries: Bahrain, Syria, Egypt, Yemen, Libya, and Tunisia. Presented in sequences, the clips show various gatherings in public spaces, ranging from cheerful moments of conviviality to violent clashes between civilians and the police. They also include personal messages by vloggers or occasional speakers who address people in their countries, urging them to support the emancipatory struggle. Significantly, the film refrains from a voice-over or intrusive editing. The various clips made by different and often anonymous people are stitched together regardless of their divergent origins. As a result, images of protesters in Bahrain connect visually with those from Egypt or Syria as if they were part of the same struggle. The only editing device used by Snowdon is the insertion of intertitles dividing the sequences into seven days. The film begins with the intertitle 'Seven Days Ago' and ends with 'Today'.

Peter Snowdon's *The Uprising* (2013) started as a project of compiling the protest videos stemming from Egypt and other countries in the region.[3] As a former resident and journalist in Cairo who subsequently embarked on practice-based research in Hasselt, Snowdon was both personally and professionally connected to the region. His video selection was screened for different publics in both Europe and Egypt. Viewing large numbers of these clips for an extended period of time – and seeing the reactions of the audiences during the screening – the film-maker decided to stitch the videos

2 I compared *The Uprising* to two films on the protests in Syria in an earlier article (Ruchel-Stockmans 2017). Drawing on this earlier research, I consider it here in the context of participatory media and the cultures of online sharing.

3 In an interview, Snowdon explains he lived in Egypt as a journalist and activist prior to the protests. When the upheaval began, he had already moved to Belgium and could only watch the events unfold live on-screen. In fact, YouTube and social media were the only spaces where his friends and colleagues from Cairo could be contacted in the heated days of the protests. Telephone lines were inaccessible and it was difficult to get in touch with the people caught in the revolution (Cohn 2014).

together in one overarching visual narrative. Interestingly, the various and mostly anonymous videos nevertheless bear many characteristics that invite such a narrative, at the same time challenging the viewer with difficult, opaque, and unsteady visual information.

Filmed with small mobile devices, these videos clearly show the possibilities and limitations of the technology involved in their making. Mobile phones become extensions of the human hand (and not so much the eye) as they are held in front of the body, often shoulder-high, and reflect the movements of the bodies that use them.[4] In clips from mass demonstrations, we often get the perspective of one of the participants who looks from behind the backs of other people and who regularly has to move forward or even run away when the situation becomes more dire. These particular conditions of production have a paradoxical effect. On the one hand, we have a sense of being directly involved in the events portrayed as we are granted the viewpoint of a participant. On the other hand, however, the videos are often blurry, shaky, or pixelated, making it difficult to discern or interpret the scene. Such ambiguous effects recall and at the same time complicate the analysis of media made by David Bolter and Richard Grusin (1999, pp. 19–50). In *Remediation*, Bolter and Grusin argue that all new media strive to provide an immediate experience of the world outside of the media. Yet the 'improvement' that each medium introduces is in fact created in an oscillation between transparency and opacity. A new medium attempts to create an experience of immediacy, yet at the same time reveals its own status as a mediation. In the case of the handheld, low-resolution, and shaky video images, which are spread by participants of sometimes geographically distant events, the immediacy is generated by the images' manifestly mediated character. It is also impossible to forget that the poor quality of these images, low in information and formally deficient, is precisely what guarantees their authenticity and directness. They are authentic and direct in the sense that they are not edited, processed, or improved. This impression of unmediated, raw visual information is particularly strong when we are confronted with images from a camera operator who is hit by a bullet and falls to the ground. The camera falls too, showing the off-kilter perspective of the pavement and close-ups of bloodstained clothing.[5]

4 In this sense, it becomes relevant to revisit the media theory of Marshall McLuhan, who conceptualized media as extensions of man (McLuhan 1964).

5 The phenomenon of protesters who continue filming even when they are in the aim of snipers has been compellingly analysed by Rabih Mroué in his lecture-performance 'The Pixelated Revolution'. See Rabih Mroué, 'The Pixelated Revolution', *Image(S), Mon Amour. Fabrications* (Madrid: CA2M Centro de Arte Dos de Mayo, 2013), pp. 378–393.

Performativity and Peer-to-Peer Communication

Apart from the videos showing unrest and violence in the streets, *The Uprising* (2013) also contains other types of images. On several occasions, we see vloggers commenting on the current situation, appealing to other citizens or analysing media coverage about the protests. These images are more static, yet even here the handheld camera is often visible and image resolution is rather low. What is clear is that these messages are a form of peer-to-peer communication eschewing the habitual channels of information circulation. Such direct address can equally be found in some street scenes, in which protagonists demonstrate the intention to reach other citizens. As the clips from the middle of the protests are made by serendipitous passers-by, we often see individuals emerging from the crowd gathered in the street and expressing their discontent in front of the camera. The presence of the person filming seems to act as a trigger for many other participants to stage an address to the potential viewers of the video. In a negative sense, the sight of a camera also prompts aggressive reactions from the police or the military, who try to prevent protesters from filming. In several clips, there is a dialogue between the camera operator and the people in front of the camera. At one point, the security forces urge the protester holding a camera to stop filming. On other occasions, voices from the crowd encourage the videographer to continue recording so the world can see what is happening. In this sense, the act of filming can be seen as performative. Both media activists and those visible in the frame enact their performances of dissent in front of and for the camera (Ruchel-Stockmans 2012; 2015). It seems as though the various originally disconnected videos were made as part of a large, overarching script for the revolutions.

The sense that there is one script designed by an unknown director was noted earlier in recent history. When the film-makers Harun Farocki and Andrei Ujică embarked on a documentary about the 1989 revolution in Romania, they discovered vast archives of amateur videos in Bucharest that were recorded by various unconnected participants of the events. Having watched parts of the footage, they decided to reconstruct the chronology of the events without much editing of the images. In that period in Romania, long before internet and mobile phones became commonplace, there were already enough video cameras available for accidental participants to cover virtually all important episodes in the violent revolution. Amateurs, students, and activists went down to the streets to record the long-awaited change of the regime. Already at this point in history the footage shows performativity of the cameras discussed earlier. The main difference with the 2011 uprisings lies in the analogue and offline character of the Romanian video archives.[6] The footage used for *Videograms of a Revolution* (1992) was largely unknown and inaccessible until the recent digitization of parts of these archives.

Image Operations

Since 1989, the conditions for viewing and accessing images of citizen journalists have changed significantly. Peter Snowdon had at his disposal the unruly and impenetrable, but principally open and accessible, archives of YouTube, Twitter, and Facebook. While Farocki and Ujică gave visibility to images habitually buried in archives, Snowdon reused clips that could generally be found in their original form by anyone connected to the internet. Many of these images also disappear, are removed, or are simply difficult to trace, but their general status is that of open access. Analogue video and film required projection or a professional apparatus to be viewed. Electronic data can be viewed and processed on common and widely used devices. Short videos posted by citizen journalists in 2011 are normally viewed on the small screens of laptops, tablets, or mobile phones. It is a distinct form of viewing which does not require black box projection, concentrated attention, and time investment in a full-length film. This distinction is analogous to the difference between a theatre performance and a broadcast, described aptly by the sound artist Lucas Derycke. Asked in an interview why he prefers a sound performance in a theatre to, for example, a radio or an online broadcast, he replies that a theatre performance takes the audience hostage for the entire duration of the performance (Derycke 2017). This mode of reception imposes, so to speak, concentrated and isolated viewing – or, for that matter, listening – of the piece. Clips found on social media invite participation by sharing, commenting, and even re-editing the audiovisual material.[7] Transferring these clips onto a big screen alters these viewing conditions. The blurry and shaky videos are enlarged on a projection screen and viewed collectively in the darkened space of the cinema. The spectators in a screening are expected to watch the whole film, while viewing small clips on the internet allows for pausing, replaying, or abandoning the viewing

6 Politically, the differences between 1989 and 2011 are abundant. See: Guehenno 2011.

7 This different mode of viewing has already been recognized by Laura Mulvey with respect to video technology (Mulvey 2009, pp. 7–16). A video player or recorder allowed for pausing, replaying, or fast-forwarding films, thus changing the modalities of moving-image reception. Online and mobile device viewing enhanced and multiplied these possibilities.

of particular scenes. Peter Snowdon's intention was to give homage to these activists and citizens who produced the videos, while he was also determined to make his film available online after it screened in festivals. The question remains as to what particular consequences this transfer onto film and its inclusion in a narrative might have on the original footage.[8]

The last sequences of the film do not show mass protests directly and are devoid of the original soundtrack. We see a distant explosion or smoke, followed by a man raising a flag or followed by a view of a burning tank. The closing clip shows a storm somewhere in North America. We see menacing dark clouds gathering in the distance. Here the sound is re-edited so that an impassioned call can be heard inviting us, the viewers, to join the revolution. The female voice belongs to the eminent vlogger Asmaa Mahfouz who helped spark the protests in Tahrir Square (Wall and El Zahed 2011). It is the only moment of the film when the diegetic sound is severed from the image. Looking at the images of the approaching storm, we hear the urge to participate in civil dissent. The film therefore imparts the idea that it is not so much a political analysis or a conventional documentary about the series of revolutions in the Arab countries, but rather a vision of an uprising that is yet to come. This utopian tone of the closing scenes might seem at odds with the palpable videos showing actual people engaging in demonstrations and often risking their lives in the name of concrete social and political change. The subtle yet persistent inscription of the clips in the visionary filmic tale generates a distance from the visceral scenes shown in the footage.

A Man with a Movie Camera

A different approach to amateur video sharing can be found in the work of Abounaddara. Founded in Syria in 2010, Abounaddara is an anonymous film collective whose members produce short videos from and about the war-torn country. Since 2011, the collective regularly released new videos on Vimeo and other social media channels. At the time of writing, there are about five hundred videos available online. In its statements and polemics, Abounaddara strives to portray a counter-image of the Syrian revolution and conflict. Against the propagandistic language of the state media and the spectacle-driven international news channels, the collective proposes what they have coined 'emergency cinema'. In content as well as in its mode of production, the emergency cinema aims to counterbalance distorted, myopic,

8 For watching the full movie see: https://theuprising.be/

and unreliable journalistic reporting. It stems from the realization that most citizen journalism falls into the same trap: activists on the ground are asked by large news channels to provide certain types of footage, mostly to show spectacular violence, the violations of ceasefires, and stereotyped images of victims.[9] These repetitive and conventional images, however, fail to convey knowledge and insight about the complex situation in Syria.

The alternative collection of short videos by Abounaddara bears many similarities with Peter Snowdon's *The Uprising* (2013). Yet the videos also differ significantly from the footage spontaneously uploaded by various anonymous users of social media. Although any details on particular members or even the number of contributors to the Abounaddara channel remain undisclosed, the collective name lends these film-makers a recognizable identity. Abounaddara has been invited to film festivals, conferences, and art events, and after winning the main prize at the Sundance Film Festival in 2014, it gained unprecedented visibility (Bayoumi 2015; Creswell 2016; Elias and Omareen 2014; Feinstein 2015; Holland 2014). Since then, it has also been awarded the Vera List Center Prize and has participated in various art exhibitions such as the Venice Biennale (2015) and documenta 14 in Kassel (2017). Significantly, it has also refused to participate in art or film events that conflicted with its commitments and views. Although relying on volunteers and working in extremely harsh conditions, the collective managed to create a distinctive and recognized body of work.

The emergency cinema proposed by Abounaddara is marked by a plurality of voices. The length of videos varies between less than a minute and over ten minutes, but some videos are arranged into series connected by the same title.[10] They often show ordinary street scenes and various occupations of ordinary citizens devoid of any commentary. A significant part of this output is observational footage of, for example, a young butcher, a street vendor, or a women's hairstyling class. Often, people filmed in close-up tell their stories. Some faces are visible, while some are made unrecognizable by backlight or a colour filter. Occasionally, a montage of found footage is also used (e.g. *Kill Them*, 24 March 2015).

These sometimes confessional or impressionistic, and at times ironic or critical, videos relay various issues in Syrian society in a multivocal way, allowing different actors in the ongoing military conflict to defend their

9 See especially Abounaddara's video *We Who Are Here* (19 August 2016).

10 Examples of such series include *The Team* (October 2015), consisting of ten parts, and *The Syrian Who Wanted A Revolution*, which is an interview consisting of five parts, made on the fifth anniversary of the Syrian revolution (March 2016).

perspectives. There is no direct political commentary or analysis present in the shorts; rather, various individuals tell their stories or are shown in their daily activities. Similar to the many anonymous citizen-journalist videos posted on the internet, there is little background knowledge given in the videos or in the commentary on the Vimeo or YouTube channels. As a rule, there is no voice-over, although occasionally there are subtitles or text-frames providing often jarring or ironic commentary to the images in the style of Russian avant-garde film. Taken from an Egyptian satirical magazine, the name Abounaddara, which is Arabic for 'a man with glasses', also alludes to the famous 'man with a film camera' from Dziga Vertov's 1929 film. Unsurprisingly, there are many parallels in their videos to the revolutionary film practices of the 1920s.

Plural Identity and Collective Struggle

The many disparate short videos produced by Abounaddara often call to mind the postulates of Dziga Vertov for what he called 'Kino-Eye' and 'Kino-Pravda' (Michelson 1984, pp. 14-21). The observational videos such as *Just Do It* (2016),[11] *The Butcher of Aleppo* (2016),[12] or *And Yet it Moves* (2016)[13] register everyday situations or activities of citizens without any visible involvement of the film-maker. In his writings and manifestos, Vertov calls for a new 'factory of film footage in which life, passing through the camera lens, does not vanish forever, leaving no trace, but does, on the contrary, leave a trace, precise and inimitable' (Michelson 1984, p. 45). Registering life as it is implies that scripts, costumes, and play-acting should be abandoned. Life should be 'caught unawares', without any pretence or masquerade (Michelson 1984, p. 41). Cinema becomes a prosthesis of the human eye, observing the chance occurrences it encounters.

Undeniably, these anonymous film-makers do more than just roll the camera in a fortuitous place and time. It is clear that the members of Abounaddara make very deliberate choices. This is especially noticeable not so much in what they film but rather in what they refuse to register on their cameras. The collective's short videos programmatically avoid showing victims of war in the way that is common for mainstream news channels. This choice comes from the perceived need to rebuild a respectful

11 https://vimeo.com/188270532
12 https://vimeo.com/196005030
13 https://vimeo.com/176779632

way of representing people, especially when it comes to images of those victimized by war. They make this point clear not only verbally, in debates and statements (Abounaddara 2015; 2016; 2017b; Kiwan 2014), but also by visual means. In the half-minute-long video entitled *Aleppo* (2016), the frame remains white from beginning to end. Only two sentences appear on screen, condemning the live broadcasting of human tragedy on television channels worldwide. This is a counter-image whose blankness must strike the viewer as an intended decontamination of the news culture. Many human tragedies are present in the various short films by Abounaddara. However, they are presented mainly through personal stories and oblique images addressing the viewer's sensitivity and imagination. The main strategy in showing the reality of war through such visual means appears to be a microhistory of individual lives and human choices. For example, in *Sons Without Fathers* (2016), a man recounts situations of what he calls 'sodomy' among youngsters and how the community tries to create a makeshift justice system to fight it. For an outside viewer, it might be hard to comprehend that a minor was executed for a sexual offence in a makeshift trial. No further details are given about the case. The complex and painful story is told from one particular perspective, without any comment or judgement. In *Hassoun* (26 June 2015), a young woman narrates a story about her younger brother, who was conscripted into the army. She talks about her trauma and, wiping away her tears, admits to being happy that he is no longer alive. Her story, filmed in a close-up and with a steady camera held by someone she knew and trusted, is also a form of counter-image to the news reports. It is a story of a premature and tragic death, but told in a personal account that reveals the conflicting emotions of the bereaved.

In several statements and articles, Abounaddara argues for film as a public good that should not become a marketable product. It therefore insists on producing films with little means, but with the uncompromised artistic freedom. The videos are available online, but nevertheless have a copyright notice at the end of each episode. An incident at the Triennale of Milan in 2017 has demonstrated that there are a host of new challenges for this open-source approach. Abounaddara was invited to participate in the exhibition *Restless Earth*, but declined the invitation because, in the group's perception, its curatorial concept exclusively served the Western narrative of the refugee crisis. However, its videos were included in the exhibition, without the collective's consent. When inquired for an explanation, the curator's spokesperson argued that the videos were not exhibited as part of the show but only made available through the online streaming on monitors placed in the exhibition (Muñoz-Alonso 2015; Ryzik 2017). In response to this violation of copyright, the collective decided to make their online channel unavailable for the duration of the exhibition (Abounaddara 2017a).[14] As a result, the videos, which were previously freely accessible online, temporarily disappeared from the internet.

Recollecting Visual Evidence

The large collection of videos produced by Abounaddara was expressly removed from the internet as a sign of protest against the unauthorized use of the footage by an art institution. Apart from the copyright issues, this event poignantly reminds us of the ephemeral nature of online data. Many videos uploaded on social platforms are either deleted, hacked, or simply get lost in the avalanche of similar visual material. It has recently been noted by investigative journalists that large amounts of potential evidence of war crimes in Syria have been removed from YouTube as a result of the decision by the commercial platform to censor possible terrorist and propaganda content (Martin 2017). More often than not, politically sensitive material is subject to cyberattacks by parties involved in the conflict. The Syrian Archive was launched in 2014 as a response to that issue. It is a freely accessible, independent database of visual material from the post-2011 conflict in Syria intended to facilitate accountability and advocacy processes. The videos it gathers pertain to human rights violations and can serve as possible judicial evidence. The founder of the project, Hadi Al Khatib, notes that many participant and activist videos either get lost or are insufficiently verifiable or contextualized to serve as evidence. The Syrian Archive is an endeavour aimed at preserving, verifying, and making available those vernacular videos which testify to human rights violations. It essentially saves visual material that appeared elsewhere or obtains videos directly from film-makers or activists on the ground. This means that the videos selected to be preserved have quite specific content. They make visible horrific events post-2011 in Syria. Unlike *The Uprising* (2013) and the Abounaddara films, there are very few joyous or 'ordinary' moments preserved in the database. More importantly, the archive differs from the other two examples in that the videos it contains are provided with metadata, including the source, author, date, and time, as well as geolocation. The events shown in a given image can thus be historically located and identified within a larger narrative.

14 1 August 2017, https://m.facebook.com/notes/abounaddara-films/our-films-are-not-available-nos-films-ne-sont-pas-disponibles/1529777670416268

The inconclusiveness and illegibility of many online videos was the main incentive for this archival approach. Videos and photographs can be uploaded by various users on different platforms, making it difficult to navigate through the abundant but scattered resources. Especially in times of turmoil, the amount of visual data disseminated on the internet can rise exponentially, making it impossible to process all of the available material. Unavoidably, viewing or following of these videos will be selective. Online sharing platforms such as YouTube as well as search engines such as Google use ranking algorithms that determine the hierarchies of search results. Their logic and effectiveness is largely determined by the commercial goals of these platforms. Some content might simply land at the bottom of the ranking because its relevance or significance was not recognized. Moreover, as has been mentioned earlier, many visual data get lost. Some files may be deleted, hacked, or purposefully removed. Cyberattacks are increasingly part and parcel of military conflict, as is abundantly clear in Syria (Xynou and Al Khatib 2015).

Even if the relevant visual material is accessible and accurate, it is often difficult to interpret what is seen. Some of the essential information often cannot be determined by only viewing the images, such as the identity of the people shown, the location of an incident, or the exact time when it happened. This issue is obviously not new. Images cannot easily be translated into words, and more often than not, there is a cleavage between what can be seen in an image and what can be said about it. W.J.T. Mitchell proposed to designate this gap between image and text as 'image/text', in which the slash is a sign of a problematic rupture (1995, p. 89). The Syrian Archive proposes to respond to that rupture by establishing a standardized scheme of metadata. Before adding new videos and photographs to the database, their credibility is checked and metadata restored or collected. The verification process is needed to establish whether the source – the maker, the organization, or individual who uploads the content to the internet – is credible as a source of information. The archive lists over three hundred credible sources who are already verified. New sources are added according to a transparent verification process which includes screening the upload history of a source and checking its credibility among the known contacts on the ground (the Syrian Archive 2017a). Subsequently, the videos or photographs from credible sources are subjected to a metadata scheme. In this process, the essential information is established about the contents of the visual material, including the actors, location, and time of the incidents. Some of this information, such as geolocation, can be gathered by comparing and cross-analysing various open-source databases. For example, satellite images available on Google

Earth can be used to establish the exact coordinates of the places shown in the visuals. The videos are subsequently categorized according to the type of violation they document. These categories include, for example, massacres and other unlawful killings, torture, hostage taking, use of illegal weapons, or sexual or gender-based violence. This might appear as a bureaucratic division of evils, but it is necessary to make the archive searchable. Inevitably, choices have to be made as to what is defined as a human rights violation. The typology of offences chosen for the archive adheres to the standards developed by the Office of United Nations High Commissioner for Human Rights. The compatibility with these international standards ensures that the archive can be used by various actors in human rights work.

An example of how this collection of visual material can be put to work is demonstrated in the journalistic investigation of an alleged mosque bombing on 16 March 2017 near al-Jinah in Syria's Aleppo governorate. The investigations were independently conducted by each Bellingcat, Forensic Architecture, and Human Rights Watch, but the organizations exchanged information and arrived at the same results.[15] All multimedia content related to this incident and the investigations is stored in the Syrian Archive. Using open-source information, videos, and witness accounts of people on the ground, compared with official reports, the investigators could determine that the official version of the incident was not correct and that innocent civilians were targeted and killed while attending a ritual at a mosque (the Syrian Archive 2017b). The investigations were published online in the form of multimedia articles, but there is also a visual component to the publication. The researchers of Forensic Architecture created a short clip summarizing their investigation in which they rely on the amateur videos and photographs and reuse them as illustration of evidence in demonstrative text and image animations (Forensic Architecture 2017). It is therefore a new form of reporting, based on collaborative open-source investigations and pursued independently of the mainstream press. At the same time, these reports are occasionally included in the news reporting of leading press and television stations in different countries.

The Syrian Archive collects, verifies, and analyses selected citizen-journalist videos to make the information they contain more accessible. The archive therefore generates knowledge on the conflict in Syria, but also facilitates knowledge production by other actors in the field. The investigation results already achieved demonstrate that this method of work

15 Bellingcat is an organization founded by Eliot Higgins with the purpose of developing open-source intelligence investigations and tools. Forensic Architecture is an interdisciplinary research agency led by Eyal Weizman at Goldsmiths, University of London.

returns to the tools and formats of investigative journalism yet avoids the mainstream media channels, creating alternative routes of communication. At the same time, the archive has its clear, self-imposed limitations. Aimed at facilitating accountability in times of prolonged military conflict, it preserves only the visual materials pertaining to perceived human rights violations. By default, it excludes the everyday, peaceful, or neutral footage that seems so significant for the message formulated in *The Uprising* (2013) and in Abounaddara's filmic output.

Apart from storing verified visual data, the Syrian Archive has published its code open-source, so that other organizations or individuals can create their own databases.[16] Not only the content, but also the methodology and computational tools developed for the archive are made available for other users. The project makes clear that new strategies and skills are increasingly needed to yield new and relevant knowledge out of the endless flow of data on the internet, especially in times of tumult such as in the post-2011 MENA region.

The Archive Project demonstrates that working with big data is one of the most acute challenges in the time of digital sharing culture. The significance of that challenge is confirmed by other researchers and artists who proposed alternative modes of gathering and presenting user-generated images of revolutions. In 2014 Lev Manovich and a team of researchers created *The Exceptional and the Everyday: 144 hours in Kiev*, a visualization of people's images uploaded on Instagram during the upheaval on Maidan Square. The large zoomable photomontage that resulted from that project consists entirely of public images that appeared on Instagram on the particular days in February 2014 in the area of central Kiev.[17] Gathering all images instead of selecting only revolution-related ones, the project demonstrated how the ordinary – the selfies, food, cats, and cityscape pics – coexisted with the images of burned trucks and agitated crowds on the main square. Manovich provocatively claims it is not artists today who make inspiring representations of a revolution, but data analysts who can rework and visualize images delivered by countless anonymous producers of images. Yet at the same time, Manovich relates his work to that of a painter. He visualizes data in a particular way but does not claim to show the truth of the revolution in Ukraine (Packer 2014). The Syrian Archive might perhaps also become a tool to transform the large mass of videos and photographs into a visually compelling, interactive narrative.

16 The Syrian Archive Toolkit can be found on the software-sharing site GitHub: https://github.com/syrianarchive/toolkit.

17 The results of the project can be viewed on its blog: www.the-everyday.net.

Conclusion

While the approaches of Snowdon, Abounaddara, and the Syrian Archive seem fundamentally divergent, they each develop methods of involvement with citizen-generated videos and photographs. Though each in a different way, they all revive the idea of 'imperfect cinema' proposed by Julio Garcia Espinosa – a cinema of little means and technical deficiencies, but a cinema that reflects the movement and energy of the collective (1979).

The found footage film proposed by Snowdon does not reconstruct the chronology or genealogy of the 2011 revolutions, and neither does it present a political or social analysis of their deplorable outcomes. It selects and subtly alters the imagery created by various first-hand witnesses of the events, saving from oblivion the chosen clips out of thousands available. By arranging them in a narrative of a promised or awaited revolution, it condenses the energy of these protests and monumentalizes the common dream of the masses. *The Uprising* (2013) can be seen as a multi-authored film, in which the anonymous contributors can find themselves in the images of their fellow videographers. Made shortly after the first wave of protests, it offers an affective response to the crisis. Abounaddara adapts the practice of anonymous citizen journalism to its own idea of 'emergency cinema'. Producing various short videos authored by diverse, unnamed film-makers, the collective enters into a dialogue with the common practices of mainstream media reporting in Syria. It gives voice to individuals from different sides of the conflict who share their often traumatic experiences. In other videos, Abounaddara comments on the political situation and ironically criticizes the one-sided and spectacle-centred mainstream news reporting on Syria. Some of its short films include found footage and photography coming from citizen journalism, documentary, or commercial videos and films. This is therefore a multifaceted online presence that builds on the advantages of user-generated videos but also formulates a response to the risks of citizen journalism. The Syrian Archive, finally, systematically selects and stores those online videos which pertain to human rights violations in Syria. It subsequently verifies this footage and provides it with metadata. Such visual material can be used in accountability processes by various organizations and individuals.

The three ways of working with citizen videos and photographs presented in this chapter demonstrate that open-source visual material produced by often anonymous users, whether witnesses or participants in conflict situations, present both an unparalleled chance and a daunting challenge to knowledge production. Writing history appears as a plural, non-linear, and multilayered process in which crowd sourced images play a leading role.

References

Abounaddara Collective, 'A Right to the Image. A Concept Paper for a Coming Revolution by The Abounaddara Collective', *The New School Vera List Center for Art and Politics*, Online Resource: http://www.veralistcenter.org/engage/exhibition/1978/abounaddara-the-right-to-the-image/ (accessed 18 November 2020).

—, 'Regarding the Spectacle', *The Nation*, Online Resource: https://www.thenation.com/article/regarding-the-spectacle/ (accessed 18 November 2020).

—, 'Our Films Are Not Available', Online Resource: https://m.facebook.com/notes/abounaddara-films/our-films-are-not-available-nos-films-ne-sont-pas-disponibles/1529777670416268 (accessed 18 November 2020).

—, 'Dignity Has Never Been Photographed', Online Resource: http://www.documenta14.de/en/notes-and-works/15348/dignity-has-never-been-photographed (accessed 18 November 2020).

Moustafa Bayoumi, 'The Civil War in Syria Is Invisible – But This Anonymous Film Collective Is Changing That', *The Nation*, Online Resource: https://www.thenation.com/article/the-civil-war-in-syria-is-invisible-but-this-anonymous-film-collective-is-changing-that/ (accessed 18 November 2020).

Cécile Boëx, 'La grammaire iconographique de la révolte en Syrie: Usages, techniques et supports', *Cultures & Conflits*, 3 (2013a), pp. 65–80.

—, 'La vidéo comme outil de l'action collective et de la lutte armée', in *Pas de printemps pour la Syrie. Les clés pour comprendre les acteurs et les défis de la crise (2011–2013)*, ed. by François Burgat and Bruno Paoli (Paris: La Découverte, 2013b), pp. 172–184.

David J. Bolter and Richard Grusin, *Remediation. Understanding New Media* (Cambridge: MIT Press, 1999).

Pamela Cohn, 'Peter Snowdon Interview', *Bomb Magazine*, Online Resource: http://bombmagazine.org/article/1000048/peter-snowdon (accessed 18 November 2020).

Robyn Creswell, 'Voices from a Different Syria', *The New York Review of Books*, Online Resource: http://www.nybooks.com/daily/2016/03/21/voices-from-different-syria-abounaddara-films/ (accessed 18 November 2020).

Lucas Derycke, interview by Nicky Aerts on *Klara.be*, 18 September 2017, Online Resource: https://klara.be/pompidou-op-maandag-18-september (accessed 18 November 2020).

Anthony Downey, *Uncommon Grounds. New Media and Critical Practices in North Africa and the Middle East* (London and New York: I.B. Tauris, 2014).

Chad Elias and Zaher Omareen, 'Syria's Imperfect Cinema', in *Syria Speaks. Art and Culture from the Frontline*, ed. by Malu Halasa and Zaher Omareen (London: Saqui Books, 2014), pp. 207–216.

Julio Garcia Espinosa, 'For an Imperfect Cinema', *Jump Cut: A Review of Contemporary Media*, 20 (1979), pp. 24–26.

Laura Feinstein, 'This Syrian Film-making Collective Shows the Banality of Life in War', *Vice*, 2015, Online Resource: https://www.vice.com/en_ca/article/mvxgap/this-syrian-art-collective-shines-an-intimate-light-on-life-in-their-war-torn-country-456 (accessed 18 November 2020).

Forensic Architecture 'Al-Jinah Mosque', Online Resource: https://www.youtube.com/watch?v=mOyihqEOfYA (accessed 18 November 2020).

Jean-Marie Guehenno, 'The Arab Spring is 2011, not 1989', *The New York Times*, Online Resource: http://www.nytimes.com/2011/04/22/opinion/22iht-edguehenno22.html?_r=0 (accessed 18 November 2020).

James Harkin, 'Good Media, Bad Politics? New Media and the Syrian Conflict', *Reuters Institute for the Study of Journalism Fellow Papers*, 2013, Online Resource: https://reutersinstitute.politics.ox.ac.uk/our-research/good-media-bad-politics-new-media-and-syrian-conflict (accessed 18 November 2020).

Ekaterina Haskins, 'Between Archive and Participation: Public Memory in a Digital Age', *Rhetoric Society Quarterly*, 37 (2007), pp. 401–422.

Simon Hawkins, 'Teargas, Flags and the Harlem Shake: Images of and for Revolution in Tunisia and the Dialectics of the Local and the Global', in *The Political Aesthetics of Global Protest. The Arab Spring and Beyond*, ed. by Pnina Werbner, Martin Webb, and Kathryn Spellman-Poots (Edinburgh: Edinburgh University Press/The Aga Khan University, 2014), pp. 31–52.

Jessica Holland, 'Syrian Video Collective Brings Glimpse of Life During Wartime', *The Guardian*, Online Resource: https://www.theguardian.com/artanddesign/2014/jul/18/syrian-video-collective-abounaddara (accessed 18 November 2020).

Henry Jenkins, 'Twitter Revolutions?', Online Resource: http://spreadablemedia.org/essays/jenkins/#.Wn2u_HkiGM8 (accessed 18 November 2020).

Henry Jenkins, Sam Ford, and Joshua Green, *Spreadable Media. Creating Value and Meaning in a Networked Culture.* (New York: NYU Press, 2013).

Lina Khatib, *Image Politics in the Middle East. The Role of the Visual in Political Struggle* (London and New York: I.B. Tauris, 2013).

Charif Kiwan, 'Cellphone War Reporting Dumbs Down the Truth', *Newsweek*, Online Resource: http://www.newsweek.com/cellphone-war-reporting-dumbs-down-truth-289763 (accessed 18 November 2020).

Jens Maier-Rothe, Dina Kafafi, and Azin Feizabadi, 'Citizens Reporting and the Fabrication of Collective Memory', in *Uncommon Grounds. New Media and Critical Practices in North Africa and the Middle East*, ed. by Anthony Downey (London and New York: I.B. Tauris, 2014), pp. 70–85.

Lev Manovich, 'The Exceptional and the Everyday: 144 Hours in Kiev', *2014 IEEE International Conference on Big Data*, 2014, pp. 72–79.

Alexander J. Martin, 'YouTube "Deleted Syrian War Crime Evidence"', *Sky News*, Online Resource: http://news.sky.com/story/youtube-deleted-syrian-war-crime-evidence-10989430 (accessed 18 November 2020).

Marshall McLuhan, *Understanding Media. The Extensions of Man* (London: Routledge, 1964).

Annette Michelson, *Kino-Eye: the Writings of Dziga Vertov* (Berkeley and Los Angeles: University of California Press, 1964).

W.J.T. Mitchell, *Picture Theory. Essays on Verbal and Visual Representation* (Chicago: University of Chicago Press, 1995).

Rabih Mroué, 'The Pixelated Revolution', in *Image(s), mon amour. Fabrications*, ed. by Rabih Mroué (Madrid: CA2M Centro de Arte Dos de Mayo, 2013), pp. 378–393.

Laura Mulvey, *Death 24x a Second. Stillness and the Moving Image* (London: Reaktion Books, 2009).

Lorena Muñoz-Alonso, 'Syrian Film Collective Abounaddara Withdraws from Venice Biennale Claiming Censorship', *Artnet*, Online Resource: https://news.artnet.com/exhibitions/syrian-film-collective-abounaddara-withdraws-venice-biennale-claiming-censorship-297525 (accessed 18 November 2020).

Randall Packer, 'Painting with Data: a Conversation with Lev Manovich', *Hyperallergic*, 2014, Online Resource: http://hyperallergic.com/159820/painting-with-data-a-conversation-with-lev-manovich/ (accessed 18 November 2020).

Alexa Robertson, 'What's Going On? Making Sense of the Role of the Media in the Arab Uprisings', *Sociology Compass*, 9(7) (2015), pp. 531–541.

Katarzyna Ruchel-Stockmans, 'Televised Image in/as History. Videograms of a Revolution and the visibility of the 1989 Changes', in *Visualisierungen des Umbruchs. Strategien und Semantiken von Bildern zum Ende der kommunistischen Herrschaft im östlichen Europa*, ed. by Ana Karaminova and Martin Jung (Bern: Peter Lang, 2012), pp. 47–68.

—, 'The Revolution Will Be Performed. Cameras and Mass Protests in the Perspective of Contemporary Art', *Acta Universitatis Sapientiae, Film and Media Studies*, 10 (2015), pp. 7–23.

—, 'Towards a Poor Cinema: The Performativity of Mobile Cameras in New Image Wars,' in *Visualizing War: Emotions, Technologies, Communities*, ed. by Kathrin Maurer and Anders Engberg-Pedersen. (London: Routledge, 2017), pp. 110-29.

Melena Ryzik, 'Videos of Syrian Life Pulled from Internet in Protest Effort', *The New York Times*. Online Resource: https://www.nytimes.com/2017/07/25/arts/design/videos-of-syrian-life-pulled-from-internet-in-protest-effort.html?mcubz=0 (accessed 18 November 2020).

The Syrian Archive, 'Methods and Tools', Online Resource: https://syrianarchive.org/en/about/methods-tools (accessed 18 November 2020).
—, 'The Al-Jinah Mosque Complex Bombing – New Information and Timeline', Online Resource: https://syrianarchive.org/p/blog/al_jinah_mosque_us_strike_timeline/ (accessed 18 November 2020).
Melissa Wall and Sahar El Zahed, 'The Arab Spring "I'll Be Waiting for You Guys"': A YouTube Call to Action in the Egyptian Revolution', *International Journal of Communication*, 5 (2011), pp. 1333–1343.
Pnina Werbner, Martin Webb, and Kathryn Spellman-Poots, *The Political Aesthetics of Global Protest. The Arab Spring and Beyond* (Edinburgh: Edinburgh University Press/The Aga Khan University, 2014).
Maria Xynou and Hadi Al Khatib, 'Syria's Digital Civila War', *openDemocracy*, Online Resource: https://www.opendemocracy.net/opensecurity/maria-xynou-hadi-al-khatib/syria%27s-digital-civil-war (accessed 18 November 2020).
Dork Zabunyan, *L'insistance des luttes. Images soulèvements contre-révolutions* (Paris: De L'Incidence, 2016).

About the Author

Katarzyna Ruchel-Stockmans is assistant professor in contemporary art, photography, and new media at the Vrije Universiteit Brussel. She graduated with a degree in philosophy and art history in Cracow, Poland, and subsequently obtained a doctorate in art history at the University of Leuven, Belgium, where she was fellow of the Lieven Gevaert Research Centre for Photography. She has served as a visiting professor at the University of Warsaw, Poland (2012–2013). Her research interests include art and photography theory, media studies, and East European cultures, history, and representations. She is the author of *Images Performing History* (2015).

7. Cinematic Spaces of 'the Arab Street': Mohamed Diab's Inverted Road Movie *Clash* (2016)

Alena Strohmaier

Abstract

This chapter examines how cinema challenges and inverts traditional spaces of social upheavals, such as streets and squares, in their capacity to be spaces of knowledge and solidarity, in conceptualizing them as enhanced media-sensible spaces. Through a close reading of Mohamed Diab's feature film *Clash* (2016), I foreground the idea of the truck as a cinematic space predicated on its ability to accommodate movement, both in a literal and a metaphorical sense. This allows for a discussion of cinematic spaces of the so-called 'Arab street', created by both mise en scène and cinematography that go against the more prevalent images of street fights and mass demonstrations as seen in documentaries about the popular upheavals in the Middle East and North Africa region since 2009.

Keywords: Cinema, inverted road movie, Egypt, the Arab street

Introduction

Films about the Egyptian Revolution 2011 and the ensuing protests in 2013 are mainly documentaries striking a balance between internet and mobile phone footage, such as videos made by citizen journalists, and first-hand material, such as interviews with talking heads.[1] They often feature subjective camera

1 Such as: *1/2 Revolution* by Omar Shargawi and Karim El Hakim (2011); *Tahrir 2011: The Good, The Bad And The Politician* by Tamer Ezzat (2011); *We Are Egypt* by Lillie Paquette (2012); *The Noise of Cairo – Art, Cairo and Revolution* by Heiko Lange (2012); *The Square* by Jehane Noujaim

Strohmaier, A. and A. Krewani (eds.), *Media and Mapping Practices in the Middle East and North Africa: Producing Space*. Taylor & Francis Group, 2021
DOI 10.5117/9789462989092_CH07

shots, clearly obtained from mobile digital cameras. In so doing, they strive to provide a comprehensive visual account of contexts and actors involved. These documentaries seem to be marked by what sociologist Ilka Eickhof calls 'the wish to support the struggle for the one, truthful narrative *of how things really were* [...] to regain firm ground in some aims at supporting a cause based on serving a hype- and interest-driven attention economy' (2016, p. 20; emphasis in original). They promise so-called real-life scenes of a revolution, which has been more complex and contradictory than the films lead us believe.

The Egyptian Revolution of 2011 consisted of mass demonstrations and the occupation of the central Tahrir Square in Cairo, which became the emblem of the revolution. Millions of protesters from a range of socio-economic and religious backgrounds demanded the overthrow of Egyptian President Hosni Mubarak. The revolution started with calls for protests from online youth groups. Initially, these included liberal, anti-capitalist, nationalist, and feminist elements, but it ultimately included Islamist elements as well. After Mubarak's downfall, the Muslim Brotherhood candidate Mohamed Mursi (1951–2019) won the election and became Egypt's new, and first democratically elected, president. However, in June 2013, marking the one-year anniversary of his inauguration, millions of Egyptians protested against him and his politics. As he refused to resign, both his opponents and his supporters took to the streets, which resulted in violent clashes.

Film-maker Mohamed Diab chose an alternative format to that of the documentary to narrate these political and social tensions of 2013. His internationally acclaimed feature film *Clash* (*Eshtebak*) (2016), officially selected by the 2016 Cannes Film Festival as the opening film of the 'Un Certain Regard' section as well as the Egyptian entry for the Best Foreign Language Film at the 89th Academy Awards, is based on a fictional set-up: it shows a group of people from different political and social backgrounds detained in a police truck, brought together by fate during the turmoil that followed the ousting of former president Mursi from power in 2013. The film's slogan states 'conflict is on the inside', emphasizing not the geographical location of the street or the square as the main public hubs of social and political struggle, but closed and coincidental spaces such as the inside of a police truck.

Through a close reading of the film, the goal of this chapter is to foreground the idea of cinematic spaces predicated on their ability to accommodate

(2013); *Cairo Drive* by Sherief Elkatsha (2013); *We Are The Giant* by Greg Barker (2014); *Art War* by Marco Wilms (2014); *The Trials of Spring* by Gini Reticker (2015); *Abdo – Coming of Age in a Revolution* by Jakob Gross (2015); *Yallah! Underground* by Farid Eslam (2015); etc.

movement, both in a literal and in a metaphorical sense. This allows for a discussion of particular cinematic spaces in relation to the notion of 'the Arab street' created by both mise en scène and cinematography in *Clash* (2016) that go against the more prevalent images of street fights and mass demonstrations as seen in documentary films. My main argument is that *Clash* (2016) challenges traditional spaces of social upheavals such as streets and squares. In so doing, I will demonstrate how this, what I would call an *inverted road movie*, shows spaces that are closed, yet mobile and fluid at the same time. Combining the film analysis with theories from political science, this chapter follows an interdisciplinary approach to offer new perspectives on cinema's possibilities to produce and negotiate mobile media spaces.

From Opinions to Movements

Laura Sharp's summary of the connection between cinema and cartography as, on the one hand, cinema's innate 'mapping impulse' (Castro 2009) and, on the other hand, a desire to re-humanize the map in cartography (Caquard and Taylor 2009) strives to encompass the multilayeredness of a rising research field. The geographer emphasizes different approaches, such as visual studies scholar Giuliana Bruno's position of cinema as a 'modern cartography' with a virtual mobile gaze (Bruno 2002), and philologist Tom Conley's broad subscription to a notion of affective, mobile geography of flânerie (Conley 2007). She therefore positions cinematic cartography as a cognitive issue in which the viewer charts experiences and emotions onto mental maps of the world (Sharp 2018). Geographer Chris Lukinbeal, on his side, reminds us of sociologist Michel de Certeau's two modes of spatial narration: the map and the tour (1988). With the former, one sees and orders places from above; with the latter, one narrates the embodied practice of going and doing. Where seeing from above is associated with maps as representations, the embodied going and doing is about practice and touring (Lukinbeal 2018).

Following these approaches, I argue in this chapter for the widely discussed notion of 'the Arab street' as not just a location or scene of the Egyptian uprisings, but as we shall see through the film analysis, a *mobile media space* that includes a variety of media and mapping practices. In Arabic, 'the Arab street' (الشارع العربي) has a double meaning. It commonly translates to 'public opinion in Arab nations' (Collins English Dictionary 2013), however refers both to 'the street' as a geographical location in urban settings as well as to 'the public' as an opposition to officials and governments. In the course of the popular uprisings of 2011, 'the Arab street' became a key symbol in its double meaning to designate both 'the street' and 'the public'. However, political scientist David Pollock criticized already before 2011 the assumption of a uniformity of *the* Arab opinion 'when in truth it varies considerably, within and among different Arab societies, and over time', and further the assumption of a static relationship between Arab governments and public opinion 'when in fact that relationship may be quite dynamic' (1992, p. 11). This paved the way for a more fluid and multidirectional concept of 'the Arab street', as taken up by political scientist Asef Bayat, who differentiates between 'street politics' and 'political streets'. For him, the 'street' designates in general

> the chief locus of politics for ordinary people, those who are structurally absent from the institutional positions of power – the unemployed, casual workers, migrants, people of the underworld, and house wives. It serves as a key medium wherein sentiments and outlooks are formed, spread, and expressed in a remarkably unique fashion. (Bayat 2010, p. 211)

However, 'street politics' does not mean the mere appropriation of the geographical, physical street as a social space of rebellion:

> *Street politics* [...] is more than just about the conflict between the authorities and the deinstitutionalized or 'informal' people over the active use of public space and the control of public order. Streets as spaces of flow and movement are not protesting, but where they extend their protest beyond their immediate circles to include the unknown, the 'strangers' who might espouse similar, real or imagined, grievances. (Bayat 2010, pp. 211–212; emphasis in original)

Ergo, 'political street' means, on the one hand, the symbolic and metaphoric manifestation of the street as a political momentum, but, on the other hand, also

> the collective sensibilities, shared feelings, and public judgment of ordinary people in their day-to-day utterances and practices, which are expressed broadly in the public squares – in taxis, buses, shops, sidewalks, or more audibly in mass street demonstrations. (Bayat 2010, p. 212)

In this context, I would like to conceptualize 'the Arab street' as an effective power that goes beyond its geographical and physical state as a location, as a mobile media space through which diverse political, social, and cultural practices, contrasting opinions and movements of and for a variety of people are negotiated. Mohamed Diab's *Clash* (2016) foregrounds this through the spatial transformation revolving around the conversion of filmed locations into cinematic spaces and the use of a genre that in itself already accommodates mobility: the road movie.

The classic road movie as a film genre dates back to the early days of cinema. Racing horses, moving cars, and incoming trains have generated film images since the beginning of film history. Therefore, it is not surprising that the road movie traditionally illustrates aesthetics and perceptions of modernity, as well as of identity quests. It is the movement between two places, which in itself appears as a way of life leading to individual or collective change. Road movies are mainly about being on the road. The first manifestations experienced the genre as a street film in the 1920s: the street as a counter-world to the ideas of bourgeois everyday life. A film such as Federico Fellini's *La Strada* (1954) takes the street as a metaphor for life, a place of probation, a restless anchor in the world. Some heroes of the road movie find themselves on the street, finding the restless movement a condition of life, such as the heroes of Carlos Diegues' film *Bye Bye Brazil* (1979). Others voluntarily choose the aimlessness of the street as a condition of life, such as the protagonists in Stephen Eliot's *The Adventures of Priscilla, Queen of the Desert* (1994); still others have to choose life on the street because they are on the run, such as the two heroines in Ridley Scott's *Thelma and Louise* (1991). Furthermore, the motif of the street is reminiscent in many ways of the travelling abilities of some frontier heroes of the Western genre. Likewise, some gangster films are road movies. The biker film has also, since Laslo Benedek's *The Wild One* (1953), designed the street as a place of special social group forms and styles. Finally, the trucker film turns life on the country road into a generic core motif, as did Sam Peckinpah's *Convoy* (1976), which interpreted the discontinuity of street life as the unpredictability of heroes – and as a condition for taking up the fight against the institutions of society.[2] Correspondingly, there are two main definitions of the road movie. The first one defines the genre through the vehicle itself; as cinema scholar Timothy Corrigan states, 'road movies are, by definition, movies about cars, trucks, motorcycles, or some other motoring self-descendants of the nineteenth century train' (1992, p. 144). The second defines it through movement, as road movies are about being on the move – about walking, driving, strolling, or racing

2 See *Lexikon der Filmbegriffe*: https://filmlexikon.uni-kiel.de/.

on the road. In the second definition, the car functions as a prop. It is functional as a motif and brings the protagonists from A to B, is new and shiny, or old and shabby.

However, this functionality of the car has been increasingly challenged in films of the Iranian New Wave, such as Abbas Kiarostami's *Taste of Cherry* (*Ta'am-e Gilas*) (1997) and *Ten* (*Dah*) (2002), as well as Jafar Panahi's *Taxi* (2015). In these films, the car as a semi-private, semi-public space moving through the streets of Tehran opened new ways for political, social, and cultural negotiations. The car transformed from the key characteristic of the road movie into a space in and of itself as a means of spatial experience, both for the protagonists and for the viewers (Strohmaier 2016b). In this space, the characters act more freely, expressing what is burning in their souls and breaking taboos; however, they concurrently stay on the inside of the closed (and secured) space of the car. From the function as a prop, the car thus becomes a space for action, which tells of changes that question the sociopolitical situations of the characters, transforming the vehicle into a space inhenrently cinematic. In these films the car is no longer an element *in* the film, but an element *of* the film itself (Strohmaier 2016a).

The specificity in *Clash*'s use of the truck – as both a semi-public vehicle, moving through the streets of Cairo, and an enclosed, semi-private space, in which opinions are pronounced and negotiated – connects to the films of the Iranian New Wave. However, none of the detained people and thus opinions can flee. At the same time, they are in constant exchange with a confusing and stirring outside. The film's main stance is, however, that 'out there' and 'in here' are increasingly indistinct. In that sense, *Clash* (2016) is an *inverted road movie* because it goes against the classical idea of the road movie as a life-changing and liberating journey, drawing attention to the capability of the truck as a mobile media space, inherently cinematic, opening up a multiplicity of dynamics of 'the Arab street'.

Shifting the Focus

Inside – truck – day. Metal-grilled windows and a narrow door. From a frontal position, the camera shows the inside of a room reminiscent of a prison. Daylight is entering the room and we hear engine noises. The room starts to move after a few seconds, revealing that we are inside a truck. While zooming in, the camera does not leave its position but grows out of focus, leaving only the silhouettes of the window and the door. The room dissolves and has no clear borders or lines anymore (Figure 7.1).

Figure 7.1: *Clash* (2016), opening (00:00:18–00:01:28).

The focus, traditionally referring to the degree to which light rays coming from any particular part of an object pass through the lens, thus creating sharp outlines and distinct textures that match the original object, is arbitrarily challenged as an optical property of the film. The variations in depth of field in the opening scene – through shallow focus, deep focus, and techniques such as racking focus – set the tone for the rest of the film.

The image becomes sharp again and the first two people, Egyptian-American journalists, are pushed into the truck. The police officers accuse them of being spies and close the door. The truck starts to move again. One of the journalists has difficulty breathing, reacting to the confined space he finds himself in. As other people are pushed into the truck without any explanation, among them men, women, and adolescents supporting different parties, and the day wears on, the heat becomes unbearable inside the truck. The people start taking turns at the windows and the police officers take pity on them, leaving the door open. However, the open door does not provide any rescue for the detainees; the only person that dares to leave, an elderly man, dies right away. Nevertheless, mobility and closeness do not contradict each other in this scenario. The closed space in the mise en scène consists of an interior location and closed setting that creates a mood of constriction and claustrophobia, and characters who are restricted in their movements. Tight shot composition, static framing, and barriers within the mise en scène and in the shots' foreground suggest this closeness. However, the camera changes its position throughout the film, constantly moving in this narrow space, changing its position in filming from above, from behind, from below, and through the windows and the door. The movement thus created is twofold: from the inside to the outside and from the outside to the inside, as the truck moves through the streets of Cairo, enabling manifold encounters.

The cinematic space, although created from supposed restriction, reveals various forms of dynamics. The image that is out of focus can mark the transition between movement and stillness, between dream and reality, between subjectivity and objectivity (Ullrich 2002; Lauenburger 2019). At the same time, these images appear, as I argue here, as intermedial references to other media and other modes of production; for example mobile phone videos. A specific form of mobile phone video that circulated through social media in 2011 were videos showing snippers on rooftops who, once they perceived the filming, shot directly at the person holding the mobile phone. The artist Rabih Mroué stated in his lecture performance *The Pixelated Revolution* at the documenta 14 in Kassel in 2014: 'one person is shooting with a camera, and the other with a rifle. One shoots for his life and the life of his citizens and the other shoots for his life and his regime' (Mroué 2014). This disturbing yet powerful analogy, recalling Paul Virilio's comparision of the camera as a weapon, is echoed in *Clash* (2016).

Inside – truck – day. As the truck moves through Cairo, it encounters protests. The camera on the inside of the truck shoots through a metal-grilled window on the side of the truck in a horizontal shot, showing the street full of protesters and police officers. The next shot goes vertically through another metal-grilled window on the roof of the truck, showing a couple standing on their balcony watching the scene of the street from above. This principal repeats several times, the camera playing on the bird's-eye view and worm's-eye view in moving from one window to the other, in alternately showing the street full of people and people witnessing the protests from above. Some choose to close their window shades; some start filming with their mobile phones from the rooftop. In both cases, the outside appears sharp, clear, and in plain sight, whereas the position of the camera on the inside remains visible through the blurry contours of the metal grill and the dirty windshield, as well as through the blurry and fragmented body parts, such as hands and faces, of the detainees in the shots' foreground. The sequence ends with a blurry worm's-eye view towards the sky. The focus then shifts slowly from a distance to nearby, sharpening the metal grill in an extreme close-up (Figure 7.2).

Blurryness and sharpness, and more precisely the shifting between the two, are two essential categories for *Clash*'s construction of mobile media spaces. They enable the viewer to get a sense of what is inside and of what is outside the truck, but, more importantly, also a sense of inside and outside fading into each other. The spatial limitations of the truck separate the visible from the invisible, the interplay of in and out of focus, however, allows the cinematic spaces to emerge as a necessary medium of perception. The truck then becomes the new means of defining the interwovenness of space and (in)visibility. In the zone between the visible and the invisible, a mapping practice takes place. By moving away

Figure 7.2: *Clash* (2016), view (00:15:21–00:17:26).

from clear, sharp, and narrow images to a stage of blurryness and thus transition, the spatial experience becomes tangible. These images mediate between being in and out of focus and share at least two spaces, two movements, two times. Hence, a density of meaning, of simultaneity, and of accessibility is established, which negotiate questions of perception, of understanding, of subject formation and spatialization. Martine Beugnet writes that 'for all its photographic objectivity, cinematographic vision allows for the indefinite to surface' – elements that 'resist stability and definition' (Beugnet 2017, pp. 1–2), such as the inherent confusion of the phenomenal world and its constant transformations. *Clash* (2016) magnifies this experience through techniques such as shallow focus, changes in speed, and other alterations of the focus of the image that pull it towards the painterly.

Outside – truck – night. The truck is still on the road, the detainees not freed. In the far distance, green laser lights illuminate the area. The green rays are visible from afar through the metal-grilled window, the truck drives closer, and the next shot shows, from an upper-view perspective out of the window, the same green rays and numerous uplifted hands. The scene looks more like a dancing or partying scene than a mass demonstration – no individual faces can be distinguished, though. The green rays enter the truck and mingle with the natural colours of the detainees' faces and clothes. The truck is shaken and pushed over, obliterating the difference between the green outside and the more colourful inside. The next shot shows the image rotated by 90 degrees, suggesting the truck is lying on its side, the detainees on the inside are fully absorbed by the green light now. The door opens and people from the outside grasp for people on the inside, dragging some of them away. The following close-up reveals that the brawl does not allow a clear differentiation between an inside and outside space anymore. Everything is green; there is only one space immersed in the same light until the greenness submerges everything, dissolving the space completely, leaving only one green spot (Figure 7.3).

Clash's ending thus perfectly demonstrates the transformative power that is made possible by the truck's ability to accommodate numerous opinions and movements. Moreover, the close investigation of mise en scène and cinematography demonstrates that the cinematic space of 'the Arab street', with its intense expressions of popular mobilities, operates as a paradoxical space of oppression and resistance. It is a lived space that shows and shapes the subjective experience of social relations and conflicts. In its ability to accommodate and advance numerous movements, exemplified by the different opinions assembled in the truck, 'the Arab street' makes manifest the assertion that 'space is not merely the setting of stories' (Konstantarakos 2001, p. 1).

This brings us back to the idea of the *inverted road movie*. In the road movie, the inside of the vehicle is equated either as a space of encounter and conversation, or as a space of physical activity. The truck, however, also represents a space of waiting and watching. Ultimately, it produces spaces that have a cinematic significance and are cinematically conditioned at the same time. 'The interior space of a car makes more dramatic possibilities of character interaction' (2006, p. 13), film scholar David Laderman says, emphasizing the car and thus the road movie as a genre for cultural critique:

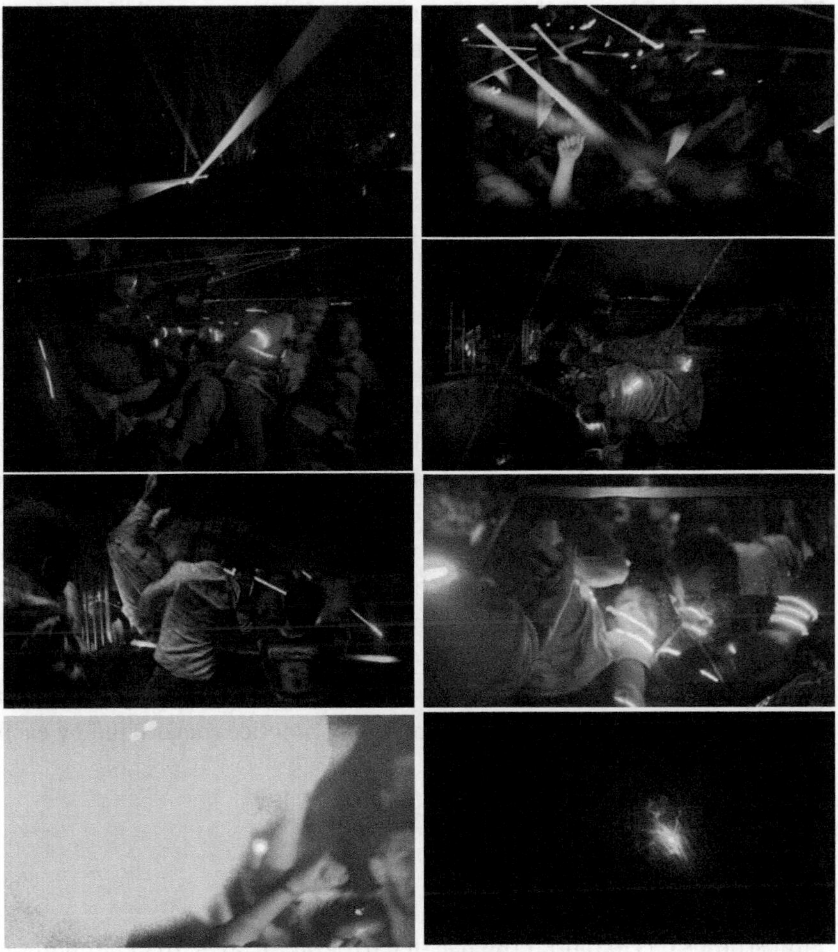

Figure 7.3 *Clash* (2016), submergence (01:30:07–01:38:00).

> The driving force propelling most road movies [...] is an embrace of the journey as a means of cultural critique. Road movies generally aim beyond the borders of cultural familiarity, seeking in unfamiliar for revelation, or at least for the thrill of the unknown. Such travelling, coded as defamiliarization, likewise suggest a mobile refuge from social circumstances felt to be lacking or oppressive in some way. This broadly conceived notion of cultural critique functions in road movies on many levels [...]. Thus the road movie celebrates subversion as a literal venturing outside of society. (2006, p. 1)

This might also be true for the *inverted road movie*; nevertheless, *Clash* (2016) does more than that: it broadens our understanding of film as map. It does so in involving locational imaging and representational practices that situate the viewer within cinematic spaces that are mobile and dynamic. Film scholar Teresa Castro uses the phrase 'mapping impulse' to reference cinema's 'particular way of seeing and looking at the world, a visual regime' (2009, p. 7). She argues that a mapping impulse envelopes cinematography through its use of various cartographic shapes. For Giuliana Bruno, on the other hand, cinema offers 'both an instrument and a route' (2002, p. 270). While cartography defines a location in absolute terms, cinema is, so I argue in this chapter, a *mobile media space*. Whereas a map requires objects in space to understand their relation to other objects, *Clash* (2016) develops a topology of image facts that behave like islands of meaning linked together by a narrative flow. Rather than dismissing 'the Arab street' as a notion that either transfers spatial knowledge or reinforces power relations, both perspectives are true when it comes to cinematic spaces. If we think of the *inverted road movie* not having any outside boundaries but always already being an interactive re-configuring of the world, then it can be constituted as an open-ended cultural practice capable of critique in itself. In this sense, *Clash* (2016) demonstrates the extent to which cinema today is marked by a complex system of media and mapping practices negotiated by each film anew.

Conclusion

Film studies has explored the use of cartographic logic to expand upon theories and methodological practices. In contrast, cartographic science has drawn on film studies to consider how to improve mapping practices related to emotion, affect, and memory. Whereas interest in cinematic cartography has focused on cross-disciplinary theoretical developments and applying spatial analysis to film, the epistemologies behind the media practices are not emphasized. Critical cartography has shown that maps as representations are social, and rather than using maps as inscriptions, the focus on understanding 'the Arab street' and its role in everyday life allows for thinking of social upheavals as media practices. This shift is a movement away from 'the Arab street' as a static product coded with power relations, ideology, and bias, to one that looks at cinematic spaces as an ongoing cultural process. The cultural critique here is bound to a specific materiality of the image (and the camera that produces it), attaching 'the

Arab street' to discursive formations and figures of (present and past) opinions.

Jean-Paul Satre's Joseph Garcin famously said, referring to the eternal entrapment with fellow damned souls in the same prison-like room of *No Exit* (1944): 'Hell is other people'. Just as there is remarkable continuity between structures, abilities, and opinions, so too are there equally remarkable differences. Member of the Muslim Brotherhood or not, police officer or not, or inside or outside the truck makes no difference. Such interstices allow insight into the fundamental role of the encounter, the connection, and separation as epistemically relevant events. This chapter situated cinematic spaces at the intersection of media and mapping practices. It did so by using the two spatial narrative modalities of 'the Arab street' and the *inverted road movie*. *Clash* (2016) shows how connections and separations work both within, beside, and in between the physicality of the street, within the aesthetic ensemble of the film, positioning the cinematic space of the truck within and amongst a topological place that resonates and confirms space as a media product.

References

Clark Arnwine and Jesse Lerner, 'Cityscapes: Introduction', *Wide Angle*, 19(4) (1997), pp. 1–7.

Asef Bayat, *Life As Politics: How Ordinary People Change the Middle East* (Stanford: Stanford University Press, 2010).

Martine Beugnet, 'Introduction', in *Indefinite Visions: Cinema and the Attractions of Uncertainty*, ed. by Martine Beugnet, Allan Cameron, and Arild Fetveit (Edinburgh: Edinburgh University Press, 2017), pp. 1–13.

Giuliana Bruno, *The Atlas of Emotion: Journeys in Art, Architecture, and Film* (New York: Verso, 2002).

Sébastien Caquard and Fraser Taylor, 'What is Cinematic Cartography?', *The Cartographic Journal*, 46(1) (2009), pp. 5–8. Online Resource: https://doi.org/10.1179/000870409X430951 (accessed 12 November 2020).

Collins English Dictionary, *Arab Street*, Online Resource: https://www.collinsdictionary.com/dictionary/english/arab-street (accessed 12 November 2020).

Timothy Corrigan, *A Cinema Without Walls. Film and Culture After Vietnam* (London: Routledge, 1992).

Teresa Castro, 'Cinema's Mapping Impulse: Questioning Visual Culture', *The Cartographic Journal*, 46(1) (2009), pp. 9–15.

Mohamed Diab, *Clash* (Arrow Films: 2016).

Ilka Eickhof, 'All That Is Banned Is Desired: "Rebel Documentaries" and the Representation of Egyptian Revolutionaries', *Middle East – Topics & Arguments*, 6 (2016), pp. 13–22. Online Resource: https://meta-journal.net/article/view/3801 (accessed 12 November 2020).

Myrto Konstantarakos, *Spaces in European Cinema* (Exeter: Intellect, 2001).

Rabih Mroué, 'The Pixelated Revolution', in *Uncommon Grounds. New Media and Ciritcal Practices in North Africa and the Middle East*, ed. by Anthony Downey (London: Tauris, 2014), pp. 247–256.

David Pollock, *"The Arab Street"? Public Opinion in the Arab World* (Washington, DC: The Washington Institute for Near East Policy, 1992).

David Laderman, *Driving Visions. Exploring the Road Movie* (Austin: University of Texas Press, 2006).

Adina Lauenburger, *Das Unschärfebild: Medientheorie einer Wissensform* (Marburg: Büchner-Verlag, 2019).

Henri Lefebvre, *The Production of Space* (Oxford and Cambridge: Blackwell, 1991).

Lexikon der Filmbegriffe, Online Resource: http://filmlexikon.uni-kiel.de/index.php?action=lexikon&tag=det&id=313 (accessed 12 November 2020).

Chris Lukinbeal, 'The Mapping of "500 Days of Summer": A Processual Approach to Cinematic Cartography', *European Journal for Media Studies* (2018), Online Resource: https://necsus-ejms.org/the-mapping-of-500-days-of-summer-a-processual-approach-to-cinematic-cartography/ (accessed 12 November 2020).

Laura Sharp, 'Embodied Cartographies of the Unscene: A Feminist Approach to (Geo) Visualising Film and Television Production', *European Journal for Media Studies* (2018), Online Resource: https://necsus-ejms.org/embodied-cartographies-of-the-unscene-a-feminist-approach-to-geovisualising-film-and-television-production/ (accessed 12 November 2020).

Alena Strohmaier, 'Bewegte Heterotopie: Zum Motiv des Autos in Filmen von Abbas Kiarostami', *Iran on the Road. Rote Fabrik – Fabrikzeitung*, (2016a), p. 5.

—, 'Why Stories Matter: Jafar Panahi and the Contours of Cinema', in *The State of Post-Cinema: Tracing the Moving Image in the Age of Digital Dissemination*, ed. by Malte Hagener, Vinzenz Hediger, and Alena Strohmaier (London: Palgrave Macmillan, 2016b), pp. 115–126.

Wolfgang Ullrich, *Die Geschichte der Unschärfe* (Berlin: Wagenbach, 2002).

About the Author

Alena Strohmaier is project leader (postdoc) of the research project "'But I'm not filming, I'm just doing a bit of video'. Cinematic appropriation processes of protest videos from the Middle East and North Africa since 2009" at Philipps-Universität Marburg (funded by the German Federal Ministry for Education and Research, 2019–2023). She did her PhD on films of the Iranian diaspora and initiated the scientific Open Media Studies Blog (Zeitschrift für Medienwissenschaft, with Sarah-Mai Dang). Her recent publications include the edited volume *Re-Configurations. Contextualizing Transformation Processes and Lasting Crises in the Middle East and North Africa*, edited with Rachid Ouaissa, and Friederike Pannewick (2021), and her first monograph *Medienraum Diaspora. Verortungen zeitgenössischer iransicher Diasporafilme* (2019). Currently she is editing a book on racism and film (*Rassismus und Film*, 2022) with Ömer Alkin and Rauf Ceylan.

8. Body-Space-Relation in Parkour: Street Practices and Visual Representations

Ines Braune

Abstract

Parkour today is a global subcultural scene that combines street with media practices. Parkour consists of a local moment, fundamentally concerned with the materiality of the street, and simultaneously of a global digital discourse, which involves millions of parkour actors. While the spatial knowledge requires a very close knowledge and tactile contact of the surface's nature of space, the media representations seem to reflect an opposite image, namely the detachedness of space. In this chapter, I will address the question of space-making and spatial practices in Morocco and the relation to parkour's visual representations.

Keywords: Morocco, parkour, body, street

Today, *parkour* is a global cultural scene that combines the street with media practices. In that sense, parkour consists of a local moment, fundamentally concerned with the materiality of the street, and simultaneously of a global digital discourse involving millions of parkour actors, called *traceurs*, negotiating and re-evaluating their navigation through both physical and digital space.

The first and often only contact with parkour, from the point of view of non-practitioners, frequently comes through its visual representations. Spectacular images and videos depict the weightlessness of the body and its detachment from space. Scholars have also weighed in here, analysing the relationship between the body and public space – that is, the interaction and engagement between body and space – while also evaluating parkour as an online culture, assessing the relevance of (digital) media practices within parkour and other youth and sport cultures such as skateboarding (Borden 2001; Buckingham 2009) and surfing (Booth 1996; 2001).

Strohmaier, A. and A. Krewani (eds.), *Media and Mapping Practices in the Middle East and North Africa: Producing Space*. Taylor & Francis Group, 2021
DOI 10.5117/9789462989092_CH08

This chapter focuses exclusively on the body-space relationship invoked by the practice of parkour in a local setting, as well as on the body-space relationship inherent in its visual representation. This focal point has been selected to link the two aforementioned discourses: the first about the engagement of the body with the public space, and the second about the relevance of parkour in digital media. The aim is also to shed light more generally on the fundamental tension between the practice itself, on the one hand, and its visualization, on the other. In the wake of the iconic/pictorial turn (Boehm 2001; Mitchell 2008; Alloa 2015) and the resulting culture of enhanced visuality, images are more easily created and distributed due to the digital transformation of the technology needed. Consequently, a renewed focus on the structure of the visual material and its relation to the depicted practices seems to be of particular importance.

Parkour is practised across the world, including in the Middle East and North Africa (MENA) region. I approach parkour from a Moroccan perspective, which does not mean that I am a Moroccan *traceuse*; I am a female German researcher taking the ideas and positions imparted to me by Moroccan traceurs during my field research (2012–2014), and using these as a lens through which the body-space dynamic of a particular global cultural practice can be observed. This chapter will address the following questions: What characterizes the body-space relationship and the spatial knowledge of the local discourse and the practice of parkour on the local ground? How is the body-space relationship visualized? How are these two approaches related, and what are the implications for (and beyond) the body-space relationship?

To answer these questions, this chapter is structured as follows: after a short introduction to parkour, the next section deals with its underlying theoretical assumptions with respect to the body and sensory ethnography. This is followed by an investigation into the body-space relationship during the practice of parkour, which will then be confronted with the body-space relationship inherent in visual and digital records of parkour (this section begins with a commentary on media and visual ethnography). Finally, this chapter offers an analysis of three parkour images from a Moroccan traceur, and some concluding thoughts.

Parkour as a Global and as a Moroccan Cultural Practice

The idea behind parkour is to find the most direct path between two points and to overcome any obstacle within that path, using only one's own physical and mental capacities (Belle 2009; Lauschke 2010; Angel 2011; Thibault 2013).

Although nothing other than one's own body is needed to practise parkour, it is mainly trained and organized in groups. Male parkour actors are called *traceurs*, while female actors are called *traceuses*, to indicate that these actors trace their way through their environment. Any barriers or hindrances on the route are perceived as a challenge, and will be integrated into the parkour movements, without in any way affecting or damaging them.

Alongside the street practice, part of the cultural practice of parkour is also the production of visual material, in the form of photos, videos, or other media. Indeed, street and media practices are interrelated constituent components of parkour, which itself is a circular process. The circulation between the performances on the street and those in the video clips is constituted through citation, repetition, and modification. Internet applications provide the virtual architecture for the media performances – that is, free platforms on which to post the clips and the possibility of easily circulating and commenting on them – while benches, buildings, and bridges provide the material architecture for the street performance.

Since its inception in the 1980s and 1990s in the suburbs of Paris, parkour – or the art of displacement (*l'art du déplacement*) – reached the wider public via media representations: parkour became internationally popular through films and documentaries with and about its founding figures, David Belle and Sébastien Foucan.[1] For many traceurs, their initial encounter with parkour occurred through films and, later, video clips on YouTube. Furthermore, music videos, commercials, and computer games began to copy and include parkour moves and, in doing so, increased parkour's international visibility and popularity.

In the early 2000s, parkour groups started to emerge in places as diverse as France, the United Kingdom, the United States, Mexico, Russia, Spain, and Japan. The story of parkour in Morocco can be traced back to the years 2005–2006, when young teenagers saw their first parkour moves in the French films *Yamakasi* (2001) by Ariel Zeitoun and *Banlieue 13* (2004) by Pierre Morel, interestingly both co-authored by Luc Besson.[2] They were attracted by these

1 David Belle is said to be the founding figure of parkour. He was born in 1973 in Fécamp (France), where his father, who trained as a soldier in Vietnam, taught him escape moves. Belle translated these movements into the concrete setting of cities. Sébastien Foucan was one of the first traceurs associated with Belle, who later followed his own ideas and founded free running. Free running accentuates the acrobatic, rather than the most efficient, mode of displacement. I am aware of these differences, which are also discussed in the community, but they are irrelevant for the purpose of this chapter. When I talk about parkour here, free running is included.

2 In the following part, I try to trace the history of parkour in Morocco based on interviews I conducted with different pioneers of the parkour community in the country, such as Soufian and Badr from MpK in Marrakech, and Adil from Accroches Toi in Casablanca. Both groups are considered the first parkour practitioners in Morocco. In addition to this, I refer to an interview

moves and were inspired to translate them into their own environment, so they began to train alone or with a handful of friends in their neighbourhoods. While searching for additional information and material about parkour on the internet, the groups came into contact with each other. These personal encounters were crucial for the further development of parkour in Morocco. Added impetus was provided by visits from international traceurs, such as a Belgian Moroccan traceur who was also in contact with David Belle, and who shared his experience during training sessions with traceurs in Morocco. As soon as Moroccan traceurs established a certain level of proficiency, they started to take pictures of their best movements, and those who had access to the technical equipment tried to produce footage for videos.

Due to the availability of technical equipment and the ongoing spread of digitization since the 2000s, digital media practices became integral to the cultural practice of parkour, alongside the global emergence of local groups. Today, Moroccan traceurs, as well as global parkour communities, organize themselves and communicate via different social media channels (mainly Facebook, YouTube, and Instagram).

Today, hundreds of traceurs are active all over Morocco.[3] A number of groups and individuals are active in metropolitan areas such as Casablanca, Marrakesh, and Tangier, but parkour is also practised by young people far from such urban centres – in Berkane and in the surroundings of Agadir, for example. Parkour is performed on the roofs of public schools, municipal buildings, or mosques in the city; on top of private houses in residential areas; in empty villas, on crowded beaches, and even at famous architectural sites. With respect to their corporeal skills, as well to their visual, technical, and digital engagement, traceurs are involved at different levels. They train as individual traceurs, but they have also established parkour groups, many of which have been active for several years, and are involved in global parkour institutions, such as Ibkano in Casablanca or Jlafet in Tangier. The different local actors are well connected via social media but also personally; they visit each other and undertake shared training sessions in different Moroccan cities.

Moroccan parkour practitioners have produced themselves. In it, Ahmad Bakkali, a brother of the founding member Badr of MpK, tells the story of parkour in Morocco: Ahmed Bakkali's Interview – PARKOUR, https://www.youtube.com/watch?v=fb9Tk_v6KGs (last accessed 18 November 2020).

3 The following description about the present-day Moroccan parkour community is the result of field research and participant observation in Morocco since 2012. In addition to this, I conducted qualitative interviews with around 30 *traceurs* and 3 *traceuses*, mainly in 2014 during field research, but also in the following years, up to today. In the course of the chapter, I refer to the interviews I conducted.

The fact that Moroccan parkour actors originate from a diverse range of social backgrounds rather contradicts the common assumption that global cultural trends are mainly the preserve of middle-class adolescents. Nothing other than one's own body is required to practise parkour, which opens up opportunities for participation, irrespective of the financial resources at one's disposal. In interviews, traceurs often mention that, through parkour, they have encountered people from different social backgrounds, and that they have established close ties with them.

Like other body-centred sport cultures (surfing, skateboarding, climbing, etc.), parkour is rather male dominated.[4] Nonetheless, it is also practised by some women, including in Morocco and other Arab countries. However, *traceuses* in Arab countries are confronted by a gamut of prejudices towards the idea of women moving expressively through public space.[5]

The Body and Sensory Ethnography

The body, in my approach, is a regulated but conscious and sensory entity. In recent decades, the body and its practices have become increasingly prominent in social theory through concepts such as Bourdieu's '*habitus*' (1977), Butler's 'gender performativity' (1988; 1990), Foucault's 'biopolitics' (1976), and Howes' 'sensing culture' (2003). Within the hyper-rationalist

4 There is a scientific debate about the role of gender in sport cultures, which also questions hegemonic white masculinities in it. See e.g. Robinson 2008 with a focus on climbing, Thorpe 2008 for snowboarding, Yochim 2009 for skateboarding, Kidder 2013 for parkour, and the important volume *Women in Action Sport Cultures: Identity, Politics, and Experience* (2016), edited by Holly Thorpe and Rebecca Olive, which includes many inspiring articles and interviews with pioneers in this field. For parkour in this edited volume, see Wheaton 2016. See also the volume entitled *New Sporting Femininities: Embodied Politics in Postfeminist Times* (2018), edited by Toffoleti et al.

5 There is sparse literature about women and gender in sport cultures in 'non-Western' settings. Wheaton writes in her article about gender and parkour: 'These insights remind us to look outside of our predominantly white Western perspectives and empirical contexts to map the differences in gender and its intersectionality with ethnicity and religion, as well as other markers of identity to expose the complex and contradictory articulations of identity in these informal but increasingly globally widespread spaces and settings in which action sports takes place' (2016, p. 130). However, here are some references for parkour: Braune (2019) on gender and parkour in the Arab World, and Thorpe and Ahmad (2015) about parkour in Gaza. There is also a quite vibrant female parkour community in Iran; you can find some journalistic snapshots about it here: 'Fast-paced parkour offers outlet for women in Iran', 2014: http://www.nydailynews.com/life-style/health/iranian-women-embrace-parkour-article-1.1731975 (last accessed 18 November 2020), and Fazekas 2018.

and hyper-intellectualized context of social theory, notions of the 'body', 'things', and 'practical knowledge' have become increasingly relevant as well (Reckwitz 2002, p. 258). A focus on practices implies centring on agency, rather than an exclusive reliance on texts, discourses, and mental processes. Reckwitz draws on the work of Bourdieu, Foucault, Latour, and Butler to describe practices as a set of 'routinized bodily performances' and 'mental activities' (2002, p. 251). Focusing on practices means 'the exploration of the embeddedness of the mental activities of understanding and knowing in a complex of doings' (Reckwitz 2002, p. 258).

In this vein, my main interest lies in the reconstruction of how parkour 'is produced by a nexus of [...] practices as body/knowledge/things-complexes' (Reckwitz 2002, p. 258). A post-structuralist approach is taken here, in that space is presented as something that is materialized through bodily practices. As de Certeau puts it:

> [I]f it is true that a spatial order organizes an ensemble of possibilities (e.g. by a place in which one can move) and interdictions (e.g. by a wall that prevents one from going further), then the walker actualizes some of these possibilities. In that way, he makes them exist as well as emerge. But he also moves them about and he invents others, since the crossing, drifting away, or improvisation of walking privilege, trans-form or abandon spatial elements. (1988, p. 98)

At the moment of the parkour encounter, the identity of the architectural site is questioned. The traceur not only confirms a distinctive architectural structure but also contests its identity. Buildings and places are perceived with alternate perspectives. For example, a windowsill is explored as a potential foothold to overcome a certain distance. This allows the sill to continue to function as intended, and at the same time, its function is changed: 'in doing so the traceur not only changes his own perception of the building but also changes the building itself as perceived object' (Fuggle 2008, p. 215).

To put the spotlight on the body allows for an incorporation of the sensorial dimension, in the sense that there is no separation between the corporeal experience and the rational mind: 'the experiencing, knowing and emplaced body is therefore central to the idea of a sensory ethnography' (Pink 2015, p. 28). Concisely, a sensory ethnography includes the senses in the process through which knowledge is produced and takes into account knowledge which is not spoken. Three topics are central in the debate on the anthropology of the senses: the relationship between senses and culture, vision and its relation to other senses, and 'reflexivity that goes beyond the interrogation of how culture is "written" to examine the sites of embodied knowing' (Pink 2015, p. 13).

Sensory ethnography takes its impulses from different disciplines such as anthropology, human geography, and sociology. Due to this chapter's focus on space, I refer to works from geography and walking studies (Wunderlich 2008; Edensor 2010; Middleton 2010; Yi'En 2013), especially with regard to practising parkour in a local setting. The reference to walking studies is made due to a basic commonality between walking and parkour – namely, that only the body, bereft of any additional instruments, moves through and shapes space. The geographer Yi-Fu Tuan has consistently argued that, to achieve a more complex understanding of objects, places, and their relation, the senses must be taken into account (Tuan 1977). In *Sensuous Geographies* (1994), Paul Rodaway proposes an approach that 'offer[s] a more integrated view of the role of the senses in geographical understanding: the sense both as a relationship to a world and the senses as themselves a kind of structuring of space and defining of place' (1994, p. 4). When moving, it is not only the feet but rather the whole body that is involved and in contact with the environment. Rodaway classifies four different ways of 'getting in touch' with the environment: the 'global touch' 'represents the body's general contact with the environment' and refers to the feeling and reaffirmation of one's own body in a certain spatial context. The 'reach touch' denotes the contact of feet and fingers, for example, with the environment. Here, the feeling of the texture comes into play. The 'extended touch' is mediated through an instrument or device, while the fourth 'imagined touch' is based on tactile memory and/or expectations (Rodaway 1994, p. 48).

In her article 'Walking and Rhythmicity: Sensing Urban Space', Wunderlich explores three modes of walking and highlights implications for urban design. Irrespective of whether the form of walking is characterized as 'purposive', 'discursive', or 'conceptual', the basic rhythmicity of walking remains inherent to it: 'the rhythm of walking is influenced by the interaction with other external rhythmical events that we came across in urban environments, such as other walking practices, or other bodily movements or activities that contrast in pace or rhythm' (Wunderlich 2008, p. 133).[6] In the case of walking, it seems that '[a] notion of place is acquired through the participatory attribute of the haptic sense' (Wunderlich 2008, p. 129), which relates to the conditions of the pavement (e.g. cobblestone, flagstones, gravel) (Gehl 1987), as well as to the responsiveness (Endensor 2010) and the

6 About rhythmicity, see also Edensor 2010; about different kinds of walking, see also Middleton 2010.

affectiveness of materialities (Yi'En 2013). Edensor includes mechanisms that regulate the style of walking, from outside weather conditions to the form of clothing (2010, p. 73). He refers to Lefebvre's notion of 'dressage' 'as a means to train the body to perform and condition it to accede to particular rhythms so that people walk "properly", do not slouch, put their shoulders back and firmly plant their heels down first' (Edensor 2010, p. 71). Middleton extends this argument to the embodiment of places (2010).

In addition to the rhythmicity and purpose of walking, the quality of surfaces, weather conditions, clothing, and the learned way of walking, there are also normative expectations which are deeply embedded in society with regard to gender-appropriate behaviour and the given architectural structures that regulate the sensory body and its movement.

Sensing Parkour while Practising It

A key component of the cultural practice of parkour is seeing the environment with parkour-eyes; the perception of one's own surroundings starts to change significantly. The priority then becomes to look for good spots to perform new manoeuvres. In what follows, I refer to this as the 'parkour gaze'. On both an intentional and unintentional level, this crucial and comprehensive change of perception structures the practitioner's movements through the city. The traceur either consciously searches for a good spot, or is en route to somewhere else, such as school or work, and coincidentally discovers a suitable setting. The parkour gaze has an explicit interest in discovering potentially interesting athletic challenges, with the attention lying primarily on the surfaces of a given structure (place or object). What happens inside a building, the historical background of a certain structure or area, is not important to the successful performance of parkour-related movements.

My argument here is that this perception represents a radical perspective on public space, because it disregards the putative symbolic meaning, social status, or historical prestige of buildings, places, or objects (see also Atkinsons 'urban deconstruction' with regard to parkour, 2009). Furthermore, it is worth highlighting that the perception of public space is primarily driven by the potential athletic challenges offered by a certain architectural structure. Different architectural structures assume importance depending on one's corporeal capability and the intended parkour manoeuvres. A wall's height or inclination for climbing over or for leaping off; the distance between benches for jumping over or from one to another; the presence of a handrail to offer balance – such features now assume a special significance.

Once the traceur has arrived at a (new) spot, the surfaces and dimensions of the location are carefully measured through a close sensory investigation. The information that is useful for practising is to be found on the surface itself, rendering a detailed and deep knowledge of it indispensable. The surface's character is experienced through close tactile contact – through touching, pushing, and scratching. By tangibly touching it, the traceur acquires the necessary information with respect to stability, firmness, and grip, for example. Furthermore, distances and conditions are evaluated through careful examination, while the smell provides further information about the place. Does it smell of fragrant flowers or of rotting rubbish? The odour offers evidence about context and situatedness, as well as about the weather. The same is true for the taste – during intensive training sessions, dust as well as sweat and water are tasted (differently).

While all of this is occurring, the traceur also hears noises which the body produces during the practice of parkour: the squeak of sneakers over different surfaces, the rush of wind while flipping in the air, the heavy breathing while performing a particularly demanding sequence of motions. Furthermore, it may be possible to hear the comments of observers, as well as a range of other background noises, as parkour sessions transpire against the soundscapes of public space, whether in the centre of a metropolitan city, at the beach, or in a lower-class residential area. For some traceurs, this sonic dimension is an important component of parkour and is by no means perceived as annoying, whereas others prefer to listen to music while training. This serves different purposes: 'During the training we can listen to music that inspires us, to make our level better and to make us feel free. It also helps us not to feel bored from training' (personal interview, 16 October 2013, Facebook).

Depending on the information collected through tasting, hearing, seeing, smelling, and palpation, the traceurs start to test aspects of their planned movements before proceeding to perform complex motion sequences. The appropriation of a new location, as well as regular play sessions, are characterized by a high level of repetition. The same motions are continuously repeated ten times, a hundred times, a thousand times. A fundamental aspect of parkour is precisely this repetition, which has two interrelated meanings: first, to exercise intensively and repetitively in preparation for a move or figure, and second, to ensure that the skill is successfully completed if the figure can be repeated again and again.

Through this process of continuous repetition, the specific location becomes embodied, in a very literal sense of the word. This embodiment is intensively related to the location insofar as it is not readily reproducible at another place with different spatial and surface conditions. Furthermore, while practising parkour, the traceur comes into bodily contact with urban materials such as dust, mud, and rubbish, inhales car exhaust gases, smells a baker's shop, hears people chatting, and gets dirty. The places inscribe themselves even deeper into the body; sometimes the body is marked by small scratches, bloody wounds, or even serious injuries. In addition, of course, the body is profoundly shaped through repetitive training, not only to perform a specific place-related move but more generally with regard to its overall constitution.

Not only the body but also the place is physically shaped by direct contact. I argued above that a given building or location changes its character and function when it is perceived as an object for the parkour encounter. In terms of the more or less visible tracks left through wear and tear, this also holds true for the physical encounter. Moreover, traceurs also restructure particular settings to suit their purposes.

The relation between body and space is characterized by a close sensorial connectedness and an embodied perception of places and their spatial relations. The body is in direct contact with the space; it gets dirty, it inhales, it touches and tastes the dust and the earth. Moreover, like the place, the body is shaped by the encounter through repetitive exercises. At the same time, the places become embodied (both inwardly and outwardly) through repetitive training, as well as through small scratches and wounds.

A deep and detailed knowledge of the structures and qualities of surfaces is acquired through tactile and haptic contact, as well as through a visual, sonic, smelled, and tasted perception of it. As the parkourian body forges paths that were not originally made for walking and jumping, the sensorial encounter provides vital information for a creative and secure performance. I argue here that it is through the very practice of parkour that spatial elements acquire new and different meanings. The body-space relation imparts this meaning in the very moment of encounter. Parkour implies new ways in which space is negotiated and appropriated that were not originally intended by city planners and architects. Parkour-related practices thus implicitly contest traditional forms of behaviour, as well as the intended functions of particular public spaces.

Media and Visual Ethnography

The visual material and its digital circulation are part of parkour and open it up to a wider audience. Approaching parkour through its media practices and products allows for varying levels of involvement. In what follows, I will briefly touch on several perspectives, which conceptualize parkour as a media practice, thereby linking it to other sport (media) cultures (such as skateboarding, snowboarding, windsurfing, and climbing). These practices all have an important visual component and are characterized by a special and fundamental relationship between cultural practices and evolving media technologies. In recent research, such practices and sports have been variously described as 'life style sports' (Gilchrist and Wheaton 2013), *Stil-Kulturen* (translated as 'style cultures', Stern 2010), or as 'action sport cultures' (Thorpe and Olive 2016), and also as 'new', 'extreme', 'alternative', or 'freestyle' sports. For the purpose of this chapter, I use the term 'cultural practice' to refer to the broader cultural context and practices in which these athletic activities are embedded.

Early subcultural studies addressed media coverage, and specifically the extent to which media coverage could constitute a 'co-opting' of the subversive potential of style-orientated cultural practices (Hebdige 1979). Scholars writing in the 1990s tended to arrive at a very different conclusion, however; namely, that different cultural communities 'use specialist or "niche" media for their own purposes – not merely to disseminate information, but also to establish (and to regulate) collective identity' (Buckingham 1999; Thornton 1995).

As in the cases of surfing (Booth 1996; 2001), skateboarding (Borden 2001; Buckingham 2009), and other sports, media has played and continues to play a crucial role in the development of parkour, on many different levels (Gilchrist and Wheaton 2013; Kidder 2012; Ladewig 2008). The history of each of these cultural practices is very much linked to medial production because through films and documentaries, these practices have become internationally visible and popular. In more recent years, parkour's international visibility and popularity have been further enhanced by YouTube, music videos, commercials, and computer games, attracting young people all over the world to imitate parkour movements and adapt them to their own surroundings. Furthermore, different parkour communities are mainly (self-) organized, on both a local and global level, through particular social media channels. Gilchrist and Wheaton highlight the importance of 'continuities with pre-Internet forms of communication' (2013, p. 170), and here I would further emphasize the role of 'internet-supplementary' forms of communication. Overall, however, Facebook, YouTube, and Instagram provide the digital architecture for parkour media practices and performance. The opportunity to participate – that is, the availability of online forums on which to post information, pictures, and videos free of charge, and the possibility of easily circulating and commenting on these clips both locally and globally – constitute structures of marginalization and exclusion.

With respect to the visual material produced by parkour practitioners – and taking into account the power of representation – the story of parkour might be told as a narrative produced firstly through Western documentaries and representations about it, until, secondly, more self-presentation by the practitioners – as expressed through their own self-made videos – became prevalent. As the technology needed to take photographs and record footage becomes cheaper and more available in Morocco, Moroccan traceurs are able to produce and spread their own visual representations. Due to their creativity and professionalism, the Moroccan traceurs are able to produce digital representations of themselves.

The value of the visual material is constantly being negotiated within the different lifestyle communities in terms of what is perceived as a 'good' picture or a 'nice' video. This is shaped by, and dependent upon, the development of filming technology and equipment. The media practices of parkour range from simple unedited photos to animated pictures, HD videos recorded with smartphones, digital reflex cameras, action cameras (such as GoPro), or camera drones. These different technologies, as well as the editing software used and perspectives chosen, are central to the pictures and videos that are ultimately produced. This also touches on sensitive questions around access to technology, and especially the highly professionalized knowledge needed to manipulate it, to produce what is perceived as a 'good video'.

In the following, I will make explicit how I work with visual material as a means of knowledge production. The engagement with visual material hints at the tense relationship between visual and perceived reality, between sight and the other senses, and between the visual and textual, as the result of academic knowledge production.

I approach visual material in several different ways. As a matter of principle, this chapter deals solely with photos, even though videos are equally important components of parkour's media culture. This focus on photos has profound implications on the sensorial encounter of the body-space relationship, because pictures rely exclusively on the visual sense (whereas videos, for example, also have a sonic component). There is a close relationship between image and video in the sense that the images are themselves usually extracted from video footage, with images effectively snipped from a movement to capture the motion more expressively.

Firstly, I use photos I have personally taken to underline the sensorial description of the spots. Even though these relate exclusively to the visual dimension, they are employed here to provide a more vivid illustration of the spatial relations of the chosen sites. Furthermore, my photos were taken without any specific technology; in other words, I used a simple camera without any specific angle or filter. I am nonetheless aware that these photos are not realistic renderings of a given place, but merely constructed moments of reality.

Secondly, I work with photos taken and produced by a Moroccan traceur to analyse the body-space relationship and its fundamental constructedness in and through the visual material. The production of parkour images is critically reliant on a profound knowledge of the movement sequences in parkour, on the ability to anticipate and accompany the movements, and on the handling and manipulation of the advanced recording technology and editing software. Furthermore, knowledge of what is perceived within the parkour community as a good and thrilling image is vital. Traceurs often told me that collaboration with otherwise talented and well-equipped cinematographers who did not have a parkour background failed because they lacked this important knowledge of parkour movements and their visual representation.

To elaborate on my argument, and to highlight the specificity of the spatial knowledge in parkour images, I contrast the sensorial description of the chosen spots with the photos I have taken, which offer some idea of the general spatial setting of the edited images of one Moroccan traceur taken at the same spot.

Therefore, out of hundreds of different spots in Marrakech, I have chosen three exemplary images, due mainly to their diversity but also to their popularity. These images are from Abderrahim Arradi, a traceur from Marrakech, who has been practising parkour since 2006 and has developed a remarkable physical proficiency. Abderrahim Arradi puts a great amount of energy into the visual representation of his talents, which has gained him widespread recognition, not just locally but also globally.

A Sensorial Description of the Spots

The following sensorial descriptions of the spots encompass the sensorial encounter of parkour while practising in different public spaces, thus offering a more intimate insight into three specific settings.

Al Anber is the name of an urban quarter where several traceurs live, including Abderrahim Arradi. Al Anber can be considered as the residential courtyard of the city's lower class. This spot is very important to the traceurs, and it is used intensively as an everyday training ground (Figure 8.1). It is soundtracked by the daily hustle and bustle of the surrounding flats and the public spaces: mothers shouting for their children to come for dinner, music, televisions, the smell of cooked food. Inhabitants usually pass through the courtyard, and children play there (and, indeed, receive their first

Figure 8.1: Al Anber. Photo: Ines Braune.

Figure 8.2: Koutoubia Mosque in Marrakech. Photo: Ines Braune.

Figure 8.3: 'The castle'. Photo: Ines Braune.

parkour lessons). Nearly every single given structure in the courtyard is perceived as a potential training surface: the branches of trees are used to swing on, the iron bars at the forefront of the picture are used to train one's balance, the different distances and heights between walls are used to jump on, off, or over. The traceurs also tend the spot and try to keep it clean, while periodically rearranging it for their own special purposes, such as by putting an old tyre in the sand to use as a springboard, while an old mattress functions as an exercise mat for secure landings.

The second spot is the famous Koutoubia Mosque in Marrakech (Figure 8.2). It is one of the most well-known and important tourist attractions in the heart of the city. It lies opposite the Jamaa El Fna, is surrounded by a network of busy central roads, and is frequented by inhabitants of Marrakech as well as by local and international tourists. The parkour sessions here take place in a quite astounding setting, amidst the vivid soundscape of a city centre alive with the noise of traffic and crowds, of passers-by and astonished spectators. The athletic challenges at this spot lie in the minaret, different-sized bases of columns, and various handrails.

The next spot is a building located on the outskirts of Marrakech known to the traceurs as 'the castle' (Figure 8.3). This is an unfinished empty villa on a hill with an amazing view over Marrakech that extends to the snow-capped Atlas Mountains. Its outer façade is complete but devoid of window glass, while the interior of the multilevel villa comprises a set of incomplete stairs and only a preliminary, foundational distribution of rooms. These structures have proven inspiring, setting the scene for several extraordinary parkour recordings. The castle is a semi-public place, in the sense that it is not on any public transport link. In addition, the whole area is sealed off by a wall at the foot of the hill. The place was discovered by one of the traceurs while working nearby as a painter. However, traceurs are not the only lords of this castle; they share it with other refuge-seekers (such as the homeless or drug addicts). Because it is far from the city, the air here is filled with a certain freshness, but it also offers a glimpse into the abysses of human existence.

Visual Analysis of the Body-Space Relation in the Images

Al Anber

In the first picture, taken in the courtyard at Al Anber, we see a human body elevated in midair in the middle of the image (Figure 8.4). The bottom third is taken up by a panoramic view of Al Anber's courtyard, while the upper two-thirds are filled with blue sky and wispy clouds. It seems to be a backlit photograph: dark, with only few gradations. Details are hardly recognizable, and it is not clear whether the person depicted is male or female. The picture is characterized by a tension between the flat body and the increased spatiality of the architecture through the curved horizon.

Figure 8.4: 'Superman'. Photo: Abderrahim Arradi.

While the body is mainly black, the colours of the background are edited and form a complementary contrast between blue and orange (against the coloured background, the dark figure does not really fit into the image, or even appears detached from it). With the help of the 'hero' or 'superman' perspective (i.e. from below), the body appears frozen in the sky, with seemingly no relation to the architecture or the ground, like a flying bird that contradicts human limitations and gravity.

Koutoubia

This picture shows the famous mosque Koutoubia in the city centre of Marrakech (Figure 8.5). In the middle of the picture, the traceur bridges a divide in a wall. The leaping body is in focus, and even the face is recognizable. As in figure 8.4, the body appears entirely framed by the sky, and the photo is also taken from the 'hero' perspective, from below. In the bottom half of

Figure 8.5: 'Jump'. Photo: Abderrahim Arradi.

the picture, we see a panorama of the paved floor, parts of the wall, and the minaret of the former mosque. The upper half of the picture is filled with cloudy sky. This picture has been reworked and the contrasts sharpened; consequently, the structure of the pavement, as well as the stone wall and the minaret, are extremely clear. The same is true of the cloudy sky. The spatiality is highlighted through the curved horizon. Against this backdrop, the body seems to be frozen and serves as a point of tranquillity. In contrast to the other pictures, the standpoint of the observer is obvious, as is the landing point of the traceur. The photographer and the observer stand on the pavement and most probably recognize the sequence of the movement, the starting and landing points. Furthermore, the body is more integrated into the picture and serves as a kind of connection across the divide in the architectural structure.

While in the other two pictures the tension is created through the unexpected, the intensely motionable body and its exclusion, in this picture the body seems to recognize the impossibility of remaining suspended in the air, and connects to the manmade architecture.

The Castle

The picture is divided into two parts (Figure 8.6). On the left-hand side of the foreground, there are close, finely structured parts of an unfinished,

Figure 8.6: 'Flying body'. Photo: Abderrahim Arradi.

flat-roofed, multilevel villa. On the right-hand side, the observer has an extensive view and can see, firstly, the incline on which the villa was built, followed by the (typically Marrakechian) red-coloured sea of houses, with the horizon composed of the spectacular snow-capped Atlas Mountains. In the upper part of the picture is a rotating human body, which, as in the other two pictures, is surrounded completely by sky. In contrast to the other two pictures, however, this has not been taken from the 'hero-perspective' – in other words, not from below. Instead, the camera is located at the same height as the screened body. Due to the architectural structure of the building, this was only possible with a drone camera, which was used to produce a video clip. The photo was extracted from the video footage and subsequently edited; the colours and their contrasts were sharpened, and the photo was mirrored.

Against the flat, monotonous sky, the 'flying body' is striking, also due to its (constructed) posture, the angled arms, offset legs, curved upper body, and falling hair. The tension in this picture is created through the spatial arrangement, the sharp contrast between extreme movement and stillness, and through the uncertain moment of contact with the ground.

With regard to the description of the image content in all three pictures, the body is surrounded by sky and has no contact at all with the architectural structures of the particular setting. The images convey the impression of a body detached from the places and from the natural law of gravitation; they seem to fly or hover above their surroundings. The frozen body in the sky conveys a certain separateness from the banality of the grounded everyday: lightness and freedom versus seriousness and problems. Nevertheless, the architectural structures of the places serve as crucial points of reference for the situatedness and proficiency of the body.

The construction of this surface of the image content is quite complex. The viewer looks at what is not actually shown, what remains invisible, and thus at the techniques of construction. All signs of physical exertion are invisible; sweat-distorted faces due to the complicated and challenging moves are hidden from view. There is no indication of the direct encounter with the surfaces, no dirt, and no dust. Furthermore, there is no contextual information about the places, or about how the video was recorded, or about how long the practitioner has trained to perform such impressive moves. The qualities of the buildings' surfaces are similarly unrecognizable.

With respect to the construction of the image, the chosen perspective plays an important role. Two of the three pictures are taken with the 'hero' or 'superman' perspective from below, which makes the body appear larger. The third picture was taken at eye level by a drone camera, but because of the specific site of the recording (the most elevated point in the location), as well as of the splitting of the image, the body is also elevated above everything else.

Furthermore, complicated graphic filters are applied to the digital image to highlight, and neglect, particular information. In all three pictures, the colours as well as the contrasts are edited. In the first two pictures, the image horizons are curved, which creates a stronger sense of spatiality of the architectural structures.

Perspective, the splitting and arrangement of the image, and different graphic filters are all important tools for producing parkour images that simultaneously reflect weightlessness, tension, and risk.

Concluding Remarks on the Sensorial, the Visual, and the Textual

Street practices require an intense knowledge about and a dense sensorial contact with the surfaces of the architectural structures of the public space.

The body exerts itself, it transpires; it is exhausted, and it is satisfied. The body has a close haptic relation to the places and feels it; the body carefully measures, maps, and evaluates the surfaces through its eyes; the body smells, tastes, listens, sees, and feels the spatial surroundings. By doing so, the body reaffirms and recreates spatial relations. The body is involved as subject and actor of the movements and is manipulated through intense training sessions to perform the movements. Simultaneously, the body is shaped by the places and in a literal sense, the places are embodied as well, and the places are shaped by the encounter. Furthermore, the regular street practices are characterized by a high level of repetition, control, and safety. The body as a tool to practise will not be risked easily or recklessly.

In contrast to this, the visualization of the body-space relation in parkour reflects a contrary approach. The images by their very nature refer exclusively to the visual sense. With regard to the body-space relation, there is no direct contact between the body and the surfaces depicted. The images are edited in a way to visualize a distance between the body and the different architectural structures, even though the different places are crucial references to give meaning to the body-space relation. Moments of exertion and strain are not visible and the body becomes a flying object (instead of actively producing the moves). The detached body as object is technically manipulated by camera settings, angles, and the editing process. Repetition, control, and safety are replaced in the pictures by extraordinary singularity, risk, and danger.

Both approaches – the body-space relation while practising parkour on the streets and the body-space relation in the visual material – seem to be driven by antagonistic momentums: connectedness versus detachedness, the routine versus the extraordinary, subject versus object, and control versus risk. Both approaches follow different ideas and construct the relationship between body and space in a distinct way. This hints at the crucial claim about the limitedness of assertions about practices while looking exclusively at the visual material and vice versa.

Working with the visual material opens up a broader access to embodied knowledge but underlines a certain predominance of the visual sense, which excludes further frames of sensorial knowledge. Furthermore, the academic knowledge production is based on textuality, meaning the knowledge gained has to be translated into a written text.

I argued in the beginning of this chapter that the exploration of the body-space relationship as practice and as representation are used to relate the two to each other. They constitute components of parkour but also of other media-intense lifestyle and sport practices, though the approaches to how body and space are related are diverse and sometimes even contradictory. Parkour can be and is practised without any recording technology, but when visual material of one's own practices is targeted, the recording becomes part of the training session. Furthermore, the images, which are edited and technically manipulated for extraordinary moments, are mainly taken during routinized training sessions and settings. The body-space relationship while practising parkour is mediated through the production of visual material in various ways. The places of the routinized training sessions are reimagined through the images as results of the visual material. Moreover, the digital recordings of the movements are evaluated directly after performing the movements on the screens of the cameras. Here again, the body-space relation is perceived through the visual material with an aesthetic but also athletic claim. The direct evaluation of the material is done through an analysis of the aesthetic composition; in other words, how to improve and accentuate certain body parts or how to improve the athletic performance. The digital recording technology enables the traceurs to slice their movements into the tiniest parts (invisible to the eye while performing) to improve, control, and shape their body knowledge. Referring to Rodaway's classification of getting in touch with the environment, the 'global touch' of parkour is intensively shaped by the parkour gaze, a gaze on the places that signals aesthetic as well as athletic challenges but disregards any historical or reputational meaning. 'Reach touch' describes the sensorial encounter of body and place with all senses. The 'extended touch' is mainly mediated through the recording technology, while the 'imagined touch' refers to memory as well as the power of imagination of the different moves performed or to be performed at a place.

References

Ahmed Bakkali's Interview – PARKOUR, Online Resource: https://www.youtube.com/watch?v=fb9Tk_v6KGs (accessed 18 November 2020).

Emmanuel Alloa, 'Iconic Turn. A Plea for Three Turns of the Screw', *Theory, Culture & Critique*, 53(3) (2015).

Julie Angel, *Ciné Parkour* (self-published, 2011).

Michael Atkinson, 'Parkour, Anarcho-Environmentalism, and Poiesis', *Journal of Sport and Social Issues*, 33 (2009), pp. 169–194.

David Belle, *Parkour* (Paris: Intervista, 2009).

Ines Braune (2019), 'Gender and Parkour in the Arab World', in *Building Bodies: Transnational Historical Approaches to Sport, Gender and Ethnicities, Yearbook of Women's History*, ed. by Saskia Bultman, pp. 193-207.

Gottfried Boehm, ed., *Was ist ein Bild?* (Munich: Fink, 2001).

Douglas Booth, 'Surfing Films and Videos. Adolescent Fun, Alternative Lifestyle, Adventure and Industry', *Journal of Sport History*, 22 (1996), pp. 313–327.

—, *Australian Beach Cultures. The History of Sand, Sun and Surf* (London: Frank Cass, 2001).

Iain Borden, *Skateboarding, Space and the City: Architecture and the Body* (Berg: Oxford, 2001).

Pierre Bourdieu, *Der Tote packt den Lebenden* (Hamburg: VSA Verlag, 1997).

—, *Outline of a Theory of Practice* (Cambridge: Cambridge University Press, 1977).

David Buckingham, 'Skate Perception: Self-Representation, Identity and Visual Style in a Youth Subculture', in *Video Cultures. Media Technology and Everyday Creativity*, ed. by David Buckingham and Rebekah Willet (Basingstoke: Palgrave Macmillan, 2009), pp. 133–151.

Video Cultures, *Media Technology and Everyday Creativity*, ed. by David Buckingham and Rebekah Willet (Basingstoke: Palgrave Macmillan, 2009).

Judith Butler, 'Performative Acts and Gender Constitution: An Essay in Phenomenology and Feminist Theory', *Theatre Journal*, 40 (1988), pp. 519–531.

—, *Gender Trouble. Feminism and the Subversion of Identity* (New York: Routledge, 1990).

Michel de Certeau, Michel, *The Practice of Everyday Life* (Berkeley: University of California Press, 1988).

Tim Edensor, 'Walking in Rhythms: Place, Regulation, Style and the Flow of Experience', *Visual Studies*, 25 (2010), pp. 69–79.

Agnes Fazekas, 'Fatimeh fliegt', *Brigitte*, 2 (2018), pp. 44–48.

Michel Foucault, *The Will to Knowledge: The History of Sexuality*, 1, trans. by R. Hurley (1976) (London: Penguin, 1998).

Sophie Fuggle, 'Discourses of Subversion: The Ethics and Aestethics of Capoeira and Parkour', *Dance Research*, 26 (2008), pp. 204–222.

Jan Gehl, *Life Between Buildings. Using Public Space* (New York: Reinhold, 1987).

Paul Gilchrist and Belinda Wheaton, 'New Media Technologies in Lifestyle Sports', in *Digital Media Sport: Technology and Power in the Network Society*, ed. by Brett Hutchins and David Rowe (Florence Routledge, 2013), pp. 169–185.

Dick Hebdige, Subculture: *The Meaning of Style* (London: Methuen, 1979).

David Howes, *Sensing Culture: Engaging the Senses in Culture and Social Theory* (Ann Arbor: The University of Michigan Press, 2003).

Emily Chivers Yochim, *Skate Life: Re-Imagining White Masculinity* (Ann Arbor: University of Michigan Press, 2009).

Jeffrey L. Kidder, 'Parkour, The Affective Appropriation of Urban Space, and the Real/Virtual Dialectic', *City & Community*, 2 (2012), pp. 229–253.

—, 'Parkour, Masculinity, and the City', *Sociology of Sport Journal*, 30 (2013), pp. 1–23.

Rebecca Ladewig, 'Play It, But Seriously. Zur medialen Inszenierung von Le Parkour', *Montage AV. Zeitschrift für Theorie und Geschichte audiovisueller Kommunikation*, 17 (2008), pp. 109–130.
Jennie Middleton, 'Sense and the City: Exploring the Embodied Geographies of Urban Walking', *Social & Cultural Geography*, 11 (2010), pp. 575–596.
William John Thomas Mitchell, *Bildtheorie* (Frankfurt: Suhrkamp, 2008).
Sarah Pink, *Doing Visual Ethnography* (Los Angeles: Sage, 2013).
—, *Doing Sensory Ethnography* (Los Angeles: Sage, 2015).
Paul Rodaway, *Sensuous Geographies: Bodies, Sense and Place* (London: Routledge, 1994).
Andreas Reckwitz, 'Toward a Theory of Social Practices. A Development in Culturalist Theorising', *European Journal of Social Theory*, 5 (2002), pp. 245–265.
Victoria Robinson, *Everyday Masculinities and Extreme Sport. Male Identity and Rock Climbing* (Oxford and New York: Berg, 2008).
Martin Stern, *Stil-Kulturen. Performative Konstellationen von Technik, Spiel und Risiko in neuen Sportpraktiken* (Bielefeld: transcript, 2010).
Vincent Thibault, *Parkour and the Art du déplacement. Strength, Dignity, Community* (Montréal: Baraka Books, 2013).
Yi-Fu Tuan, *Space and Place: The Perspective of Experience* (Minneapolis: University of Minnesota Press, 1977).
Holly Thorpe and Nida Ahmad, 'Youth, Action Sports and Political Agency in the Middle East: Lessons from a Grassroots Parkour Group in Gaza', *International Review for the Sociology of Sport*, 50 (2015), pp. 678–704.
Holly Thorpe and Rebecca Olive, ed., *Women in Action Sport Cultures. Identity, Politics and Experience* (London: Palgrave, 2016).
Holly Thorpe, 'Foucault, Technologies of Self, and the Media: Discourses of Femininity in Snowboarding Culture', *Journal of Sport and Social Issues*, 32 (2008), pp. 199–229.
Kim Toffoletti, Jessica Francomb-Webb, and Holly Thorpe, ed., *New Sporting Femininities. Embodied Politics in Postfeminist Times* (London: Palgrave, 2018).
John Urry, 'The City Life and the Senses', in *A Companion to the City*, ed. by Gary Bridge and Sophie Watson (Chichester: Wiley-Blackwell, 2012), pp. 388–397.
Belinda Wheaton, 'Parkour, Gendered Power and the Politics of Identity', in *Women in Action Sport Cultures. Identity, Politics and Experience*, ed. by Holly Thorpe and Rebecca Olive (London: Palgrave, 2016), pp. 111–132.
Mato Filipa Wunderlich, 'Walking and Rhythmicity: Sensing Urban Space', *Journal of Urban Design*, 13 (2008), pp. 125–139.
Cheng Yi'En, 'Telling Stories of the City: Walking Ethnography, Affective Materialities, and Mobile Encounters', *Space and Culture* (2013), pp. 1–13.

About the Author

Ines Braune has a background in Middle Eastern and media studies and has been researching culture and media changes from a cultural studies perspective. Her research interests include youth, inequalities, and cultural practices in the Middle East. Having focused on everyday internet practices of young Moroccans, her PhD dissertation dealt with local appropriations of globally available resources. Her current research is on parkour as a mediatized cultural practice in the Arab world. She was co-editor of the open access *META Journal – Middle East Topics and Arguments* and is currently the managing director of the Marburg University Research Academy (MARA).

9. Mediated Narratives of Syrian Refugees: Mapping Victim–Threat Correlations in Turkish Newspapers

Ayça Tunç Cox

Abstract

Turkey has become the first and main transition hub for Syrian refugees. Furthermore, Turkey is spatially as well as culturally simultaneously referred to as European and Asian or Middle Eastern depending the point of view. Therefore, the representation of refugees in the Turkish press proves significant for the knowledge produced about refugees. Accordingly, this chapter strives to investigate the coverage of Syrian refugees in newspapers, which constitutes only one aspect of the overall reception of the issue in Turkey, and therefore does not claim to be exhaustive. Yet, because daily newspapers are still among the most important media sectors in Turkey, they constitute a special case of knowledge production worth investigating.

Keywords: Turkey, refugees, print media, narratives

One of the biggest current global issues is arguably the so-called refugee crisis that is likely to have an impact on every possible aspect of daily life – from employment to housing, education, and health – especially in Europe, for the coming decades. This should be evaluated in light of major debates about transnational mobility, migration, ethnic and racial discrimination, and the concomitant image and knowledge production. Arjun Appadurai argues that 'mediascapes provide large and complex repertoires of images, narratives and ethnoscapes to viewers throughout the world' (2003, p. 35). In other words, mediascapes help to create and then to consolidate transnational 'imagined communities'. In this context, mediated narratives about migration and about refugees play an important role in ethnic and cultural identification processes.

Strohmaier, A. and A. Krewani (eds.), *Media and Mapping Practices in the Middle East and North Africa: Producing Space*. Taylor & Francis Group, 2021
DOI 10.5117/9789462989092_CH09

Turkey has become the first and main transition hub for Syrian refugees, and furthermore, the country is spatially as well as culturally simultaneously referred to as European and Asian/Middle Eastern. Therefore, the representation of refugees in the daily Turkish press is significant for the knowledge produced about them. Accordingly, this study strives to investigate how different media groups in Turkey deal with the issue from different perspectives based on political stance, affiliations, and agenda. Focusing on the coverage of Syrian refugees in newspapers comprises only one aspect of the overall reception of the issue in Turkey, and therefore, this study does not claim to be exhaustive. Yet, because the daily newspapers are still among the most important media sectors, especially owing to increased global accessibility via the internet, they constitute a special case of knowledge production worth investigating.[1]

The Turkish government's official welcoming attitude notwithstanding, it appears newspapers with varied ideological affiliations subscribe to differing interpretations of the issue. To this end, the coverage in *Hürriyet*, a mainstream conservative paper; *Milliyet*, a right-wing pro-government paper; and *Cumhuriyet*, the nationalist Kemalist representative of the Turkish press, are analysed in the following. Rather than employing random sampling, certain periods that raised momentous interest among the public have been chosen: 1) the three days following the outburst of the Syrian crisis (17–19 June 2011); 2) the three days following the discovery of baby Ailan's body washed ashore (3–5 September 2015); and 3) the three days following the European Union's three billion-euro fund for Turkey (29 November–1 December 2015).

It is important to locate the critical discourse analysis of Turkish newspaper coverage of Syrian refugees in the appropriate context. It therefore seems crucial to first illustrate Turkey's position with regard to the Syrian war, and second, to disclose the ownership structures and ideological affiliations of the newspapers in Turkey that would presumably affect their framing of the issue, and thus, the ultimate portrayal of Syrian refugees.

1 As explained by Sőzeri and Güney, based on an official report by the Press Audience Research Committee (BİAK), there were 11.4 million newspaper readers in cities alone in Turkey in 2010. This indicates a decrease in readership compared to previous years; however, such a drop appears to be a common trait across the world, according to an OECD Directorate for Science, Technology and Industry Committee for Information, Computer and Communications Policy report. And yet, Sőzeri and Güney argue, the drop ratio in Turkey seems to be less in comparison with other countries, which might owe to the widespread promotional campaigns employed by Turkish newspapers (2011, p. 58).

Is the Door Really Open?

The war in Syria, which has been going on for seven years, is now an issue of international importance even though 'it was not a problem until it hit Europe', as argued by Kirişci (2016, p. 84). Millions of Syrians have escaped the country due to increasing violence. Human rights abuses such as rape, torture, and the use of incendiary and chemical weapons have been extensively reported.[2] People have fled from the war-torn area to several neighbouring countries such as Jordan, Lebanon, and Turkey, and some have even managed to take refuge further west despite extreme difficulties on migratory routes, including being forced to travel illegally, often ending up in the hands of human traffickers. It is estimated that so far around 150,000 people have been killed during flight from Syria; over twelve million have been displace, either internally or externally (Kaya 2017, p. 365).

Among all the neighbouring countries, Turkey arguably harbours the highest number of refugees from Syria. Since the beginning, Turkey officially acknowledged an 'open-door' policy regarding Syrian refugees, albeit calling them 'guests' rather than refugees, with the hope that they would soon leave.[3] This allowed an unprecedented flux of people to be admitted into the country, first in the contained refugee camps in border cities, and then across the country. According to a recent report by the European Commission (2017), the current number of refugees in Turkey is over 3.4 million.[4] This means Turkey is the country with the highest number of refugees in the world right now. By January 2017, the number of refugee babies born in Turkey was around 230,000.[5] It can easily be claimed that, although Turkey has received migrants throughout its history, the sheer scale of this recent human mobility, as well as its duration, struck the country unexpectedly; Turkey was not ready, neither legally nor socially. As rightly put by Erder, 'it did not have the legal arrangements, institutions, and resources that would cater to the needs of such a large and diverse refugee group with differing expectations' (2016, p. 126).

Initially, while complying with the 1951 Geneva Convention, to deal with this unforeseen number of refugees, Turkey appropriated a temporary

2 For further information, see: https://www.hrw.org/world-report/2017/country-chapters/syria

3 'This is partly because the geographical limitation clause of the 1951 Geneva Convention relating to the Status of Refugees maintained by Turkey restricts the refugee status only to those coming from Europe' (Toğral Koca 2015, p. 209).

4 Among those, 85 per cent appear to live outside of specifically designated refugee camps. Refer to ORSAM Report 2015 for details.

5 https://www.milliyet.com.tr/, 7 January 2017.

protection regime.[6] In addition, in February 2016, Syrian refugees were granted the right to work legally in Turkey. These implementations undeniably provided certain assurances for Syrian refugees: 'An open-door policy for all Syrians; no forced returns to Syria (non-refoulement); and unlimited duration of stay in Turkey' (Kaya 2017, p. 370). Nonetheless, as Toğral Koca explains, 'if asylum seekers are recognized as refugees, they are resettled in a third country with the support of the UNHCR. However, said latest Directive prevents refugees [...] from being considered for resettlement into third countries as *recognised refugees*' (2015, p. 211; emphasis in original). That is to say, they have no other option than to settle in Turkey as somewhat 'unwanted/overstayed' guests or to resort to dangerous, illegal means to migrate to Europe. The latter has been made even harder, if not impossible, by the highly criticized deal signed between Turkey and the European Union in March 2016 to stop movement from Turkey to European countries.[7]

For those who opt to stay in Turkey, the prospect does not seem any brighter either, as the local people's attacks against Syrians in 2014 showed. Such incidents underline the fact that the more ill-conceived, stereotypical images prevail in the media, the more prejudices are formed against the refugees among the Turkish public. Furthermore, research conducted in areas with a high population of refugees demonstrate the effects of these prejudices. Even though most of the Syrian interviewees express their gratitude towards the Turkish state and society for accepting them, among their complaints are exploitation in the labour market, bigotry and discrimination they are exposed to on a daily basis, limited access to health and education facilities, housing problems, and integration problems – further made difficult not only because of linguistic and cultural differences but also by their perceived temporariness (Dinçer et al. 2013; Erdoğan 2014; ORSAM Report 2015; Toğral Koca 2015; Kaya 2016). On the other side of the coin is the impact of the increasing number of refugees on Turkey's socio-economic structure. As addressed by the ORSAM Report in 2015, forecasted problems are an increase in child labour, an increase in polygamy and the resultant divorce cases, a potential risk of ethnic and religious polarization, a change in demographic

6 With the additional introduction of the Temporary Protection Directive in 2014, the scope of temporary protection was expanded (Toğral Koca 2015, p. 211).

7 'The so-called one-to-one initiative, as part of the deal, stipulates that for every Syrian who has travelled without authorization to Greece and been returned to Turkey, EU member states will resettle one Syrian from Turkey' (Rygiel et al. 2016, p. 316). See: http://www.consilium.europa.eu/en/press/press-releases/2016/03/18-eu-turkey-statement/ for the details of the deal. Also refer to: https://www.amnesty.org/en/latest/news/2016/06/eus-reckless-refugee-returns-to-turkey-illegal/ for the reasons for criticism of this deal.

structure, rising levels of poverty and unemployment, an increase in cases of human trafficking, an expected increase in crime levels in the long term, unplanned urbanization, inflation in border regions, undocumented workers causing resentment among Turkish residents, and so on.

In light of these issues, Turkey's official 'open-door' policy needs to be further questioned not only because this policy was suspended several times over the years but also because of the increased securitization (Huysmans 2000) of the borders via militarized and technology-driven practices (Toğral Koca 2015, p. 212). This implies at the same time exclusionary rather than standardized practices on the borders:

> The various groups arriving from Syria – Kurds, Assyrians, Turkmens, Armenians, and Yezidis – carry with them varied contentious political histories. Thus, refugees may be initially classified as Syrian, but may then subsequently be grouped as either *friend* or *enemy* along lines of ethnicity and/or religion. (Erder 2016, p. 119; emphasis added)

In other words, not every refugee from Syria receives the same treatment from the Turkish authorities while crossing the border, nor in the society, provided that they manage to enter the country in the first place.

Free Press or Whose Press?[8]

Common public perception of Syrian refugees is inevitably shaped by the knowledge produced about them in the media. The end product – that is, the news reports, narratives, and mediated images – eventually serves to create a popular imaginary that inscribes itself in Turkish collective memory. Every text is a product of a strongly connected net of social, cultural, and historical determinants, and thus, each text has a certain meaning depending on the particular context in which it is produced. In this respect, revealing the principles of the Turkish press will provide a basis for understanding the infrastructure of the news media, and consequently, the specific modes of knowledge production.

8 This section of the study provides a modified version of a previously published work: Ayça Tunç Cox. 'Hyphenated Identities: The Reception of Turkish-German Film-makers in the Daily Turkish Press', in *Turkish German Cinema in the New Millennium: Sites, Sounds, and Screens*, ed. by Barbara Mennel and Sabine Hake (New York: Berghahn, 2012), pp. 161–172.

The Turkish press has never been truly independent, either economically or politically. This is mainly due to particular structures of ownership and legislative measurements that determine the context in which the Turkish press operates. In Turkey, as elsewhere in the world, newspapers evolved from individual enterprises to national and then international monopolies or oligopolies that are subject to the rules of capitalist industrial regulations.[9] However, they heavily relied on the financial support of the state. The political and economic interventions of the government have been justified by claiming that they are to protect the public, whereas in fact the legislative measures have functioned as a censorship tool to silence the press when required (Ünlüer 2006; Erdoğan 2007). As a result, there has been an ongoing and intrinsic financial dependency upon the state and a resultant ideological conformity of the Turkish press with successive governments.

In the mid 1990s, an amendment was made to liberate the mass media industry from the state monopoly (Darendeli 2007, p. 7). However, despite their seeming financial independence, the media remained under the control of the state by the enactment of several articles such as 3984 and 301, which made it possible to imprison journalists for articulating their opinions on controversial matters, as the latter presents a vague definition of what constitutes insulting Turkishness and the Turkish state (Turkish Penal Code-TCK 2004). The political conformism enhanced during this period has become one of the inherent characteristics of the mainstream Turkish press. That is, the newspapers have customarily acted in line with official state policies, approximating the dominant discourse, if not totally reproducing it.[10]

State control over the press has increased dramatically from 2013 onwards, following Gezi Park Protests in 2013. The failed coup d'état in July 2016

9 The first newspapers in Turkey can be categorized as official and semi-official on the basis of their institutional structure. The first newspaper in the Ottoman Empire, *Bulletin de Nouvelles*, was printed in the eighteenth century by the French Embassy. This was followed by more private newspapers that were published in other foreign languages in Izmir and Istanbul. They were subsidized by the government, as they had pro-empire policies. The first newspaper in Ottoman, *Takvim-i Vekayi*, was printed in 1831 as an official journal under the full control of the government. For a detailed historiography of the Turkish press, see: Erdoğan 2007.

10 An example from the 2000s shows the continuing state influence on the Turkish media: the current governmental party in Turkey, the Justice and Development Party (AKP), charged Doğan Media Holding with a 2.5 billion-dollar fine for alleged improper business dealings on 8 September 2009 following *Hürriyet* and *Milliyet* coverage of a scandalous court case in Germany involving top AKP bureaucrats (Cagaptay). Following this fine has been a period of serious suppression of the press either via acquisitions, fines, or legislation, which has never been done so manifestly before.

exacerbated the situation, as Turkey was governed under the state of emergency for a while. Subsequently:

> Turkey jailed more journalists than any other country had in any year since the Committee to Protect Journalists began keeping records in the early 1990s [...]. The Turkish news media were hollowed out. More than 100 outlets were closed. Journalists were jailed or pushed into exile to avoid retaliation for their work. Hundreds of media workers were left unemployed. Pro-Kurdish media was almost entirely decimated. Any media outlet accused of any affiliation with the movement of exiled preacher Fethullah Gülen – whom the Turkish government accuses of being the mastermind of the coup attempt – was closed. Small leftist publications were closed. (Öğret and Ognionova 2017)[11]

Today the media industry in Turkey is an oligopoly in which a few giant media groups compete at the centre, exercise excessive power, and demonstrate a high level of political influence.[12] The newspapers constitute one leg of a vertically integrated industry, which has also expanded towards other sectors such as energy, telecommunications, banking, insurance, and finance, resulting in the formation of formidable conglomerates in the field (Figure 9.1).[13] The concentrated ownership structure and the expansive financial interests of the patrons suggest they have interests other than creating a democratic media environment. Power and business relations as such have an impact on the attitudes of newspapers towards issues that would determine the success of their wider corporate businesses.[14] Put

11 For lists of closed media institutions by statutory decrees under the circumstances of state of emergency, see: http://www.haberturk.com/gundem/haber/1272968-ohal-kararnamesiyle-16-televizyon-3-haber-ajansi-45-gazete-kapatildi; http://www.tgrthaber.com.tr/gundem/iki-yeni-khk-yayinlandi-18-vakif-14-dernek-1-gazete-ve-1-dergi-kapatildi-176205; and http://www.didf.at/index.php/tr/17-haberler/14-yeni-khk-ile-10-gazete-2-ajans-ve-3-dergi-kapat-ld.

12 It should also be highlighted here that the structure of the Turkish press has been changing rapidly. For instance, *Milliyet* used to belong to Doğan Group, owned by the giant media mogul Aydın Doğan; now it is under the Demirőren Group, which is openly pro-government. Similarly, the newspaper with the highest circulation, *Zaman*, which was alleged to have ties with Fethullah Gülen, was closed by a statutory decree. Such transformations inevitably changed the dynamics and power relations in the Turkish press. For an insightful analysis of these changes, see: Adaklı 2006.

13 For detailed historical analyses of the Turkish press's ownership structure, see: Tunç 2004; Adaklı 2006; Erdoğan 2007; Sőzeri and Güney 2011; Sőzeri 2015.

14 'This is particularly visible in the distribution of public procurements (biddings) in various industries; those who maintain close relationships with the government seem to secure the deals. This entire process was made possible by legal changes in pertinent legislations, allowing media moguls to have investments in other areas, first in 2001 and then 2011' (Sőzeri 2015, p. 14).

	Media Industry	Other Sectors
Doğan Holding	**Newspapers:** HÜRRİYET, Posta, Fanatik, Hürriyet Daily News, TME **Periodics:** Doğan Burda Dergi **Publishing (Book):** Doğan Egmont, Doğan Kitap **News Agency:** DHA TV Channels: Kanal D, CNN Türk, tv2, Dream TV, Kanal D Romania, EuroD **Production:** D Productions, InDHouse, Kanal D Home Video, Doğan Music Company Digital TV: D-Smart **Online Media:** hurriyet.com.tr, posta.com.tr, kanald.com.tr, netd.com, hurriyetaile.com, mahmure.com, bigpara.com Online Advertising and Marketing: MedyaNet, arabam.com, hurriyetemlak.com, hurriyetoto.com, yenibiris.com, ekolay.ne, yakala.com Other: Doğan Factoring, Doğan Foreign Commerce Social Responsibility: Aydın Doğan Vakfı	**Energy:** Doğan Enerji, Boyabat Elektrik Üretim ve Ticaret Ltd. Şti, Giresun Aslancık Barajı, Mersin RES, Şah RES, D-Tes Elektrik Enerjisi Toptan Satış A.S., Gas Plus Erbil Aytemiz Petrol **Retail:** D&R **Property:** Milpa **Finance:** DD Konut Finansmanı **Tourism:** Milta
Demirören Group	**Newspaper:** MİLLİYET, Vatan **Online:** milliyet.com.tr, gazetevatan.com, milliyet.tv, skorer.com, milliyetemlak.com	**Energy:** Milangaz, Oto Milangaz, Moil, Demirören-AXPO Enerji **Inudstry:** Demirören Ağır Metal Sanayi A.Ş., Parsat Piston, MS Motor Servis, D-Mermer, İstanbul Zeytinburnu Zeyport **Property:** Kemer Golf&Country Club, Kemer Golf Resort Hotel, Taksim Gayrımenkul Yatırım Geliştirme ve İşletmecilik A.Ş. Construction: Lidya Yapı, Demirören Eskidji İnşaat **Education:** Ata Koleji

Figure 9.1: Vertical integration of the surveyed media groups. Source: Sőzeri *Türkiye'de Medya-İktidar İlişkileri*, pp. 30–34

briefly, the economy politics of the Turkish press is crucial in its ideological reproduction and in the creation of knowledge (Sönmez 2003; Bek 2004; Adaklı 2006).

Partisan Papers (pro-AKP/ Government)	**Oppositional Papers** (Dissident)	**Indefinable Papers** (Government sees them as rebels while some opposition sees them as conformist/mainstream)	**Flip-floppers** (From pro-government to oppositional)	**Flip-floppers** (From oppositional to pro-government due to being seized)
MİLLİYET	CUMHURİYET	HÜRRİYET	Taraf	Zaman
Habertürk	Evrensel	Posta	Özgür Düşünce	Bugün
Sabah	Birgün	Milli Gazete	Meydan	Millet
Vatan	Yurt	Aydınlık	Yarına Bakış	
Türkiye	Sözcü		Yeni Asya	
Takvim	Özgür Gündem			
Güneş	Ortadoğu			
Akşam	Hürses			
Star	Korkusuz			

Figure 9.2: Political affiliations of Turkish newspapers. Source: http://www.aynahaber.org/yandas-akp-yanlisi-ve-muhalif-gazete-ve-televizyonlar/517/

Accordingly, Nick Ludington once identified three main ideological trajectories among Turkish newspapers: 1) private-sector liberals that support pro-Western policies; 2) nationalist Kemalist papers; and 3) political Islamist papers (cited in Kışlalı 2009).[15] Yet, with the long-lasting Justice and Development Party (AKP) reign in Turkey, we have seen a new type of categorization, which appears to be more accurate and convenient in the current Turkish context (Figure 9.2) (Köktürk 2016).[16] This classification in particular will guide our mapping in the following section.

Victims or Threats?

As addressed by Abid et al., among numerous studies across the globe that analysed the media coverage of minority groups in general and refugees in

15 Being a Kemalist basically means supporting and following the ideas of Mustafa Kemal Atatürk, and cherishing the founding principles and constituents of the Turkish Republic, which were conceived and introduced by the founder Atatürk. I believe this is the reason why Ludington uses the term 'private sector' to describe the liberal pro-Western papers, because *Cumhuriyet* in essence is a private enterprise, too. Also see Kahraman and Keyman 1998 for a further comprehensive discussion of Kemalism as an ideology and a discourse.

16 Most of these newspapers have now been closed based on statutory decrees after the failed coup d'état in July 2016. Still, the newspapers that constitute the research sample for this study continue to exist in the defined categories above; hence, the table remains.

	HÜRRİYET	MİLLİYET	CUMHURİYET
Current Ownership	Doğan Holding (Aydın Doğan)	Demirören Group (Erdoğan Demirören)	Cumhuriyet Foundation
Circulation (roughly)	340,000	142,000	51,000

Figure 9.3: Ownership and circulation figures of the surveyed newspapers. Source: Table created by the author based on gathered data.

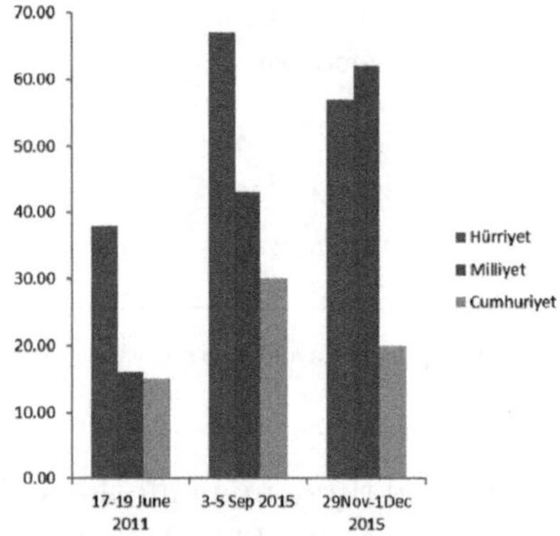

Figure 9.4: Total number of news reports analysed. Source: Graphic created by the author based on the survey results.

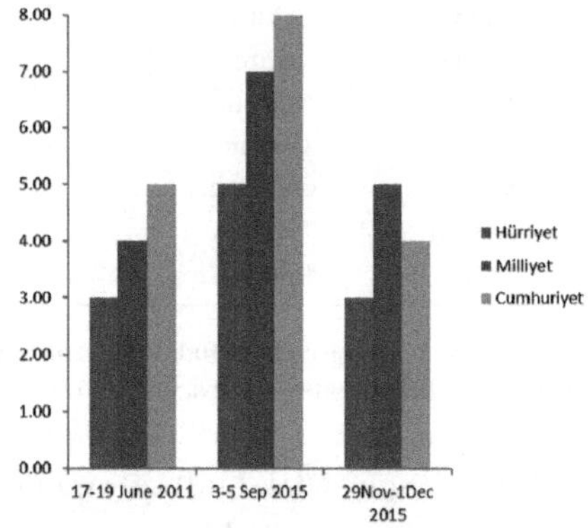

Figure 9.5: Front page coverage in the newspapers analysed. Source: Graphic created by the author based on the survey results.

particular, the majority of them 'indicates that these minority groups are systematically discriminated in media and deemed as a security, economic and hygiene threat to the majority groups' (2017, p. 122). To investigate how and to what extent the Turkish newspapers' coverage of Syrian refugees corresponds to or differentiates from this general trend, three national newspapers with online archives have been surveyed (Figure 9.3).[17] The analysis has been conducted bearing in mind Van Dijk's leading approach to critical discourse analysis, which ultimately attempts to connect the microstructure of language to the macrostructure of society (Kintsch and Van Dijk 1978; Van Dijk 2001). The focus has been news reports only, excluding any other articles and newspapers' additives. Using the keywords 'Syria', 'refugee', and 'Syrian refugees', 348 news reports in total have been surveyed (Figures 9.4 and 9.5).

17 *Hürriyet*, founded on 1 May 1948 by Sedat Simavi with the motto '*Türkiye Türklerindir*' ('Turkey belongs to Turks'), is a daily newspaper and has the highest circulation figures among the three. *Milliyet*, founded on 3 May 1950 by Ali Naci Karacan with the motto '*Basında güven*' ('Trust in Press'), is also a daily newspaper. *Cumhuriyet*, founded on 7 May 1924 by Yunus Nadi, together with his partners Nebizâde Hamdi and Zekeriya Sertel, is a daily newspaper which can be described as a low-circulation, high-impact paper.

The first three-day period is dominated by the international star Angelina Jolie's visit to a refugee camp in Hatay. That is, the front-page coverage mostly consists of various accounts of her visit. Her visit appears to have been utilized to raise a sense of national pride as well as to draw attention to the issue (Tunç Cox 2012). In this context, news reports such as *Hatay'da Angelina heyecanı* ('Excitement for Angelina in Hatay') (*Milliyet*, 17 June 2011); *Hatay'dan bir melek geçti* ('An angel has been to Hatay') (*Milliyet*, 18 June 2011); *Duydukları karşısında kahroldu* ('She was heartbroken by what she heard') (*Hürriyet*, 18 June 2011); and *İlk kez gülen çocuk gördüm* ('This is the first time I saw smiling kids') (*Hürriyet*, 19 June 2011) frame her visit as the success of the Turkish state that has welcomed Syrian refugees and offered them a decent place to live and favourable conditions compared to other refugee camps across the world. Both *Milliyet* and *Hürriyet* report Jolie's comments praising the Turkish government and the Turkish Red Cross, underlining Turkish authorities' management of the issue, their organizational skills, and their generosity. This complies with the suggested categorizations of the two papers as 'partisan' and 'indefinable', respectively (Figure 9.2), especially when combined with other news reports published during this period. For instance, news reports titled *Evden aldılar tecavüz ettiler asit döktüler* ('They took them away, raped them, and threw acid on them') (*Milliyet*, 17 June 2011); *Dünya Erdoğan'ı konuşuyor* ('The world talks about Erdoğan') (*Milliyet*, 19 June 2011); and *Erdoğan'ı alkışlıyoruz* ('We are applauding Erdoğan') (*Hürriyet*, 19 June 2011) contribute to the prevailing 'victimhood discourse' while presenting Turkey and the Turkish prime minister as the saviour of these 'wretched' Syrian refugees. It is particularly emphasized that Turkey's efforts have been highly appreciated by the international community as the reported foreign newspapers' coverages indicate.

Since the Turkish government tried hard to convince the public that Turkey's involvement in the matter was necessary, these papers make the most of an international star's visit to highlight the humanitarian aspect of the issue. Graphic narrations of Syrian refugees' horrifying experiences are used to justify Turkey's interference and its 'open-door' policy, which has been increasingly contested in the public sphere. It is worth mentioning here that both papers accentuate the then prime minister's name as the sole architect of this success story. This is probably one of the reasons why Turkey's 'Syrian refugee problem' is widely associated with the Justice and Development Party (Adelet ve Kalkinma Partisi, AKP) rather than the Turkish state's foreign policy. It is also why opposition parties such as the Republican People's Party (*Cumhuriyet Halk Partisi*, CHP) and the Nationalist Movement Party (*Milliyetçi Hareket Partisi*, MHP) 'have instrumentalised Syrian refugees to express their critics against the AKP, which they blame for deepening the Syrian crisis in the first place, thus leading to the massive migration of Syrians to Turkey at the expense of Turkish citizens' (Kaya 2017, p. 373). The more the pro-government newspapers have framed this phenomenon as the AKP's triumph; the more the oppositional media have used it as a point of critique.

In this respect, *Cumhuriyet*, as the 'dissident' representative of the Turkish press, covers Jolie's visit from a rather different perspective. As seen in the title *Angelina Jolie izdihamı* ('Stampede during Angelina Jolie's visit') (*Cumhuriyet*, 18 June 2011), somewhat negative words such as 'stampede' are employed to describe her visit. Also, Turkish authorities' superficial efforts, such as doing engineering work on the roads Jolie would use or spending extra hours cleaning, are diligently emphasized. In fact, the paper uses Jolie's visit two days in a row to attract attention to more critical issues. Accordingly, the news report *Musul-Hayfa boru hattı* ('Musul-Hayfa petrol pipeline') contains the CHP MP Mehmet Ali Edipoğlu's commentary. He argues that an ethnic conflict is being deliberately created by the parties that are after the petrol in the area. Among Edipoğlu's claims are secret meetings organized by the local authorities in Hatay, who are banned from speaking publicly, and the government's attempts to cover up the fact that the actual number of refugees is much higher (around 350,000) than what is revealed in the media (mere thousands). Furthermore, the official public declarations such as the 'refugees would be Turkmens from the region' does not reflect the truth since 'the majority of the refugees are the wives and the children of rebel fighters in Syria' (*Cumhuriyet*, 19 June 2011). This implies that Turkey sides with the rebels instead of being unbiased, as the government publicly claims to be. Through Edipoğlu's commentary, the paper argues that it was a mistake to accept refugees; instead there should have been a buffer zone for these people to take refuge, and it should have been under the protection of the United Nations. Such negative remarks are presented in bold even if given in quotation marks, assigning these parts of the report extra significance. *Cumhuriyet* has always been associated with the CHP and also has a generally nationalist attitude; therefore, it is not surprising the paper refers to the CHP members as the voice of authority in the discussion.

Similarly, the news report titled *Türkiye sınırından silah geldi* ('Weapons arrived from the Turkish border') disputes the arguments raised in the other two papers by stating that 'the Prime Minister Erdoğan receives intense criticism regarding his Syrian policy' (*Cumhuriyet*, 17 June 2011). Based on the comments of an authority figure, Allepo's mayor Ali Ahmet Mansura, who is naturally pro-Assad, Turkey is alleged to have provided weapons for the war in Syria. These news reports demonstrate that while *Milliyet* and *Hürriyet* represent

Syrian refugees as 'victims' in need of Turkey's help, thus supporting the government's decision for an 'open-door' policy, *Cumhuriyet* renders them as a 'problem', if not yet a complete 'threat', in order to criticize the government.

The second three-day period is marked by the tragic death of a refugee child, Ailan Kurdi. *Milliyet* in its coverage of the issue, including headlines such as *Utan dünya* ('Shame on you world') (*Milliyet*, 3 September 2015), which states that Turkey is the 'only' country who welcomes these refugees, cites lengthy speeches of both President Erdoğan and Prime Minister Ahmet Davutoğlu.[18] Two particular news reports, *Cumhurbaşkanı Erdoğan: O fotoğrafı görünce yıkıldım* ('President Erdoğan: I was devastated when I saw the photo') (*Milliyet*, 4 September 2015) and *Davutoğlu: O cansız beden bir uyarı sinyali* ('Davutoğlu: That inanimate body is a warning') (*Milliyet*, 5 September 2015), specifically stress how European countries have been indifferent to the issue until this tragedy. Both the president and the prime minister present a very emotional and highly dramatized account of the incident, whereby they blame 'the EU, the modern world and the international community for turning a blind eye to the ongoing crisis'. Turkey, on the other hand, is portrayed as the responsible, merciful, dutiful, hospitable, and selfless country. Such juxtapositions, reinvigorating age-old dichotomies such as 'us' versus 'them', serve to revive and consolidate Turkish national pride in the eyes of the public. These political figures' attempts to engage the Turkish public in the matter by praising them for their hospitality seems especially important when it is considered that this was the time when the highly controversial Syrian and Northern Iraq note was being discussed in the parliament.[19]

Hürriyet, too, uses these two speeches for its coverage of the tragic incident, in addition to headlines such as *Dünyayı sarstı* ('He shook the world') (*Hürriyet*, 3 September 2015); *Tokat gibi* ('Like a smack') (*Hürriyet*, 4 September 2015); *Başbakan Davutoğlu'ndan Aylan açıklaması* ('Prime Minister Davutoğlu's declaration on Ailan') (*Hürriyet*, 3 September 2015); and *Cumhurbaşkanı Erdoğan'dan önemli açıklamalar* ('Important comments from President Erdoğan') (*Hürriyet*, 3 September 2015). All in all, both papers appear to have a similar attitude again, aligning with the official discourse that portrays Syrian refugees as victims to garner public support and legitimize the government's actions.

18 It should be noted here that in August 2014, having been elected in a referendum, Recep Tayyip Erdoğan became the president of the country, and subsequently, Ahmet Davutoğlu became the leader of the AKP and the prime minister.

19 This is the note that allowed Turkey's military intervention in the region, and it was accepted despite heated debates between the government and its opposition, both in the media and in the parliament.

Subscribing to an opposing interpretation of the issue, *Cumhuriyet* develops a critical stance, albeit accepting the humanitarian aspect of the tragedy. Consequently, the news report *Elimden kayıp gitti* ('He just slipped away from my hands') (*Cumhuriyet*, 4 September 2015), narrating Ailan's father's story, alludes to President Erdoğan's comments such as 'what has drowned in the Mediterranean is our humanity'. However, the paper also reveals information that is not mentioned by the other two papers at all: a relative of the family who lives in Canada argues that Ailan's family's official application for refugee visas to Canada was refused due to complications and mistakes in the procedures in Turkey, where the family was treated badly. In a systematic attempt to undermine the government's official discourse of glory, *Cumhuriyet* highlights contradictory statements. Accordingly, the news report *Küçük Aylan artık vatanında* ('Little Ailan is in his homeland now') (*Cumhuriyet*, 5 September 2015) includes comments of some relevant parties on the matter. For instance, Human Rights Association Şanlıurfa leader Atilla Yazar claims that 'the AKP is showing off, they are not sincere [...]. They do politics over dead bodies'. Similarly, the CHP MP Sezgin Tanrıkulu argues that 'the government is responsible for the interests and lives of refugees'. The People's Democratic Party (*Halkların Demokratik Partisi*, HDP) MP İbrahim Ayhan also reinforces the criticism by stating that 'the reason for all these deaths is pro-war mentality. AKP's support for DAESH is manifest now'. It is clearly detectable here that political difference plays an important role in newspapers' coverage of the issue. So much so that *Cumhuriyet* does approve a HDP MP's statement, which it would customarily be against since the HDP is primarily a Kurdish party, not compliant with the paper's nationalist ideals and sensitivities.

When it comes to the last three-day period examined, Turkish domestic politics shadow the coverage of the European Union's three billion-euro fund for Turkey.[20] In this last period, alongside the Turkey–EU negotiations, the straining Turkish–Russian relations are covered with regard to Turkey's Syrian policy by all the papers that constitute the research sample for this study. Not surprisingly, they advocate varied aspects of the issue.

In compliance with Ludington's aforementioned categorization as papers with pro-Western policies, both *Milliyet* and *Hürriyet* celebrate the outcome of the Turkey–EU negotiations. News reports such as *AB ile 2016 umudu* ('Hope for 2016 with the EU') (*Hürriyet*, 29 November 2015), *Türkiye Almanya'nın partneridir* ('Turkey is Germany's partner') (*Hürriyet*, 1 December 2015), and *AB-Türkiye*

20 Since Tahir Elçi, the head of Diyarbakır Branch for the Body of Lawyers, was killed on 28 November 2015, *Cumhuriyet* does not cover the issue on the front page at all. Likewise, the other two papers devote much of their attention to the assassination of Elçi.

zirvesi başladı ('EU-Turkey summit has started') (*Milliyet*, 29 November 2015) present the negotiations and the resultant agreement as the turning point in Turkey–EU relations. In particular, the potential visa exemption for Turkish citizens in the Schengen area is highlighted as the government's achievement.

This should certainly be considered in the context of Turkey's long-standing attempts to become a member of the European Union. The AKP, which has been in power since 2002, paid extra attention to deepening the relationship with the European Union. 'During the AKP's first legislative term (2002–2007), its EU policy focused on using the pursuit of membership to support its legitimacy, maintain its public support, and expand Turkey's economy and its role as a regional and international actor' (Goff-Taylor 2017, pp. 1–3).[21] In this context, it can be argued that the issue of Syrian refugees is used as leverage in the Turkey–EU negotiations by both sides. For Turkey, it meant the revival of the relationship. 'Turkish identity is perpetually imagined and constructed in relation to Europe, leading to an ambivalent sense of self because Turkey has long been denied any proximity by its everlasting object of desire' (Tunç Cox 2012, p. 167). Hence, the focus on this possible outcome increases the sense of gain and victory in the Turkish reader's mind. The supportive coverage of *Milliyet* and *Hürriyet*, therefore, corresponds to their classification as 'partisan' and 'indefinable', respectively (Figure 9.2).

Cumhuriyet, on the other hand, covers the same deal as a sign of corruption. The news report *AB'den Türkiye'ye Sığınmacı rüşveti* ('The EU bribes Turkey on refugees' (*Cumhuriyet*, 30 November 2015) interprets the three billion-euro fund from the European Union to Turkey as an enticement in return for Turkey becoming a refugee camp. The paper touches upon the anxiety in the Turkish public concerning the ever-increasing number of Syrian refugees by arguing that

> the deal means the number of refugees which is currently 2.2 million will increase even more, and the promise that Turkish citizens would not need visa to enter the EU countries is not a definite but just a weak possibility.

In other words, the paper's attitude revolving around a discourse that addresses Syrian refugees as the scapegoats persists.

21 This pursuit was so determinant for the party's continued success that it even established the Ministry for EU Affairs. However, this positive attitude, which was addressed as 'Islamic Europhilia' by Ioannis Grigoriadis, was gradually replaced by a strong Euroscepticism (Tezcan and Aras 2015). Beginning with the Cyprus conflict in 2005, followed by the Syrian refugee crisis in 2011, and then the repercussions of the failed coup d'état in 2016, Turkey–EU relations reached an all-time low.

In terms of the diplomatic crisis between Turkey and Russia following the downing of a Russian plane by the Turkish army at the border, *Milliyet* and *Hürriyet* again side with the government while *Cumhuriyet* increases its criticism. News reports such as *Birçok Rus Türkiye'yle yaşanan krizden Putin'i sorumlu tutuyor* ('Many Russians hold Putin responsible for the crisis with Turkey') (*Hürriyet*, 29 November 2015), *Rus medyasından Putin'e eleştiri* ('Russian media's criticism of Putin') (*Milliyet*, 1 December 2015), and *DAEŞ'in en önemli petrol müşterisi* ('The most important oil customer for DAESH') (*Milliyet*, 1 December 2015) antagonize President Vladimir Putin and Russia for supporting Assad and present the Turkish government as the strongest side in the conflict. On the other hand, *Cumhuriyet*, in the news report *Kremlin'den Ankara'ya sert çıkış* ('The Kremlin tells off Ankara') (*Cumhuriyet*, 29 November 2015), implies that the government's restless attitude is asking for a war with Russia, putting Turkey in danger. In a similar fashion, the news report *Işid'in Petrol trafiği* ('ISIS's oil traffic') (*Cumhuriyet*, 29 November 2015) points out Turkey's involvement in the distribution of ISIS's oil as one of the biggest revenue sources for the organization. The paper claims that ISIS oil is sold to Israel through Turkey and indicates the Iraqi Kurdish Regional Government's connection in the process. Taking into account the history of Turkey–Israel relations as well as the Kurdish conflict in Turkey, it is clear that the paper depicts a negative picture to cater to the nationalist sentiment of the Turkish public.

In brief, Turkish newspapers appear to selectively highlight different aspects of the refugee phenomenon, reflecting their political affiliations and ideological standpoints. Guided by corporate interests, *Milliyet* and *Hürriyet* act as mouthpieces for the government, turning every newsworthy incident into a showcase that justifies the official Syrian refugee policy. In contrast, *Cumhuriyet*, which is categorized as 'centre left' in the Turkish context by some scholars (Kőse and Yılmaz 2012), does not refrain from creating a discourse that feeds the sense of fear in the public in order to criticize the AKP government. In other words, having been framed in the context of Turkish domestic concerns, Syrian refugees are either depicted as the victims of the global system as well as the cruel Assad regime, or as threats to the well-being and stability of Turkey and the Turkish nation.

Conclusion

In the age of new media, democratized and interactive modes of knowledge production are possible when anyone can potentially make use of advanced technologies. This increasingly includes news-making practices by ordinary/

non-professional individuals across the world. Nonetheless, conventional media, such as newspapers, still appear to play an important role in terms of knowledge production, distribution, and consumption. In other words, corporate media constitute the norm despite the celebrated interference of citizen-generated content.

Accordingly, this chapter has explored alternative approaches predominant in the Turkish press that have evidently shaped the pertinent newspaper coverage of Syrian refugees. In doing so, it has drawn attention to the intrinsic connectedness between institutions, ownership structures, and media practices, which ultimately determine the production of knowledge in the public sphere. Locating the analysis of the news coverage in the wider contexts of statutory refugee policies and the news media structure has made it possible to accentuate not only what knowledge is produced, but also how it is produced and why it is produced in that specific way. Demonstrated entanglement of press and politics has revealed the dynamics of official knowledge production.

This captured, aggregated knowledge produced about Syrian refugees both shapes and is shaped by the public discourse in a circular fashion, as exemplified by the analysis above. The fact that the Turkish news media do not differ much from the global media in the handling of refugees as the 'marginalised others' indicates a need to seek a plurality of knowledge structures. Therefore, to achieve true diversity, the hegemony in knowledge production needs to be challenged, and alternative modes of knowledge production should be introduced.

References

Preben Aamann, 'EU-Turkey Statement, 18 March 2016', *European Council of the European Union*, Online Resource: http://www.consilium.europa.eu/en/press/press-releases/2016/03/18-eu-turkey-statement/ (accessed 12 November 2020.

Raith Z. Abid, Shakila A. Manan, and Zuhair A. A. Abdulrahman, '"A Flood of Syrians Has Slowed to a Trickle": The Use of Metaphors in the Representation of Syrian Refugees in the Online Media News Reports of Host and Non-Host Countries', *Discourse & Communication*, 11(2) (2017), pp. 121–140.

Gülseren Adaklı, 'Türkiye de Medya Sahipliği Kamuya Açık Bir Bilgi Alanı mı?', Online Resource: https://m.bianet.org/bianet/medya/185011-turkiye-de-medya-sahipligi-kamuya-acik-bir-bilgi-alani-mi (accessed 12 November 2020).

–, 'Yeni Hegemonya Köşe Yazarlarına Çok Şey Borçlu', *Sol Dergisi*, Online Resource: http://arsiv.sol.org.tr/?yazino=884 (accessed 12 November 2020).

—, 'Türk Basınından Türk Medyasına: Hakim Medya Gruplarının Kısa Tarihi', Online Resource: http://www.barikat-lar.de/gorusler/medya20.htm (accessed 12 November 2020).

Amnesty International, 'EU's Reckless Refugee Returns to Turkey Illegal', Online Resource: https://www.amnesty.org/en/latest/news/2016/06/eus-reckless-refugee-returns-to-turkey-illegal/ (accessed 12 November 2020).

Benedict Anderson, *Imagined Communities* (New York: Verso Books, 2006).

Anonymous (No author given), 'Erdoğan'ı alkışlıyoruz' ['We Are Applauding Erdoğan'], *Hürriyet*, 19 June 2011.

Anonymous (No author given), 'Utan dünya' ['Shame on You, World'], *Milliyet*, 3 September 2015.

Anonymous (No author given), 'Cumhurbaşkanı Erdoğan: O fotoğrafı görünce yıkıldım' ['President Erdoğan: I Was Devastated When I Saw the Photo'], *Milliyet*, 4 September 2015.

Anonymous (No author given), 'Hatay'da Angelina heyecanı' ['Excitement for Angelina in Hatay'], *Milliyet*.

Anonymous (No author given), 'AB-Türkiye zirvesi başladı' ['EU-Turkey Summit Has Started'], *Milliyet*, 29 November 2015.

Anonymous (No author given), 'AB'den Türkiye'ye Sığınmacı rüşveti' ['The EU Bribes Turkey On Refugees'], *Cumhuriyet*, 30 November 2015.

Anonymous (No author given), 'Birçok Rus Türkiye'yle yaşanan krizden Putin'i sorumlu tutuyor' ['Many Russians Hold Putin Responsible for the Crisis With Turkey'], *Hürriyet*, 29 November 2015.

Anonymous (No author given), 'Rus medyasından Putin'e eleştiri' ['Russian Media's Criticism of Putin'], *Milliyet*, 1 December 2015.

Anonymous (No author given), 'DAEŞ'in en önemli petrol müşterisi' ['The Most Important Oil Customer for DAESH'], *Milliyet*, 1 December 2015.

Anonymous (No author given), 'Kremlin'den Ankara'ya sert çıkış' ['The Kremlin Tells Off Ankara'], *Cumhuriyet*, 29 November 2015.

Anonymous (No author given), 'Işid'in Petrol trafiği' ['Isis' Oil Traffic'], *Cumhuriyet*, 30 November 2015.

Anonymous (No author given), 'Dünya Erdoğan'ı konuşuyor' ['The World Talks About Erdoğan'], *Milliyet*, 19 June 2011.

Anonymous (No author given), 'Tokat gibi' ['Like a Smack'], *Hürriyet*, 4 September 2015.

Anonymous (No author given), 'İki yeni KHK yayınlandı: 18 vakıf, 14 dernek, 1 gazete ve 1 dergi kapatıldı' ['Two New Decree-Laws Announced: 18 foundations, 14 associations, 1 newspaper and 1 periodic closed down'] 29 April 2017, Online Resource: http://www.tgrthaber.com.tr/gundem/iki-yeni-khk-yayinlandi-18-vakif-14-dernek-1-gazete-ve-1-dergi-kapatildi-176205 (accessed 12 December 2020).

Arjun Appadurai, *Modernity at Large* (Minneapolis: University of Minnesota Press, 2003).

Mine G. Bek, 'Research Note: Tabloidisation of News Media', *European Journal of Communication*, 19(3) (2004), pp. 371–386.

Soner Cagaptay, 'Too Hot for Turkish TV', *Foreign Policy*, Online Resource: http://www.foreignpolicy.com/articles/2009/10/01/too_hot_for_turkish_TV (accessed 12 November 2020).

A. Vahap Darendeli, 'Medya Yoğunlaşması, Tekelleşmenin Denetimi ve Çoğulculuğun Kurulması' ['Media Concentration, Its Regulation and Establishing Pluralism'], *Türkiye Barolar Birliği Dergisi,* 68 (2007), Online Resource: www.rtuk.org.tr/sayfalar/IcerikGoster.aspx?icerik_id=0977101d-4e81-4e87a3bb-6f2e10388fd5" (accessed 12 November 2020).

Prime Minister Davutoğlu's declaration on Ailan, 'Başbakan Davutoğlu'ndan Aylan açıklaması' ['Prime Minister Davutoğlu's Press Conference on Aylan'], *Hürriyet*, 3 September 2015.

Nilüfer Demir, 'Dünyayı sarstı' ['He Shook the World'], *Hürriyet*, 3 September 2015.

Osman B. Dinçer et al., *Turkey and Syrian Refugees: The Limits of Hospitality* (Usak: USAK Press, 2013).

Emre Döker, 'Elimden kayıp gitti' ['He Just Slipped Away from my Hands'], *Cumhuriyet*, 4 September 2015.

Uğur Ergan, 'İlk kez gülen çocuk gőrdüm' ['This is the First Time I Saw Smiling Kids'], *Hürriyet*, 19 June 2011.

İrfan Erdoğan, *Türkiye'de Gazetecilik ve Bilim İletişimi* (Ankara: Pozitif Matbaacılık, 2007).

M. Murat Erdoğan, 'Perceptions of Syrians in Turkey', *Insight Turkey*, 16(4) (2014), pp. 65–75.

Sema Erder, 'Preliminary Thoughts on the Syrian Refugee Movement', *New Perspectives on Turkey*, 54 (2016), pp. 119–130.

Föderation demokratischer Arveitervereinigungen, 'Yeni KHK ile 10 gazete, 2 ajans ve 3 dergi kapatıldı' ['With the New Decree-Law 10 newspapers, 2 agencies and 3 periodics closed down'], Online Resource: http://www.didf.at/index.php/tr/17-haberler/14-yeni-khk-ile-10-gazete-2-ajans-ve-3-dergi-kapat-ld (accessed 12 November 2020).

Moria Goff-Taylor, 'The Shifting Drivers of the AKP's EU Policy', *The Middle East Program Occasional Paper Series*, Online Resource: https://www.wilsoncenter.org/sites/default/files/shifting_drivers_of_akps_eu_policy_0.pdf (accessed 12 November 2020).

Ioannis N. Grigoriadis, 'AKP and the Paradox of Islamic Europhilia', presented at the 2nd METU Conference of International Relations (Ankara: June 2003), Online Resource: http://turkishpolicy.com/images/stories/2004-01-evasivecrescent/TPQ2004-1-grigoriadis.pdf (accessed 12 November 2020).

Servivi Haberler, 'Evden aldılar tecavüz ettiler asit döktüler' ['They Took them Away, Raped them and Threw Acid on them']. *Milliyet*, 17 June 2011.

Haberturk, 'OHAL Kararnamesi'yle 16 televizyon, 3 haber ajansı, 45 gazete kapatıldı', Online Resource: http://www.haberturk.com/gundem/haber/1272968-ohal-kararnamesiyle-16-televizyon-3-haber-ajansi-45-gazete-kapatildi (accessed 5 December 2017).

Randall Halle, *The Europeanization of Cinema: Interzones and Imaginative Communities* (Chicago: University of Illinois Press, 2014).

Jef Huysmans, 'The European Union and the Securitization of Migration', *Journal of Common Market Studies*, 38(5) (2000), pp. 751–777.

Hasan B. Kahraman and Emin F. Keyman, 'Kemalizm, Oryantalizm ve Modernite' ['Kemalizm, Orientalizm and Modernity'], *Doğu Batı*, 1(2) (1998), pp. 65–77.

Ayhan Kaya, 'A Tale of Two Cities: Aleppo and Istanbul', *European Review*, 25(3) (2017), pp. 365–387.

Walter Kintsch and Teun A. van Dijk. 'Toward a Model of Text Comprehension and Production', *Psychological Review*, 85(5) (1978), pp. 363–394.

Mehmet A. Kışlalı, 'Yabancı Gözle Gazeteler' ['Newspapers from an Outsider's View'], *Radikal*, Online Resource: http://www.radikal.com.tr/haber.php?haberno=81556&tarih=15/07/2003 (accessed 12 November 2020).

Kemal Kirişci, 'Turkey's Role in the Syrian Refugee Crisis', *Georgetown Journal of International Affairs*, 17(2) (2016), pp. 80–85.

Abdullah Köktürk, 'Yandaş, AKP Yanlısı ve Muhalif Gazete ve Televizyonlar' ['Partisan, pro-AKP and Oppositional Newspapers and Televisions'], Online Resources: http://www.aynahaber.org/yandas-akp-yanlisi-ve-muhalif-gazete-ve-televizyonlar/517/ (accessed 12 November 2020).

–, Humanitarian Information Unit, Online Resource: https://hiu.state.gov/products/ (accessed 12 November 2020).

Avrupa Komsiyonu-Avrupa Sivil Koruma ve İnsani Yardım ECHO Tanıtım Belgesi Türkiye, *Mülteci Krizi* [*'Refugee Crisis'*], 2017, Online Resource: http://echo.info.tr/ (accessed 12 November 2020).

Aynur Köse, and Mustafa Yılmaz, 'Flagging Turkishness: the Production of Banal Nationalism in the Turkish Press', *Nationalities Papers*, 40(6) (2012), pp. 909–925.

Özgür Öğret and Nina Ognionova, 'A Year After Attempted Coup in Turkey, Media Landscape Purged of Critical Voices', Online Resource: https://cpj.org/blog/2017/07/a-year-after-attempted-coup-in-turkey-media-landsc.php (accessed 12 November 2020).

Orhan Onur Gemici, 'Davutoğlu: O cansız beden bir uyarı sinyali' ['Davutoğlu: That Inanimate Body is a Warning'], *Milliyet*, 5 September 2015.

Mahmut Oral and Fırat Kozok, 'Küçük Aylan artık vatanında' ['Little Ailan is in his Homeland Now'], *Cumhuriyet*, 5 September 2015.

ORSAM Report Orsam Yayınları, 2015, Online Resource: http://orsam.org.tr/orsam/rapor/10429?dil=tr (accessed 12 November 2020).

Güven Özalp, 'AB ile 2016 umudu' ['Hope for 2016 with the EU'], *Hürriyet*, 29 November 2015.

Cengiz Özbek, 'Duydukları karşısında kahroldu' ['She was heartbroken by what she heard'], *Hürriyet*, 18 June 2011.

Kim Rygiel, Feyzi Baban, and Suzan Ilcan, 'The Syrian Refugee Crisis: The EU-Turkey "Deal" and Temporary Protection', *Global Social Policy*, 16(3) (2016), pp. 315–320.

Mehmet A. Solak, and Akın Bodur, 'Angelina Jolie izdihamı' ['Stampede During Angelina Jolie's Visit'], *Cumhuriyet*, 18 June 2011.

—, 'Musul-Hayfa boru hattı' ['Musul-Hayfa Petrol Pipeline'], *Cumhuriyet*, 19 June 2011.

Mehmet A. Solak, 'Türkiye sınırından silah geldi' ['Weapons Arrived from the Turkish Border'], *Cumhuriyet*, 17 June 2011.

Mustafa Sönmez, *Filler ve Çimenler* (Istanbul: İletişim Yayınları, 2003).

Selahattin Sönmez, 'Cumhurbaşkanı Erdoğan'dan önemli açıklamalar' ['Important Comments from President Erdoğan'], *Hürriyet*, 3 September 2015.

Ceren Sözeri, *Türkiye'de Medya-İktidar İlişkileri* (Istanbul: İstanbul Enstitüsü Yayınları, 2015).

Ceren Sözeri and Zeynep Güney, *Türkiye'de Medyanın Ekonomi Politiği: Sektőr Analizi* (Istanbul: TESEV Yayınları, 2011).

Ercüment Tezcan and İlhan Aras, 'Adalet ve Kalkınma Partisi'nde Euroseptisizm: Avrupa Birliği Desteğinin Eleştiriye Dönüşümü', *Uluslararası Hukuk ve Politika*, 11(41) (2015), pp. 1–35.

Burcu Toğral Koca, 'Syrian Refugees in Turkey: from 'Guests' to 'Enemies'?', *New Perspectives on Turkey*, 54 (2016), pp. 55–75.

—, 'Deconstructing Turkey's "Open-Door" Policy Towards Refugees from Syria', *Migration Letters*, 3 (2015), pp. 209–225.

Aslı Tunç, 'Faustian Acts in Turkish Style: Structural Change in National Newspapers as an Obstacle to Quality Journalism in 1990–2003', in *Quality Press in Southeastern Europe*, ed. by Orlin Spassov (Sofia: SOEM2, 2004), pp. 306–323. Online Resource: http://www.interactproject.org/content/docs/reading_list/Faustian_Acts_in_Turkish_Style_Structural_Change_in_National_Newspapers_as_an_Obstacle_to_Quality_Journalism_in_1990-2003.pdf (accessed 12 November 2020).

Ayça Tunç Cox, 'Hyphenated Identities: The Reception of Turkish-German Film-makers in the Daily Turkish Press', in *Turkish German Cinema in the New Millennium: Sites, Sounds, and Screens*, ed. by Barbara Mennel and Sabine Hake (New York: Berghahn, 2012), pp. 161–172.

Ayhan O. Ünlüer, 'Medyada Etik Söylem ve Medya Etiğini Biçimleyen Temel Belirleyiciler Üzerine Bir Değerlendirme', *Küresel İletişim Dergisi*, 2 (2006), Online Resource: http://globalmedia-tr.emu.edu.tr/ (accessed 12 November 2020).

Teun A. van Dijk, 'Critical Discourse Analysis', in *The Handbook of Discourse Analysis*, ed. by Deborah Schiffrin, Deborah Tannen, and Heidi E. Hamilton (Oxford: Blackwell Publishers, 2001), pp. 352–371.

World Report 2017: 'Syria', Online Resource: https://www.hrw.org/world-report/2017/country-chapters/syria (accessed 12 November 2020).

Bünyamin Yil-Murat KİBRİTOĞLU-Ümit Kozan, 'Hatay'dan bir melek geçti' ['An Angel Has Been to Hatay'], *Milliyet*, 18 June 2011.

Begüm Zorlu, 'Suriye İç Savaşında Dört Yılda Neler Oldu?', Online Resource: http://bianet.org/bianet/siyaset/163049-suriye-ic-savasinda-dort-yilda-neler-oldu (accessed 12 November 2020).

About the Author

Ayça Tunç Cox is an assistant professor at the İzmir Institute of Technology. She completed her PhD at Royal Holloway College, University of London and received her MSc from Ege University in İzmir Turkey. She has published extensively on Turkish cinema and Turkish-German diasporic cinema. Among her research interests are transnational cinema, diasporic cinema, national cinemas, alternative cinema, and the new Turkish cinema. In 2018, she completed a research period as a Fulbright Visiting Scholar at University of Pittsburgh with a project entitled 'Images of Displacement: The Mediated Turkish Diaspora in Turkish and Turkish-German Films since the 1960s.' She is currently conducting a research project, funded by the Scientific and Technological Research Council of Turkey, entitled 'Transforming the Social Perception of the Roma in Turkey via Auto-Ethnographic Narratives.'

Part III

Agencies

10. Documenting Social Change and Political Unrest through Mobile Spaces and Locative Media

Angela Krewani

Abstract

In this chapter, I explore the media coverage of the Arab Spring and the reactions of Western media communities. Focusing on interactive documentaries and websites, this chapter clearly demonstrates to what extent media bring about individualized coverage to major events. Digital media especially have merged with cartographic competencies to provide topical information. Compared to the informational range of classic print media and television, these digital platforms and digitally distributed art forms create new and interactive forms of media participation.

Keywords: Locative media, Arab television, i-docs, interactive websites

Up to now, the majority of research on mobile spaces and locative media has proceeded with regard to European and American urban spaces – always taking for granted that a young, well-educated middle class is using these technologies for entertainment (Farman 2013; Hemmersam and Aspen 2015; De Souza e Silva 2015). However, there is a new awareness of media agency in the Middle East and North Africa (MENA) region. Digital media have become important factors within the changes and upheavals in this region as well as agents of social developments (Mellor and et al. 2011; Mellor and Rinnawi 2016; Sakr 2007; Kraidy 2016; Richter 2018). To keep track of the changing media landscapes, this chapter sets out to explore the practices and social dimensions of mobile media in the MENA region in contexts of political and humanitarian crises. Before focusing on the impact of social and mobile media systems, this chapter provides a short overview on television, since the quality and proliferation of television systems define the use of mobile and social media.

Strohmaier, A. and A. Krewani (eds.), *Media and Mapping Practices in the Middle East and North Africa: Producing Space*. Taylor & Francis Group, 2021

DOI 10.5117/9789462989092_CH10

Television in the MENA Region

Similarly to the Global North, the Global South has also been deeply affected by media changes from controlled broadcast systems to individualized, non-censored communication. Although broadcast and satellite television are a well-known factor in the MENA region, their organizational forms differ from European or North American television systems. Contrary to their democratic or commercial institutionalizations, the MENA region's television systems follow a different timeline. According to Mellor et al., until the 1980s television was a 'local, government-monopolized terrestrial operation with limited reach' (2011, p. 85), thus completely controlled by the state. The first commercial Arab television station was aired from broadcast facilities in London in 1991, a circumstance that, in some regard, matches the frequent cartographic situation of marking a region from foreign places. Since, Arab television has been on the rise:

> To the present date, television broadcasting in the Arab world has experienced a real revolution: not only in terms of global reach, channel diversity, and program quality, but also in terms of the increased involvement of the private sector in broadcast operations. Arab States Broadcasting Union (ASBU) statistics show that over 500 free-air-television channels are available to Arab viewers across the region, in addition to scores of channels furnished by cable and digital satellite TV systems. (Mellor and et al. 2011, p. 85)

The Arab States Broadcasting Union (ASBU) also conveys that over 65 per cent of television channels in mid 2009 were 'privately controlled and operated' (Mellor and et al. 2011, p. 85). Mellor et al. come to the conclusion that, despite the rapid surge of television programmes and stations, television has not supported democratization or cultural advances in the region: 'The emergence of a private broadcasting sector was seen as a purely commercial development, simply because it was not matched by an emergence of viable pluralistic and independent political structures with adversarial functions' (2011, p. 85).

Accordingly, it can be concluded that television does not offer a democratic, discursive media space. Liberal political communication is delegated to individualized social networks. Facing the individuality and variety of social media, there is not one common media space, but the construction of space as an individually embodied act – usually with the help of a smartphone:

> Because space, as Henri Lefebvre has famously noted, is social (and cannot be considered outside of the social), we must ask whose space is considered when looking at how locative media projects impact space (and how the space impacts these locative interfaces). (Farman 2015, p. 85)

Moreover, the turn towards networked digital cultures brings about a transformation of political powers and enables individuals to participate in political movements (Lim 2018), and, returning to this chapter's focus, in media spaces. Digital media bring about a reformulation of spatial dimensions by bloggers, artists, film-makers, and interested individuals. Contrary to the classical broadcast media such as television, mobile media activate diverse, sometimes small, groups before their knowledge infiltrates into a general media public. Lim terms this 'brokerage' (2018, p. 19). The term also refers to the change in the idea of media publics and audiences, which have to be re-conceptualized in accordance with the communication forms and audiences. This reformulation of audiences also affects the spatial dimensions of media, the media constitutions of space, and the reflections within media. Parallel to the re-conceptualization of regimes of knowledge and their respective media practices, the construction of media spaces plays an important role and has to be integrated into research agendas: the concept has to be triangulated between regimes of knowledge, media methods, and spatial practices, whereby the media and spatial practices usually merge. In the following, media examples will be understood along these lines. Additionally, there will be a close look at selfie cultures and artistic projects with media spaces, interactive documentaries, and investigative journalism websites. All these media forms generate a variety of media practices that accompany specific configurations of media spaces and political knowledge.

Smartphones and Geotagged Selfies

In the context of what has been called 'locative media', one of the most typical media habits is the use of smartphones by refugees. Although the smartphone is considered a luxury item, it has come to be the most valuable communication and orientation device, since it is used for communication, self-representation, and spatial orientation (Literat 2017, p. 4). Although it is an indispensable tool for communication, anti-refugee propaganda commented on refugees' possession of smartphones, which presumably turns them into 'economic migrants' (Literat 2017, p. 2). More than four million Syrians have fled their country, in what the United Nations has labelled the 'the worst refugee crisis in 25 years' (quoted in Literat 2017, p. 1). The debate about financial support and immigration options for refugees has been transported into European media. The 'refugee selfie' is especially a point of debate and a marker for the supposed cheating of refugees. As Literat remarks,

> anti-refugee commentators on social media have been especially vitriolic on one particular way refugees use smartphones: taking selfies. These images of refugees taking selfies – and significantly, not the selfies themselves – have been published in various media outlets, from *Time Magazine* to *The Daily Mail* and *The Daily Mirror*. Following their publication online, the images are then picked up and circulated via social media (especially Twitter, Facebook and Reddit), where they are fodder for further discussions and, most often, outrage. (2017, p. 2)

Besides the possession of the luxury item smartphone, the practice of selfie-taking seems to contradict the general concept of being a refugee. In a politically vitriolic and adverse treatment of refugees, the selfie represents a forbidden media practice, which is attributed to advertising and leisure time. The desperate refuge from a war-torn country 'fundamentally conflicts with the image of the refugee in our cultural imaginary, in which refugees are presumed to be sad, disconnected, and uprooted' (Literat 2017, p. 4).

However, taking a selfie is not just a technologically advanced form of self-portrait. The selfie's technology and social practice implies the sharing of the image on social websites such as Flickr, Instagram, or Facebook. Exiled from the original, located home community, the selfie thus offers participation in a virtual, social space. Moreover, selfies are usually geotagged, which connects them to a specific place. Eszter Zimanyi experiments with the aspects of the geotagged selfie by looking up the specific places of refugees on Instagram. Thus she points at the geotagged aspects of the selfie, which is added to the digital social communities (Zimanyi 2017, p. 7), insofar as the 'geotagged selfie offers an opportunity for the refugee to re-inscribe himself/herself into place after becoming stateless' (Zimanyi 2017, p. 10).

By taking selfies, refugees inscribe themselves into a new geographical space as part of the embodiment structures Jason Farman describes as connected to the phenomenological debate on film and digital media, as it is conceived by – among others – Vivian Sobchack (1988) and Mark Hansen (2004). The latter notably discusses the affective impact of digital art on the body by exploring what he calls the 'affective dimension' (Hansen 2004, p. 101) or the 'pre-sensory moments of vision' (Hansen 2004, p. 27). Affection especially takes on a dominant role in the connection between experience and medium, as Hansen argues with reference to Henri Bergson: 'There is no perception without affection. Affection is, then, that part or aspect of the inside of our body which we mix with the image of external bodies [...]' (Bergson quoted in Hansen 2004, p. 100). Whereas Hansen speaks of digital art, Farman accounts for the smartphone and its affective structures, which he considers as embodied media that help to connect to material spaces and thus shape individual, spatial embodiments, since the spaces are a hybrid of 'material' and 'medial' spaces, which can be experienced through virtual media (2012, pp. 16–34).

With this theoretical background in mind, I understand the refugee selfie as an individual inscription into space and also as an individual negotiation of the topic's visual politics. One of the first acts is to inform relatives and friends of the safe advent on an island and about the local whereabouts. As Katazyrna Marciniak observes, the visual discourse – 'viscourse'[1] – of the refugee is dominated by opposing patterns, which includes the refugee having safely arrived at the shore, like Tom Stoddard's famous photo of a father with his child celebrating their safe arrival on the Greek island of Lesbos.[2]

However, there is also the absolute counterpoint of this narrative: the body of dead child, Alan Kurdi, washed up on the Turkish shore.[3] In all these images, the shore also denotes a mythological shelter against the waves and the empty spaces of the sea. Thus the images of the living or dead bodies on the shores bring about intertextual references to narratives of shipwreck and rescue. This contextualization of shipwreck and European mythology explains why refugee selfies have been subject to artistic considerations. As Frosh resumes, 'the selfie is the production of the mediated phatic body as a visible vehicle for sociable communication with distant others who are expected to respond' (2015, p. 1623). The Chinese artist Ai Weiwei reacts to these contrasting images with a photograph of himself lying on the shore, which expresses the tragic end of a flight in a paradoxical turn: the body has arrived on the shore, not safely and triumphantly, but as a lifeless dead body.[4] In a further turn, this image could be read as commentary on refugees' situation in host countries. Sometimes the islands' shores figure as natural

1 Karin Knorr-Cetina has coined the German term *Viskurs*, referring to a combination of textual and visual discourses. I relate to this concept and translate it as 'viscourse'.

2 https://www.theannenberg.org/person/tom-stoddart/.

3 Helena Smith, 'Shocking images of drowned Syrian boy show tragic plight of refugees', https://www.theguardian.com/world/2015/sep/02/shocking-image-of-drowned-syrian-boy-shows-tragic-plight-of-refugees.

4 Monica Tan, 'Ai Weiwei poses as drowned Syrian infant refugee in "haunting" photo', https://www.theguardian.com/artanddesign/2016/feb/01/ai-weiwei-poses-as-drowned-syrian-infant-refugee-in-haunting-photo.

borders for the refugees, who are sometimes confronted with the stereotypical antipathy of the alt-right's biologically orientated rejection of their identity.

A further artistic reaction to the refugee's situation is offered by the Sri Lankan rapper M.I.A., who, having arrived in London as a refugee herself, produced and directed a video titled *Borders*, in which she restages iconic images of the refugee situation as tableaux vivants with dark-skinned actors. In this video, she visually re-enacts the iconic media images of refugee situations, such as the wired fence (which has to be overcome), the boat, and the advent on the shore. Placing herself centrally within the image, she sings 'Borders: What's up with that?/ Politics: What's up with that?/ Your privilege: What's Up with that?/ Broke people: what's up with that?/ Boat people: what's up with that?'. Through choosing exclusively men, she avoids the feeling of empathy in humanitarian crises, and can easily meet and undercut anti-migrant sentiments. Concerning aspects of media embodiment, the video's mise en scène offers the most radical relationship between body and image. By constructing iconic images by staging human bodies, the term 'embodiment' is taken more than literally, since the space is the space of the individual body, culminated in a perverse 'ornament of the masses' (Kracauer 1995, pp. 75–88). These images evoke the Busby Berkeley musical films in the use of bodily ornamentation, but they turn them into a perverse demonstration of the refugee's body, which is aligned with the land. Contrary to the manifold images of individual refugees, this underlines the fact that refugees come ashore in great numbers and that they are politically conceived of as an anonymous mass.

Interactive Documentaries

If we perceive the selfie and its geotagged possibilities as a new form of media space, the interactive documentary (i-doc) can be understood as a sampled collection of these images. A relatively new and collective form, the i-doc steps back from the construct of the documenting subject in the form of the film-maker in favour of documentary production as a collective event. In this context, the idea of documentary work as active mapping is pointed to, as Brian Winston's foreword to Judy Ashton's publication on i-docs clearly voices:

> Any project that starts with the intention to engage with the 'real' and does so by using digital technology. Unlike the traditional linear 'closed', Barthesian 'readerly' documentaries, with i-docs, at the intersection between the linear and the digital, the audience become active agents within a 'writerly', more open, documentary text. (2017, p. xv)

This development from the authorizing documentary subject towards the documenting collective, using a variety of digital media, parallels the development within cartography from institutionalized cartography towards collective geo-webbing, proceeding based on mobile digital technologies. Both cartographic and documentary activities have established a collective turn within these media practices. One of the questions arising from these changes points to the authority of speech, or who is allowed to speak for whom. Canadian film-maker Kat Cizek makes the demand for collective agency explicit when she says:

> Co-creation for me is a method or an intention to make media with people that aren't media-makers. As I go into different disciplines and worlds I see that there are very specific definitions of co-creation as being type of, for example, social entrepeneural approach. So the term is causing as many problems as it's solving. But for me co-creation is a very broad term that implies a thoughtful process, which involves a collaboration with the intent to make quality media *with* partners instead of just *about* them [...]. Technology allows us to be collaborative in new ways. (2017, p. 39; emphasis in original)

Thus, contrary to geo-webbing, which can be applied from outside the respective region, i-docs seem to rely on the actual events being displayed, since they are usually shot with a smartphone or a video camera by active participants from inside, guaranteeing their participation. These recordings can be post-produced or collected on a website and made accessible over the internet. These conditions apply to the website of the Egyptian media collective Mosireen presenting an online archive of the Egyptian revolution and the events around Tahrir Square. The collection '858.ma – an archive of resistance' provides raw footage of the revolutionary events. The number 858 refers to the number of individual videos presented by the collective, which informs the site's visitors: '858 is, of course, just one archive of the revolution. It is not, and can never be, *the archive*' (emphasis in original). Before giving access to the videos, the website informs visitors about the group's history, intent, and media practice. The 858 video clips are organized by place, topic, date, month, and keywords. Thus, the mapping aspect becomes one of the determinations for events, since users can change a certain place and are presented with the respective videos. Through this presentational form, the drop-down menus map the different places of the uprising, providing a spatial dimension to the revolutionary action. Although these videos cannot be checked for the truthful documentation of events, it is nonetheless true that they offer proof of hidden crimes and human rights violations in closed societies with media censorship (see also Dang and Strohmaier 2018). Through the open access, they offer material for further documentary and aesthetic film production, as the Syrian film-maker Ossama Mohammed has done in his documentary *Silvered Water, Syria Self Portrait* (2014), which uses the material of 1001 Syrian citizens.

Another project documenting flight from Syria is the Abounaddara Group collection, which was displayed as part of an exhibition on Arab Art in the New York City New Museum in September 2017. As the review mentions, the Abounaddara Group offers more than 250 videos online; most of the film-makers are anonymous. The group's intent is to offer alternative media coverage to the mainstream media, as one of their members recounts:

> If we accept the mainstream representation of the conflict in Syria we just have two actors. A gentleman who is a dictator but he is very polite, he smiles, a very attractive guy, but he is a criminal – and the ugly jihadists. But we don't see the society. We don't see ordinary people. So we try to show those people who are not victims, not heroes, they just try to struggle for freedom and to live with dignity. (cited in Krüger 2018)

According to this principle, the group offers a documentarian display of everyday life in Syria, avoiding direct depictions of war. Contrary to the Egyptian Mosireen collective, Abounaddara does not present names or places; it instead presents an anonymous, nameless Syria as a virtual space. On Facebook, the group rejects the clichéd media images of a war-torn Syria on account of their reproduction of

> old Orientalist clichés constraining Syrians ever since their first appearance on screen in a film by the Lumière brothers. […] Syrians are subject to a regime of representation that treats them in a way that violates the principle of dignity inscribed in stone on the pediments of global institutions. […] It also threatens Syrian society, whose fate rests on the production of images that so little reflect reality. (Abounaddara Films)

The film collective's rejection of mainstream media images clearly corresponds to the structure of the site. Contrary to the Egyptian Mosireen Group and their effort to establish a political discourse and symbolic value to the Egyptian revolution, Abounaddara expressly refuses to become a part of the journalistic-documentarian intent to inform about topical situations. The Facebook site instead connects the group's work to a European film culture and specifically to the visual and aural history of the French Resistance, with a link to an interview with Alain Resnais on *Nuit et Brouillard* (*Night and Fog*) from 1955 and a link to Léo Ferré chansons, which are underlined by images from the resistance. There are references to the French 1968 revolt and a recent post, in French and German: 'La mort n'éblouit pas les yeux des Partisans/Der Tod blendet die Augen der Partisanen nicht' (Abounaddara Films). The collective's Facebook presentation makes clear that they want to bypass the actual Syrian situation in order to address wider issues of resistance. However, this intent shapes their grasp on the spatial aspects of their website. As just mentioned above, contrary to the approaches of activists, who eagerly locate their videos within real situations and places, Abounaddara employs opposing strategies. Guided by the intellectual and theoretical critique on the 'Orientalism' of media images, they clearly dis-locate their videos and re-locate them within the European romantic discourse of historical revolutions and resistances. This is cleverly done not so much on the group's website, but through networking on social media, which shifts the focus away from the Syrian situation to a general discourse on revolution and resistance.

The organization of collective material is also undertaken by the German documentary project *My Escape* (2016),[5] which edits a collection of individual smartphone videos into a narratively coherent documentary on the topic of escape. Diverging from the collective websites, the documentary points to the refugee situation, independent of the topical situation; the film conflates refugees' stories from Syria, Afghanistan, and Iran into a coherent documentary narrative of their dire situation. Although the film is constructed from a compilation of individual videos, it is edited into a narrative on flight and escape. Correspondingly, the film starts with a sequence displaying a bomber dropping its deadly weapons on a residential area, while off-screen a voice shouts 'Allahu akhbar', accompanying the sound of explosions. The sequence ends with a zoom in on the bomb exploding into apartment buildings and people running away from the scene. In terms of the construction of filmic space, this documentary seems at first to display a conventional setting and subverts it at the same time. Although the narrative starts with a clear image of the residential situation and Arabic sounds, it does not specify the place, and it is undercut by the following stories of flight and escape from different countries. By dispersing the narrative subjects and their origins, the film dislocates stories and characters by constantly shifting the point of

5 https://www1.wdr.de/mediathek/video/sendungen/wdr-dok/video-my-escape--meine-flucht-100.html.

view. Characters tell individual stories of flight from a variety of countries, and they often present their smartphones with videos and photos. Through the constant shifting in character, place, and story, the film accentuates the situation of flight and escape with the iconic places and situations: driving in a car, walking through a desert area, waiting in a refugee camp. Thus, *My Escape* (2016) transfers the individual, locative smartphone videos into a traditional documentary without narration or a filmic subject. In this way, the film functions on a double level as multidimensional collection of individual clips and traditional documentary at the same time. Additionally, the film sports a Facebook site, which allows for communication with the film-makers, and it offers further information on connected activities.

Investigative Jounalism Websites

Contrary to the locative features of 858.ma, the investigative site bell¿ngcat uses open-source and social media investigation to explore a variety of subjects, and it 'brings together contributors who specialize in open-source and social media investigation', and creates 'case studies so others may learn to do the same'.[6] bell¿ngcat was founded by the journalist Eliot Higgins, who also ran the *Brown Moses Blog* up to 2016. Besides Eliot Higgins, the site lists several international contributors with short biographical notes and a picture, which makes it a collective project, but contrary to the locative aspects, international reporters who seek to reflect and examine international media coverage inform the site. The home page links to different international regions; after Africa and America, the MENA region comes third. The reports from the MENA region include a close inspection of the events in Syria. Under the headline 'Schrödingers's Strike: Who Hit These Buildings in Raqqa, Syria', Jeanne Sanya and Christiaan Triebert offer a detailed investigation into the news coverage of the bombing of a civilian block of buildings in Raqqa. The story starts with a summary of media events, which poses relevant questions and underlines the uncertainty of international journalism and the usage of unspecific imagery:

> The Russian Ministry of Defense accused American-led Coalition of bombing a block of civilian buildings in Raqqa, Syria. However, a year earlier, the Russians said it was *them* who targeted leader of the Islamic State in those exact same buildings. What is going on here? (2018; emphasis in original)

6 https://www.bellingcat.com/about/

To analyse the factual events behind the bombings, Sanya and Triebert refer to satellite images of the buildings before and after the bombings, which had been tweeted by the official Twitter account of the Russian Ministry of Defence (MoD). According to the authors, a similar post appeared on the Russian MoD's official Facebook page as well. In the course of their analysis, Sanya and Triebert compare satellite images from 2017 and 2018 and come to the conclusion that the images 'show exact the same location in Raqqa, Syria'.[7] Because of these images, the authors trace the exact location of the buildings near the Al-Ala mosque, which is documented by an image of the respective mosque as 'shown on Wikimapia with Bing satellite imagery' (Sanya and Triebert 2018). A photo published in 2015 shows the mosque's interior. Having established the specific place with the attached symbolic meaning of a mosque, the article offers two images of the destroyed mosque published by local activists on social media, showing the perspective from the ground of the location shown in the aerial photos of the Russian Ministry of Defence. According to the website, the local activists of *Raqqa is Being Slaughtered Silently* are convinced that the mosque was destroyed by the US-led coalition in 2017. The photos had been published on Facebook and Twitter, and they have been geolocated to the 'same site as shown in the Russian MoD satellite imagery' (Sanya and Triebert 2018). At the same time, there is a video to be found on Facebook, which shows the damaged but not completely destroyed mosque.

Further research on the photos documents that the local activists used screenshots from a video from the A'maq New Agency, 'which functions as IS's auxiliary media wing', and minor damage to the mosque can be observed between 1 August and 13 August. As a next step, Christiaan Triebert uses satellite images from TerraServer/DigitalGlobe to show that the mosque must have been destroyed between 13 and 25 August 2017. They present a territorial control map of Raqqa, Syria from 25 August, from Liveuamap (Live Universal Awareness Map), which displays the respective area circled in red; the subtitle informs us that 'the area is being encircled by the Syrian Democratic Forces, supported by the U.S.-led Coalition' (Sanya and Triebert 2018). The investigation concludes that it cannot be decided who was responsible for the bombing of this area of Raqqa. Of importance here is the insight into interconnected informational media, which are condensed within the article on the bell¿ngcat website.

The article starts with information on the author Christiaan Triebert, a freelance journalist specializing on the Syrian war. Following on general information about the bombing, Triebert relates to the official Twitter and

7 https://www.bellingcat.com/news/mena/2018/07/13/schrodingers-strike-hit-buildings-raqqa-syria/.

Facebook profiles of the Russian MoD, which accuse the coalition forces of the bombing. This information is supported by a screenshot of the respective tweet; a link connects to the Facebook site. These are followed by Russian satellite images before and after the respective bombing. These images are countered by Bing satellite imagery on Wikimapia in which the area is still intact. Next, the article offers images from the inside of the mosque, displaying men at prayer – the most peaceful images amongst the imprints of war. The next images are activists' photos of the destroyed area, with the mosque displaying some damage, but still intact. These images are supported by an activist video also documenting the mosque's dome. These images are followed by satellite images tweeted by Triebert, and finally images from Liveuamap about this area. To sum up, the impressive variety and complexity of digital images, video proofs, and cartographic information distributed over the social networks provide a collective cartographic knowledge. Comparing this article with the regular German news show *Tagesschau* and the respective correspondents' networks, the deficiency in the German news coverage becomes striking, especially in comparing the thin network of MENA correspondents with the collective activity of freelance journalists. Compared to the multifarious activities of videographers, bloggers, and journalists in the MENA region, the German television channel ARD (*Arbeitsgemeinschaft der öffentlich-rechtlichen Rundfunkanstalten der Bundesrepublik Deutschland*) foreign correspondent's network looks relatively meagre. The network of media correspondents is displayed on OpenStreetMap, which proves the evident lack of media correspondents in the MENA region: one correspondent in Morocco and another in Cairo. Compared to the political influence and the evident instability of this area, the number of correspondents is more than insufficient.

A further point for discussion is the presence of collective mapping sites. Wikimapia is a website that combines satellite cartographies with a limited wiki system, relying on the maps of Google Maps but with the chance to switch to maps from Bing, OpenStreetMap, or Yahoo. Wikimapia is run by Alexander Korjakin and Jewgeni Sawelew and is registered in Moscow. There is no impressum on the website and no further information on the project's general intent; the user can add remarks and symbols within each place on the map. There is more information provided with Liveuamap, which, according to the website's organizers,

> is a leading independent global news and information site, dedicated to factual reporting of a variety of important topics including conflicts, human rights issues, protests, terrorism, weapons deployment, health matters, natural disasters. [...] Our innovative map-centric approach to organization

> of information will allow our viewers to quickly find the stories most relevant to them, and in geographies of their interest. [...] Liveuamap was founded in 2014 by a team of devoted software developers and journalists who wished to inform the world about the Ukrainian conflict.[8]

The site also offers a list of additional links to Facebook, Twitter, and other social networks, and it informs about the founders and the legal situation. Compared to the rather clandestine site Wikimapia, Liveuamap appears to be a well-organized and well-funded enterprise: the pro version offers Google Satellite maps. Similar to Wikimapia, the free version offers open maps, and users can integrate message and information via a set of icons, including to another information sources, usually social networks, that expands on the message. The site advertises as follows: 'Liveuamaps is opendata-driven [sic] media platform that change [sic] the way you receive latest news. Explore a map, messages, pictures and videos from the conflict zone'. This cartographic news website somehow offers a continuation and expansion of the Egyptian 858.ma with its collection of videos from the Egyptian revolution. Liveuamaps connects individual messages to mapping systems – alongside the map, the news can be viewed and followed up through individual videos and further information on social media.

In conclusion, these diverse means of connecting locative media to cartographic places bring about a variety of geographical and media configurations. These forms of collective journalism and information systems focus on nations and events, which lack proper and democratically organized media systems and are shaken by situations of violence and war that are not covered by internationally established journalists and media correspondents. Thus, these areas are withdrawn from official mapping activities – our knowledge of these war-torn regions is brought about by the manifold efforts of media activists.

References

Abounaddara Films, 'The Revolting Animals', Online Resource: https://www.facebook.com/notes/abounaddara-films/the-revolting-animals-de-revolterande-djuren-la-révolte-des-animaux/1745373475523352/ (accessed 17 November 2020).

An Archive of Resistance, Online Resource: https://858.ma (accessed 17 November 2020).

8 https://liveuamap.com/about/

bell¿ngcat, Online Resource: https://www.bellingcat.com (accessed 17 November 2020).
Manuel Castells, *Networks of Outrage and Hope: Social Movements in the Internet Age* (Cambridge: Polity Press, 2015).
Lili Chouliaraki, 'Symbolic Bordering: The Self-Representation of Migrants and Refugees', *Digital News. Popular Communication*, 15(2) (2017), pp. 78–94.
Katharina Cizek, 'Documentary as Co-Creative Practice. From Challenge for Change to Highrise: Cat Cizek in Converstation with Mandy Rose', in *I-Docs. The Evolving Practices of Interactive Documentary*, ed. by Judith Aston, Sandra Gaudenzi, and Mandy Rose (New York and Chichester: Columbia University Press, 2017), pp. 38–65.
Jason Farman, 'Map Interfaces and the Production of Locative Media Space', in *Locative Media*, ed. by Rowan Wlken and Gerad Gogin (New York and Oxon: Routledge, 2015), pp. 83–93.
–, *Mobile Interface Theory: Embodied Space and Locative Media* (New York: Routledge, 2012).
–, *The Mobile Story: Narrative Practices with Locative Technologies* (Hoboken: Taylor and Francis, 2013).
Paul Frosh, 'The Gestural Image: The Selfie; Photography Theory, and Kinesthetic Sociability', *International Journal of ommunication*, 9 (2015), pp. 1607–1628.
Elisa Giaccardi, 'Reframing Heritage in a Participatory Culture', in *Heritage and Social Media: Understanding Heritage in a Participatory Culture*, ed. by Elisa Giaccardi (New York: Routledge, 2012), pp. 5–10.
Mark B.N. Hansen, *New Philosophy for a New Media* (Cambridge: MIT Press, 2004).
Peter Hemmersamand, et al., 'Exploring Locative Media for Cultural Mapping', in *Mobility and Locative Media. Mobile Communication in Hybrid Spaces*, ed. by Adriane e Silva de Souza (London and New York: Routledge, 2015), pp. 167–187.
Jessica Holland, 'Syrian Video Collective Brings Glimpse of Life During Wartime', *The Guardian*, Online Resource: https://www.theguardian.com/artanddesign/2014/jul/18/syrian-video-collective-abounaddara (accessed 17 November 2020).
Harold A. Innis, *The Fur Trade in Canada; an Introduction to Canadian Economic History* (Toronto: University of Toronto Press, 1956).
Henry Jenkins, *Convergence Culture. Where Old and New Media Collide* (New York and London: New York University, 2006).
Karin Knorr-Cetina, 'Viskurse der Physik: Konsensbildung und Visuelle Darstellungen', in *Mit dem Auge denken. Strategien der Sichtbarmachung in wissenschaftlichen und virtuellen Welten*, ed. by Heintz, J. Huber (Zürich: Voldemeer, 2001), pp. 304–320.
Siegfried Kracauer, *The Mass Ornament. Weimar Essays*, ed. by Thomas Y. Levin (Cambridge: Harvard University Press, 1995), pp. 75–88.
Marwan N. Kraidy, *The Naked Blogger of Cairo* (Cambridge: Harvard University Press, 2016).

Paul Anton Krüger, 'Roh wie die Revolution', *Süddeutsche Zeitung*, HF2 (2018), p. 11.

Bruno Latour, 'Drawing Things Together', in *Representation in Scientific Practice*, ed. by Steve Woolgar and Michael Lynch (Cambridge: MIT Press, 1990), pp. 16–68.

Henri Lefebvre, *The Production of Space* [1974] (Malden: Blackwell, 2000).

Merlyna Lim, 'Sticks and Stones, Clicks and Phones: Contextualizing the Role of Digital Media in the Politics of Transformation', in *Digital Media and the Politics of Transformation in the Arab World*, ed. by Carola Richter, Anna Antonakis, and Cilja Harders (Wiesbaden: Springer, 2018), pp. 9–35.

Ioana Literat, 'Refugee Selfies and the (Self-)Representation of Disenfranchised Social Groups', *Media Fields Journal*, 12 (2017), pp. 1–9.

Live Universal Awareness Map, Online Resource: https://liveuamap.com/about (accessed 17 November 2020).

Lev Manovich, 'Die Poetik des erweiterten Raums. Von Prada lernen', in *Black Box White Cube*, ed. by Lev Manovich, trans. by Ronald Voullié (Berlin: Merve, 2005), pp. 105–144.

Noha Mellor, Khalil Rinnawi, Nabil Dajani, and Mohammad I. Ayish, *Arab Media: Globalization and Emerging Media Industries* (Cambridge: Polity, 2011).

Noha Mellor and Khalil Rinnawi, *Political Islam and Global Media* (Oxon and New York: Routledge, 2016).

M.I.A., *Borders*, Online Resource: https://www.youtube.com/watch?v=r-Nw7H-baeWY (accessed 17 November 2020).

My Escape (Germany 2016), Online Resource: https://www.ardmediathek.de/ard/player/Y3JpZDovL3dkci5kZS9CZWlocmFnLTA1ZjJjMGE2LWZjMDktND-M1Yy05NTViLWZkZGUoYjgyMDZhYg/ (accessed 17 November 2020).

Carola Richter, *Antonakis. Introduction: Digital Media and the Politics of Transformation – a Dialectical and Multi-Scalar Reading* (Wiesbaden: Springer VS, 2018).

Jeanne Sanya and Christiaan Triebert, 'Schrödingers's Strike: Who Hit These Buildings in Raqqa, Syria', *Bellincat*, 13 July 2018, Online Resource: https://www.bellingcat.com/news/mena/2018/07/13/schrodingers-strike-hit-buildings-raqqa-syria/ (accessed 20 January 2021).

Naomi Sakr, *Arab Television Today* (London: I.B. Tauris, 2007).

Irmela Schneider,'Zur Archäologie der Mediennutzung. Zum Zusammenhang von Wissen, Macht und Medien', in *Kulturindustrie Reviewed. Ansätze Zur Kritischen Reflexion der Mediengesellschaft*, ed. by Barbara Becker and Josef Wehner (Bielefeld: Transcript, 2006), pp. 83–100.

Bernhard Siegert, *Relais. Geschicke der Literatur als Epoche der Post 1751–1913* (Berlin: Brinkmann & Bose, 1993).

Vivian Sobchack, 'The Scene of the Screen. Beitrag zu einer Phänomenologie der "Gegenwärtigkeit" im Film und in den elektronischen Medien', in *Materialität der Kommunikation*, ed. by Hans Ulrich Gumbrecht and K. Ludwig Pfeiffer (Frankfurt: Suhrkamp Insel, 1988), pp. 416–427.

Adriana de Souza e Silva, *Mobility and Locative Media* (London and New York: Routledge, 2015).

—, 'From Cyber to Hybrid. Mobile Technologies as Interfaces of Hybrid Spaces', *Space and Culture*, 3 (2006), pp. 261–278.

Wikimapia, Online Resource: http://wikimapia.org/#lang=de&lat=51.215400&lon=6.663100&z=12&m=w (accessed 17 November 2020).

Christian Wille and Rachel Rickinger, 'Exploring Constructions of Space and Identity in Border Regions', in *Spaces and Identities in Border Regions*, ed. by Christian Wille, Rachel Reckinger, Sonja Kmec, and Markus Hesse (Bielefeld: transcript, 2016), pp. 9–14.

Hartmut Winkler, 'Landschaft und Architektur als Modell für den Datenraum', in *Hyperkult II. Geschichte, Theorie und Kontext digitaler Medien*, ed. by Wolfgang Coy (Frankfurt and Basel: Stroemfeld, 1997), pp. 227–239.

Brian Winston, 'Foreword', in *I-Docs. The Evolving Practices of Interactive Documentary*, ed. by Judith Aston, Sandra Gaudenzi, and Mandy Rose (New York: Columbia University Press, 2017), pp. xv–xvi.

Eszter Zimanyi, 'Digital Transcience: Emplacement and Authorship in Refugee Selfies', *Media Fields Journal*, 12 (2017), pp. 1–16.

About the Author

Angela Krewani is a full professor of media studies at Philipps-Universität Marburg, with a special focus on the impact of digital media on social practices and art. She is interested in the epistemic, aesthetic, and cultural configurations of digital media. She has widely published on these topics; her publication *Medienkunst. Theorie, Praxis, Ästhetik* (2016) examines the practices and the developments within media art. She recently co-edited a book on *Canadian Ecologies* (2020) (with Alessandra Boller and Martin Kuester) and another one on contemporary authorship *Constructions of Media Authorship* (2021) (with Christiane Heibach and Irene Schuetze). Currently she is preparing a publication on *Corona and Media* (with Peter Zimmermann).

11. Reframing the Arab Spring: On Data Mining and the Field of Arab Internet Studies

Laila Shereen Sakr

Abstract
This chapter presents a basic genealogy of the existing literature on the 2010–2012 era of the Arab Spring within Middle East studies and media studies more broadly. The aim is to map the plurality of patterns, stories, ideas, and analyses networked across media forms. This chapter is designed to refute the possibility of a monolithic narrative about contemporary Egypt. It also situates the study within the growing scholarship on social media and political change and introduces key concepts in design research methodology.

Keywords: Egypt, glitch, internet studies, revolution

Introduction

When activists took to Twitter and Facebook in the Middle East and North Africa (MENA) region to push for political change, digital media captured the imagination of scholars, interest groups, and multiple publics. Pundits coined the uprisings of 2011 'Facebook and Twitter revolutions'. Arab activists' use of social media became a model for other social movements. These included Occupy Wall Street, From Ferguson to Palestine, the Umbrella Revolution in Hong Kong, and 12M and 15M in Spain. These movements quickly adopted similar tactics, enabling people around the world to participate virtually as witnesses to contemporary events. In an act of transnational solidarity, in February 2011 an Egyptian activist ordered pizza for labour union protesters in Madison, Wisconsin. The international phone call from Egypt was just

Strohmaier, A. and A. Krewani (eds.), *Media and Mapping Practices in the Middle East and North Africa: Producing Space.* Taylor & Francis Group, 2021
DOI 10.5117/9789462989092_CH11

one of many messages of solidarity streaming into Wisconsin from all over the world on mobile devices and social media.

The dramatic increase of virtual and on-the-ground revolutionary activism also inspired corporate interests and government entities such as the National Security Agency (NSA) to hijack or monitor social media for a variety of purposes. These included surveillance as well as monitoring consumer behaviour. In sum, movements, activists, and governments alike have radically transformed social media platforms. In many ways, the MENA region has been ground zero for some of the most dramatic and influential changes in the ever-widening spectrum of digital possibilities. This chapter explores our understanding of and approach to digital media in relation to human rights, political speech, and social change. It forms part of a larger study on glitch and resistance that traces the impact of social media on popular social movements in the early 21st century, focusing specifically on Egypt. By centring on Egypt, rather than on the United States or Europe, and by drawing on both design research and humanistic inquiry, this chapter aims to show how digital media and politics are co-creative.

The concept of 'glitch' comes from the mechanical term for a sudden interruption, noise, error, or malfunction. It broadens to what I call 'technical semiotics' as it expands the logic or language of algorithms within networked societies by introducing messiness and mistakes. Although the glitch has been closely associated with technology, it nevertheless has the potential to illuminate other areas of experience and understanding. As a form of noise, for instance, it brings a political dimension to its practice. Noise becomes political when it translates into voice – and succeeds in entering and interrupting the ordered political world of policy. Noise is produced by anyone who acts from outside the order and, hence, can be returned to no one in particular. Building on postcolonial theoretical approaches, I argue that glitches provide an important critical lens for understanding the operational logic of systems of power: through its cracks and failures, the dynamics of systems become visible.

Hence, the following question araise: Why do programmers situated in a particular context, at a particular moment, come up with new modalities for thinking about resistance? A glitch is not a thing but a command, a script, an algorithm, a procedural set of actions. Glitch is a queer modality that processes, redirects, and opens new pathways. The genealogy of my use of the term 'glitch resistance' is associated with sound studies, queer studies, and radical feminist studies. If 21st century capitalism is driven by the logic of maximum performance (Lyotard 1973), then the glitch or failure of performance undermines the fetishizing of efficiency, accuracy, and predictability of producers, consumers, and drivers of capital. Glitch may sound undesirable in the parlance of late capitalism; however, it also serves as a point of entry for activists, hackers, or artists to resist.

A review of scholarship on media and the Middle East reveals a lack of engagement with digital media content, whether as primary sources or in critically questioning the tools and analytics provided. While many humanities scholars offer rich insights in history and culture, their analyses are often constrained by limited tools for understanding the ontology and syntax of digital production and social media networks. Arguments by writers such as Malcolm Gladwell, who writes that 'high risk social activism requires deep roots and strong ties' (2011), can likely be verified. Thus, literature that is more technologically literate helps revisiting those studies portraying the Arab Spring as a 'Facebook revolution' and broadens our understanding of networks, social media tools of counter-revolutionary bots, and fake news. Therefore, glitch theory provides an alternative to the celebration of social media in the 2010–2012 era and the demonization of the same technological platform.

When Mark Zuckerberg testified before US senators[1] in the first of two hearings about Facebook's role in the Cambridge Analytica scandal in April 2018 (i.e. the influence of Russian hacking on the 2016 US election), the gap between Silicon Valley technology giants and policymakers in Washington, DC became once more obvious. As the social media memes suggest, the gap itself was untranslatable. The Senate hearings displayed fundamental confusion in the public understanding of the meaning and implications of these technologies. Scientific and humanistic knowledge have long been recognized as dominating social, cultural, and educational systems. However, social media websites and apps can be seen as a legacy of the modernist Bauhaus movement. The expression 'form follows function' defined the European modernist design aesthetics, also known as the Bauhaus movement, originating specifically from Germany and then the United Kingdom. However, as Mohamed Elshahed has argued, there is significant difference between the European and the Arab modernist experience. While in Europe economic hardship, socialism, world wars, and the housing of soldiers drove the modern movement's development, in Egypt the wealthy capitalist elite embraced modernist design in their

1 C-SPAN, 'Facebook CEO Mark Zuckerberg Hearing on Data Privacy and Protection', 2018, Online Resource: https://www.c-span.org/video/?443543-1/facebook-ceo-mark-zuckerberg-testifies-data-protection&live

mid 20th-century grand apartments and aristocratic villas. To understand how a complex system such as Facebook might influence political change requires, however, literacy in design research (Cross 1982).

First, I will present a basic genealogy of the existing literature on the 2010–2012 era of the Arab Spring within Middle Eastern media studies. The aim of this is to map the plurality of patterns, stories, ideas, and analyses networked across media forms. It aims to refute the possibility of a monolithic narrative about contemporary Egypt. It also situates the study within the growing scholarship on social media and political change and introduces key concepts in design research methodology, which works specifically at intersections – between cultural and technical analytics, theory and practice, virtual and embodied practices, and Eastern and Western languages. These ideas are based in a multidimensional approach and thrive in situations in which differences between modalities of knowledge production, culture, and languages require one to translate and straddle these differences to create a cohesive form. Second, I will provide some context for the growing interest in the analytic properties of social media content. The stock of primary data is important, but is not determinant in creating nuanced cultural analyses. I will consider the work of data visualizations, especially in relationship to the stock of primary data (actual tweets, status updates, etc.). Third, I will argue for the importance of paying attention to the specific language of the stock of primary social media data. I focus specifically on the use of Arabic language online. This is critical for a number of reasons, one of which is the moment that has been dubbed the Arab Spring. However, most analyses of this so-called social media revolution have not taken into consideration the analysis of the meaning of actual Arabic language use. After harvesting and analysing Twitter posts for more than eight years (2010–2018), I became aware that the use of Arabic online was steadily rising. Hence, the critical analysis of digital information on the MENA region has to identify and address existing gaps in how the use of Arabic is perceived.

The medium of digital substance is computer code; the manner of its composition is through words and images that can be interactive and animated; and its form is the posted digital composition itself. Thus, to understand these factors and their influence on ideas such as political mobilization, scholars must take into account a host of variables. However, studying and interacting in a networked culture requires fluency in interfaces and digital software. Such literacy understands computational media at the level of its 'rhetoric, aesthetics, and poetics encoded in any work' (Mateas 2007). This language provides the conceptual framework for the research presented in the following.

Arabic Language Analysis and the Glitch

Before 2010, much of the scholarly research published on new media in the Middle East dealt with television politics and big broadcasters such as Al Jazeera (El-Nawawy and Iskandar 2003; Seib 2008; Kraidy 2009; Kraidy and Khalil 2010). Social media, blogs, and other online publications became part of the socio-digital convergence of technology, cultural transformations, and national narratives enabling the mobilization of the popular uprisings in the MENA region after 2010. These events raised questions about the role of communication technology in the development of social movements: Do the political uprisings in Egypt, Spain, and on Wall Street confirm the horizontal networks paradigm? What kind of hierarchical relations are undermined and which are confirmed in the stream of social media patterns that emerged?

The ongoing Egyptian uprising offers a unique case to research how political agency is mediated and how these voices are connected, creating – thanks to the characteristics of digital media – more transnational political identities and political subjectivities. An important shift is that contemporary mediation blurred the online–offline and digital–paper separation. Merged, they may represent a convergence culture (Jenkins 2008), produsage (Bruns 2007), or media synchronization. The political changes and outcomes in turn raise new questions about the efficacy of prior empirical methods and tools. Further obscuring the difficulty of analysing complex developments are the differences between the distinct theoretical and methodological foci among academic disciplines.

Recent studies about the influences of technology on social movements such as the Arab uprisings have rarely incorporated rigorous Arabic language analyses on a large scale in social media (billions of tweets and millions of website articles over several years). This is partly due to the lack of natural language processing tools to conduct computational linguistic research in languages other than English but also because of the lack of access to appropriate large-scale data collections. The growth of big data, particularly collections of social media trends, has enabled more in-depth examination of the role of digital networks, allowing scientific and policy communities to direct their attention to aggregated imprints of online and social media activity, without engaging in critical methodological analyses that take into account the historical and cultural basis to inform meaningful critique. This fetishization of data over meaning has produced much research that examines the scale, speed, and directional influences in digital networks and social media.

However, complementary approaches engaging in historical, cultural, and textual analysis can reveal richer insights to explain the wider context of such phenomena.

For example, in 2009, Harvard University's Berkman Center for Internet and Society conducted one of the first research initiatives on political dissent and digital content. The resulting publication, *Mapping the Arabic Blogosphere: Politics, Culture, and Dissent*, focused on node-to-node interaction and general demographics of bloggers. While the report marked a new direction in this field of research, it lacked explanation about the data modelling and sampling methods that were used. Moreover, the model focused on defining groups of people, rather than combining the resulting empirical data with historical and cultural understandings of the region or the transmission of ideas across multilingual discussions in the blogosphere.

In the case of the Egyptian revolution, the challenge scholars face when examining contemporary digital media and political change is how to analyse information quantitatively about groups of people in a region where, historically, data science has been used to support forms of colonialism. As Timothy Mitchell explains in *Colonising Egypt*, the practice of science and systems of ordering national standards are modern projects that enable governments to maintain discipline and surveillance. A cog in the colonial project, the science of documenting every political act reflected a 'tendency of disciplinary mechanisms, as Michel Foucault has called these modern strategies of control, not to expect and dissipate as before, but to infiltrate, and colonise' (Mitchell 1991, p. 38). Indeed, before and even after the uprisings of 2011, there was a clear divide between the qualitative, humanistic research Middle East media scholars produced and the quantitative approach of traditional communications scholars. For example, in the articles of Volume 5 of *Middle East Journal of Culture and Communication* (2012), the authors do not employ methods to examine large data sets of digital media content specifically, yet they address important sociopolitical concerns about Middle East media.

On the other hand, in his 2013 book *The Arab Uprising: The Unfinished Revolutions of the Middle East*, Marc Lynch works towards defining a political history of the Arab public sphere. Though also engaged in the subject of 'the Arab public sphere', like Anderson and Eickelman in 2003, Lynch writes post-2011 and specifically takes a stance on the role of social media in the uprisings early on in his book.

The role of social media and the internet in the Arab uprisings has often been exaggerated, with too much focus on Facebook or Twitter rather than the underlying political struggles. However, Timothy Mitchell's concern about the disciplinary and colonizing purposes of data science (1988) must be carefully employed in these empirical critiques on the popular uprisings in the MENA region. Many of the research projects, conferences, and publications that have conducted in-depth data analytics from social media and other internet content about the popular uprisings in the MENA region are sponsored by think tanks such as RAND and political organizations such as the National Endowment for Democracy or the Dubai School of Governance.

Two years after *Mapping the Arabic Blogosphere: Politics, Culture and Dissent Online* (Etling et al. 2009), and following the uprisings in Tunisia and Egypt, the same core group of scholars organized a round-table conference entitled 'Blogs and Bullets II'. Following the conference, the *Journal of Communications* published a special edition on 'Social Media and Political Change: Capacity, Constraint, and Consequence'. A year later, Rob Lever interviewed the same group of scholars who participated in the 'Blogs and Bullets II' conference for his article 'Debate Flares on "Twitter Revolutions" Arab Spring' (2013). The common thread in all these discussions has been the focus on debating the role of social media in the revolutionary praxis of the MENA region, perpetuating the debate on the causational impact social media supposedly has on social movements.

One comprehensive research effort to study the impact of social media on the MENA region is a three-part series co-authored by Racha Mourtada and Fadi Salem. Their intention was to make sense of Twitter by recoding it into geolocation/geo-coordinates; this, however, is a way to estimate, not an actual statistical method (Salem and Mourtada 2011).[2] It is important to note the effort put into translating the numerical estimate that could represent the data in terms of nation states and populations, rather than simply exploring what is in the data imprints. Geolocation is not a common data imprint. This top-down approach to data mining allowed the researchers to find answers about Facebook usage based on gender, age, and national origin. However, the categories chosen left no space for including hybrid or shifting identities. In the second report of the series, the authors built upon their previous research on Facebook and introduced Twitter analytics. By

2 They explain their statistical methods as, 'with a sample of about 10,000,000 tweets among 190,000 Twitter users, estimating the size of a Twitter population was a simple two-step process: capture a number of samples (or "sweeps") of users from each country, and use a mark-recapture based technique to compute a population estimate' (Salem and Mourtada 2011, p. 63).

the time Mourtada and Salem authored the third part of the Arab Social Media Report, 'The Role of Social Media in Arab Women's Empowerment', they had started to discuss 'netizens' (online communities) in more detail. Each of their charts provides a demographic breakdown by country (Salem and Mourtada 2011). However, their approach remained unable to account for transnational netizens, or any individuals, for that matter, who did not fit neatly into predetermined categories.

While technology enthusiasts are being heard, so are voices that are more conservative. In an article that appeared in *The Atlantic*, Megan Garber warned, 'a year after the Egyptian revolution, 10 per cent of its social media is already gone' (2011). Earlier, Bassem Youssef argued 'one of the difficulties with efforts to understand the role of social media in the successful North Africa uprising (Tunisia and Egypt) is a paucity of theory' (2011). He builds his argument on highlighting three problems in internet studies: (1) internet studies theorize new media using theories that have been shown to be unsophisticated even when applied to old media, seeing media primarily as a vehicle for the more or less passive receipt of information; (2) internet studies fail to theorize the specific elements that make new media 'new', and (3) internet studies ignore the active ingenuity, creativity, and agency of users, who not only receive but also write new narratives, and generate such innovations as the Bassem Youssef Show and Pigipedia. Furthermore, only about 20 per cent of people in Egypt are online, and they fall into specific age groups, income divides, and education demographics; yet the resistance movement itself clearly extends into many other walks of society (Youssef 2011, p. 4). An example of the kind of writing on new media that Youssef is criticizing might be the 2012 report by the Center for International Media Assistance (CIMA) and the National Endowment for Democracy (NED), 'Digital Media in the Arab World: One Year After the Revolutions'. According to its author, Jeffrey Ghannam, the research was conducted through a series of '35 interviews in person, by telephone, email, and Skype: primary and secondary documents; commentaries; websites; blogs, and other sources' (2012, p. 3). This research provided an overly optimistic conclusion:

> Social media's potential represents the brightest hope for the greater freedom of expression in the Arab region, enabling tens of millions of people, and ultimately many more, to actively pursue civic engagement, free and fair elections, political accountability, the eradication of corruption, as well as free, independent, and pluralistic media in a rapidly changing media environment. (Ghannam 2012, p. 3)

Most striking in this investigation into digital media is the absence of any analysis of primary digital information. One might appropriately ask whether publishing on digital humanities without speaking digital languages or engaging primary digital media is akin to publishing on France without knowing French or engaging with any French primary material. Furthermore, the choice to interview '35 key individuals' does not speak to the scale of the research question the report poses. It also negatively reinforces the top-down authority structures that these very upheavals have been resisting. Outside of the MENA region, Twitter data have been used for forecasting or measurement in a variety of domains, such as stock markets, politics, and social movements. In 2011, Murphy Choy's Twitter sentiment failed to predict Singapore's 2011 presidential election (Choy et al. 2011). Several research projects on Twitter sentiment analysis in particular have emerged during the last few years, including one 80-page report by the RAND Corporation that uses tweets by Iranians to measure public opinion in the aftermath of the 2009 presidential election (Elson and et al. 2011).

This research was one of the first projects to use natural language processing and text analysis to determine sentiment in tweets. Without much of a review of the historical scholarship on public opinion in Iran, this research focuses on a linguistic approach to understanding how people communicate and the functions of the words they use. In practical terms, this approach suggests that it is possible to gain insights into the emotional states and social dispositions of users, regardless of the content of the writing. It is important to note that all of this analysis to measure a Persian/Farsi-speaking public was conducted on English-language tweets, blind to tweets in any other language. Despite RAND's efforts, their computer-driven analytics recently transitioned from manual analysis to applied simple computing techniques using a free Twitter archive service that is no longer in service (TwapperKeeper.com). RAND's analysis relies on other sources to provide the technical analysis. However, these secondary sources and their methodologies are not examined in their larger research.

Few monographs provide critical analysis at the intersection of media studies and the popular uprisings in the MENA region. Zeynep Tufekci, a sociologist and associate professor in the School of Information and Library Science at the University of North Carolina, writes to the public while maintaining scholarly rigour in her first single-authored monograph, *Twitter and Tear Gas: The Power and Fragility of Networked Protest* (2017). Within Middle Eastern studies, there are a number of important works, such Paul

Amar's *The Security Archipelago: Human-Security States, Sexual Politics, and the End of Neoliberalism* (2013), in which he charts the rise of what he calls the 'human-security state', and immerses his readers in urban laboratories of human security in both Cairo and Rio de Janeiro. Vijay Prashad also provides a global comparative analysis in *The Death of the Nation and the Future of the Arab Revolution* (2016); his underlying premise, which is central to my approach as well, is that revolutions are ongoing.[3]

All the above efforts, while paving the way for new grounds in research, have highlighted a growing and very tangible need for more research to interpret aggregated mediated interactions, as well as a different set of tools and languages that take seriously the digital as an archive of primary material. By the end of the first decade of the 21st century, we clearly have moved from the world of 'new' media to a world of 'more' media. When we reached 2011 and the Egyptian uprisings – which many hailed as the revolution brought about by Facebook – the ubiquity of computers, digital media software, and computer networks had led to an exponential rise in the number of cultural producers worldwide. No longer simply a matter of the rise of new media production in new global contexts, these social media platforms served as the database architectures for the accumulation of data on a scale heretofore unknown. This over-proliferation of data challenges research methodologies – the impossibility of knowing or representing such a mass of information requires new ways of investigation and interpretation. In just a few years, a plethora of articles and books on the popular uprisings in the MENA region were published, and trillions of Twitter and Facebook postings were micro-blogged in multiple languages. Though social media is often considered co-authored by the public, much research on the Arab public has focused on political histories, public opinion through traditional polling, or ethnographic investigations. In this regard, they have tended to undercut the mechanisms for verification and authority within the domain of the digital. Digital knowledge production is not like pop culture or traditional public texts, such as newspapers or legal documents. Rather, it offers a redefinition of 'the public' as societies systematically and increasingly engaged in the internet, open-source transactions, and mobile devices, for example, worthy of rigorous study.

3 Several excellent studies on contemporary resistance include Joel Beinin and Frédéric Vairel's *Social Movements, Mobilization, and Contestation in the Middle East and North Africa* (2011), in which the contributors forsake the cultural and political exceptionalism of the Middle East, and *Revolution Without Revolutionaries: Making Sense of the Arab Spring* (2017), in which Asef Bayat laments the lack of revolutionary ideology.

Network, not Geography

On 20 March 2004, Alaa Abd El Fattah and Manal Bahey El-Din Hassan, in Cairo, Egypt, first went live with their blog, *Manal and Alaa's bit bucket: free speech from the bletches*.[4] They explored and experimented with web publishing platforms that would help other groups and small initiatives have web presence. Alaa and Manal's 'bit bucket' was one of the earliest blogs to come out of Egypt and the region. Its content was simply born of the public event of the day: 20 March 2004 was also the first anniversary of the massive global protests against the war in Iraq.

At the time, I was a fellow open-source designer building Arabic and English websites for academic and activist communities in Washington, DC. I met Alaa and Manal in 2007 through the Arabic network of designers and programmers on Drupal while building the beta version of the R-Shief software system.[5] This informal network of Arab techies grew over the years as we forged new relationships and initiatives, including my own Twitter archiving project in 2008. By 2011, the US Secretary of State invited me to Washington, DC to explain how my software enabled me to predict within six hours of accuracy when Muammar Qaddafi would fall from power. This research on global, emergent media analytics became my doctoral scholarship and a guiding focus of my career as a scholar: understanding digital communication as an indispensable component of living in the 21st century, and the importance of accessing information across borders and languages.

Two trends – one organized by activists and one by academics – seemed to run parallel to each other rather than converging. The set of research practices that emerge at the intersection of both cultural and technical analytics in the fields of media studies is born out of the contemporary socio-historical moment as we face new scales of information that demand machinated computation and political influences on our cultural practices that are systematic, transnational, and mobile. Similar to the non-linear narrative form of a NoSQL database, data-driven narratives are also mapped relationally and within larger systems and orders (Figure 11.1).

This problem in scholarship continues today. In an edited volume that came out in late 2014, *Media Evolution on the Eve of the Arab Spring* (Hudson et al.), the original 2009 Berkman Center article was republished with no critical reflection on its lack of textual analysis. This brings into urgent question the standards in this field. How can academics repeatedly engage

4 https://manalaa.net/.

5 r-shief.org.

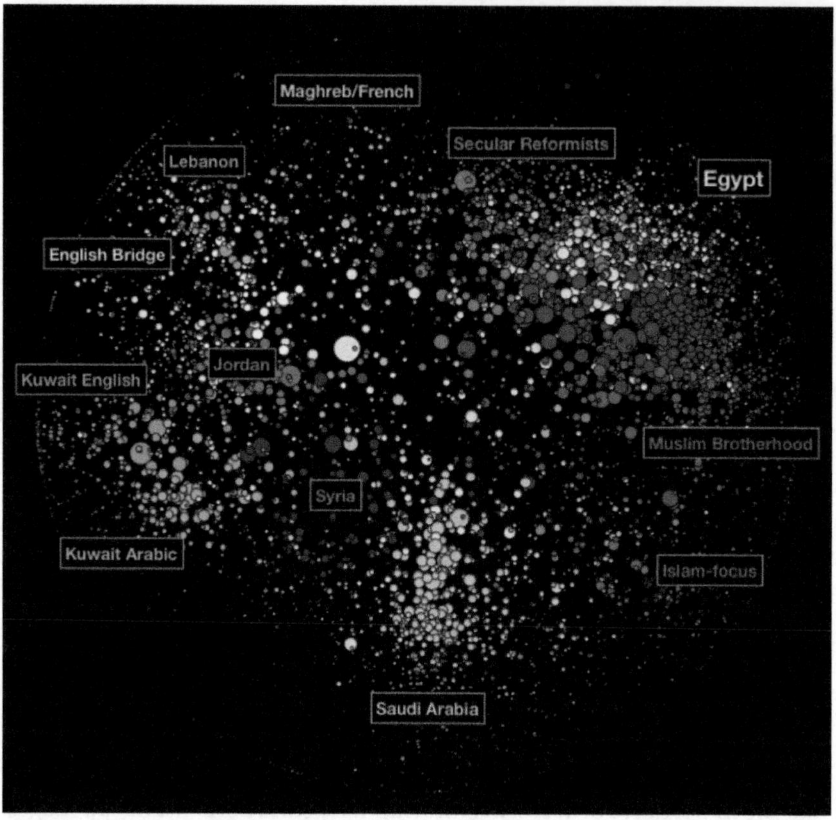

Figure 11.1: Network graph of Arab blogosphere, published in *Mapping the Arab Blogosphere* article by Etling *et al.* the Harvard Berkman Center (2009).

with the Berkman Center's article on the 'Arab Blogosphere' (which even includes the word 'culture' in its subtitle) as a scholarly work, when the writers of the article did not discuss the contents of a single primary source?

Perhaps one reason they did not analyse or read the blogs is that most of the blogs are in Arabic. More specifically, it was crucial to overcome problems of web accessibility inherent in Arabic fonts. There was a time, right before the Arab blogosphere developed, when programmers could not encode the Arabic glyph for web interfaces. In addition, we could not get those interfaces to read from right to left. These two basic limitations presented quite a challenge from both technological and visual design perspectives and took some time to resolve. One problem was that the Arabic glyph appeared much smaller than the Roman glyph. In developing a bilingual site, it was very difficult to design the Arabic legibly without the English looking dramatically large. That relationship between the Arabic and English text is reminiscent of the

English language's power in the internet's early years. Later, open-source projects such as Wikipedia and Drupal worked to change that. This is an interesting friction point between technology and scholarship.

Accessing primary materials in their native language is essential to rigorous scholarship; it allows researchers to raise the standards of what we think about social media, and elevates the scholarship and expectations from people who study social media to a set of standards. Arab digital activists addressed this issue in an interesting email thread on the Arab Techies email listserv. Sami Ben Gharbia, Slim Amamou, and Alaa Abd El Fattah expressed grievances to the University of Toronto's Citizen Lab when they approached activists on the listserv for data about open and collaborative research:

> It seems like there is a new field in academia that uses the 'newness' of the Internet as an excuse to get rid of the guidelines, ethics and traditions developed in the humanities, so results are published without using the tools, algorithms or data, bringing into question how peer review is possible. The problem would be mitigated if local academia were consulted. Why seek stewards among practitioners and not engage with local universities and researchers? Sure our brown universities are as messed up as our brown lives here in the brown south but I assure you the difficulties in working with them are nothing compared to the difficulty of doing accurate research without their understanding of the context. (2012)

Ben Gharbia, Amamou, and Abd El Fattah suggest here that academia needs to shift from building tools that further reinforce hegemonic divides between cultures and peoples, to building tools that extend the work and agency of local digital activists themselves. The set of research practices that emerge at the intersection of both cultural and technical analytics in the fields of new media studies[6] is born out of the contemporary socio-historical moment as we face new scales of information that demand machinated computation and political influences on our cultural practices that are systematic, transnational, and mobile. Similar to the non-linear narrative form of a NoSQL database, data-driven narratives are also mapped relationally and within larger systems and orders.

6 Outside of the sciences, studying technology must also instil fear among many academics. As Franklin and Rodriguez introduce their argument, 'Hypertext. Hypermedia. High Performance Computing. It's enough to make a humanities scholar hyperventilate. A debate has raged in the last decade (at least) about whether or not the Digital Age will see the death of The Book, The Library and perhaps, The Humanities more broadly' (2008).

Alaa, Manal, and I were among the group of Arab Techies who gathered on the American University of Cairo's Tahrir campus in July 2011. At the time, we thought we would be able to go to Syria and support activists. We had no way of imagining the devastation that would engulf the region after the disruptive, destabilizing glitch that shook the MENA region. We also never imagined that Alaa would end up serving a five-year sentence in Tora prison for protesting or that we would mourn the deaths of Aaron Swartz, Bassel Safadi, Ali Shaath, and many more. We were preoccupied celebrating our faults, failures, malfunctions, disturbances, anomalies, bugs, errors, and noises.

Conclusion

By 2004, two important movements in Egypt had emerged: the open-source software and localization movement, and the Kefaya political party and movement. Kefaya made its first public appearance in December 2004, when more than 50 people gathered outside the attorney general's office, making demands far beyond the established boundaries of free expression at the time. Their banners read 'Down with Hosni Mubarak', and 'The Egyptian Movement for Change'. Also founded in 2004, Open Craft is one of Egypt's prominent web-building companies that, according to their website, supports 'the values of the open-source movement to provide Drupal Web application development: implementation, integration, and internationalization'.[7] Open Craft served as a network hub of progressive, feminist programmers and designers: founder of SuperMama, Zeinab Samir, and activists Manal Bahey El-Din Hassan, Maggie Osama, and Mira Loutfi also worked at Open Craft. This network built on the online forums built by developers at Drupal. As recently as 2006, developers were still building applications to enable Arabic characters on a keyboard. Several open-source projects to develop software for Arabic on Drupal, Yamli, Kalimaat, GNOME, and other platforms enabled Arabic-language content to grow dramatically in the years to follow. There are activist platforms such as Katib and Take Back the Tech based in Lebanon. In addition, a great deal of work has gone into Arabic language analyses, from e-Space's Arabic Morphological Analyser, to Taha Zerrouki's Arabic conjugator project, Qutrub.[8]

I conceived R-Shief to fill the gap. I built the prototype in Drupal because of its free Arabic localization modules. The use of creative commons licensing

7 http://open-craft.com.

8 *Arabeyes*, Online Resource: qutrub.arabeyes.org.

and praxis of free/libre/open-source software such as MediaWiki, Ning, and WordPress emphasizes R-Shief's commitment to a collaborative, participatory model of archiving. However, Drupal had proven Arabic-language programming abilities. There are all sorts of issues concerning right-to-left non-roman script languages. Drupal's Arabic-language developers have been building modules to fix the various problems. For example, Drupal developer @Amr recently released the FireFox Yamli[9] extension and spent some time packaging a Drupal module to integrate the Arabic comma as a separator for Drupal tags.

Drupal development in Arabic[10] (right-to-left programming and UTF-8 encoding for Arabic script) grew tremendously in the years that followed, with over 50 software developers contributing to Drupal's open-source platform. One of the Arabic team administrators for Drupal is blogger and activist Alaa Abd El Fattah. He posted the following biography on his profile in 2004:

> From his work with children using Facebook to ridicule their teachers in the Arab digital expression camps, to his work with pro-democracy activists using blogs to mobilize thousands of Egyptians against the government in the Kefaya movement, Alaa just loves helping people use ICTs to stick it to the man. By day he works as a Free/Open Source Software developer, by night he dons his mask and cape and patrols the streets of Cairo, jumping from campaign to campaign, building websites, providing support and training, looking out for activists in need. He likes to pretend that his work on the Egyptian Blogs Aggregator helped bring in a new era of citizen journalism and usher in a new generation of digital activists, while the rest of the world acts as if his blog is relevant.[11]

One thing Alaa was mistaken about was that his work did not 'pretend' to 'bring in a new era of citizen journalism'. In fact, this work *did* usher in a new generation of digital activists. Between 2004 and 2011, the rise of the Arab blogosphere demanded rigorous attention. Alaa's blog posts swung like a pendulum between RSS aggregators and the Egyptian constitution. He was first jailed in 2006, along with other bloggers, while calling for an independent judiciary in Egypt. At the same time, citizen journalist

9 https://www.yamli.com/.

10 Drupal Translations, *Arabic Overview*, Online Resource: https://localize.drupal.org/translate/languages/ar.

11 Drupal Translations, *Alaa's profile*, Online Resource: https://www.drupal.org/user/8991.

Noha Atef began a crucial website and blog called Torture in Egypt,[12] Mark Zuckerberg launched Facebook in 2004, and Twitter launched in 2006. At the same time, the largest forum on the World Wide Web was a community for Egyptian women called Fatakat. From 2008 to 2010, bloggers continued to report on more violence by the state and growing social unrest.

I have argued that the scientific project of Arabic software localization (adapting computer software to different languages and regional differences) requires humanistic critical thinking (Sakr 2013). At an even deeper level, the shift in programming from C++ (a highly mathematical computer language) to Java (a computer language that contains lots of English-based vocabulary) has meant a shift into what is culturally and hegemonically English-based. The budding academic field of internet studies (i.e. the Association of Internet Researchers) is so young, however, that its ethical standards and traditions are still in formation. Scholarship on internet studies is often published without providing public access to the data or tools used in the analysis. Critiques of race, power, and colonialism are rarely brought into studies on how activists from the MENA region have used Silicon Valley tools to design their own virtualities specific to local culture and history.

References

Paul Amar, *The Security Archipelago. Human-Security States, Sexuality Politics, and the End of Neoliberalism* (Durham: Duke University Press, 2013).

Jon W. Anderson and Dale F. Eickelman, *New Media and the Muslim World: The EmergentPublic Sphere* (Bloomington: Indiana University Press, 2003).

Miriyam Aouragh and Anne Alexander, 'The Egyptian Experience: Sense and Nonsense of the Internet Revolution', *InternationalJournal of Communication*, 5 (2010), pp. 1344–1358.

Asef Bayat, *Revolution Without Revolutionaries: Making Sense of the Arab Spring* (Stanford: Stanford University Press, 2017).

Joel Beinin and Frédéric Vairel, eds., *Social Movements, Mobilization, and Contestation in the Middle East and North Africa* (Stanford: Stanford University Press, 2011).

Axel Bruns, 'Produsage: Towards a Broader Framework for User-Led Content Creation', *Proceedings Creativity & Cognition*, 6 (2007).

—, 'From Reader to Writer: Citizen Journalism as News Produsage', in *Internet Research Handbook*, ed. by Jeremy Hunsinger, Lisbeth Klastrup, and Matthew Allen (Dordrecht: Springer, 2010), pp. 119–134.

12 Tourture in Egypt, Online Resource: http://tortureinegypt.net/.

Manuel Castells, *The Rise of the Networked Society* (Oxford: Blackwell Publishers, 1996).
Murphy Choy, Michelle L.F. Cheong, Ma Nang Laik, and Koo Ping Shung, *A Sentiment Analysis of Singapore Presidential Election 2011 using Twitter Data with Census Correction*, 2011 (eprint arXiv:1108.5520).
George Dunbar, *Computers and Translation*, 3(4) (1989), pp. 263–266.
Mona El-Ghobashy, 'The Praxis of the Egyptian Revolution', *Middle East Report*, 2011, p. 258, Online Resource: http://www.merip.org/mer/mer258/praxis-egyptian-revolution (accessed 18 November 2020).
Mohammed El-Nawawy and Adel Iskander, *Al-Jazeera: The Story of the Network that is Rattling Governments and Redefining Modern Journalism* (New York: Basic Books, 2003).
Mohammed Elshahed, 'The "White City" and Egypt's modernist History', *Cairobserver*, 2012, Online Resource: http://cairobserver.com/post/21938918144/the-white-city-and-egypts-modernist-heritage#.W7_-uhNKhTY (accessed 18 November 2020).
Sarah Beth Elson, Douglas Yeung, Parisa Roshan, S.R.Bohandy, and Alireza Nader, *Using Social Media to Gauge Iranian Public Opinion and Mood After the 2009 Election* (Santa Monica: The RAND Corporation, 2011).
Bruce Etling, John Kelly, Robert Faris, and John Palfrey, 'Mapping the Arabic Blogosphere: Politics and Dissent Online', *New Media & Society*, 12(8) (2010), pp. 1225–1243.
Kevin D. Franklin and Karen Rodriguez, 'The Next Big Thing in Humanities, Arts and Social Science Computing: Cultural Analytics', *HPC Wire* (2008).
Alexander R. Galloway, *Protocol: How Control Exists after Decentralization* (Cambridge: MIT Press, 2004).
Alexander R. Galloway and Eugene Thacker, *The Exploit: A Theory of Networks* (Minneapolis: University of Minnesota Press, 2007).
Megan Garber, 'A Year After the Egyptian Revolution, 10% of Its Social Media is Already Gone', *The Atlantic*, Online Resource: http://www.theatlantic.com/technology/archive/2012/02/a-year-after-the-egyptian-revolution-10-of-its-social-media-documentation-is-already-gone/253163/ (accessed 18 November 2020).
Jeffrey Ghannam, 'Digital Media in the Arab World: One Year After the Revolutions', *A Report to the Center for International Media Assistance*, Online Resource: http://cima.ned.org/publications/digital-media-arab-world-one-year-after-revolutions (accessed 18 November 2020).
Tarleton Gillespie, *Custodians of the Internet: Platforms, Content Moderation, and the Hidden Decisions that Shape Social Media* (New Haven: Yale University Press, 2018).
Malcom Gladwell, 'Does Egypt Need Twitter?', *The New Yorker*, Online Resource: http://newyorker.com/online/blogs/newsdesk/2011/02/does-egypt-need-twitter (accessed 18 November 2020).

Jack Halberstam, *The Queer Art of Failure* (Durham: Duke University Press, 2011).

Linda Herrera, *Wired Citizenship* (New York: Routledge Press, 2014).

Charles Hirschkind, 'Uprising in Egypt: The Road to Tahrir', *The Immanent Frame: Blog of Social Science Research Council*, Online Resource: http://blogs.ssrc.org/tif/2011/02/09/the-road-to-tahrir/ (accessed 18 November 2020).

Philip N. Howard, *The Digital Origins of Dictatorship and Democracy: Information Technology and Political Islam* (Oxford: Oxford University Press, 2010).

Leila Hudson, Adel Iskandar, and Mimi Kirk, *Media Evolution on the Eve of the Arab Spring* (London: Palgrave MacMillan, 2014).

Henry Jenkins, *Convergence Culture: Where Old and New Media Collide* (New York: New York University Press, 2008).

John Kelly and Bruce Etling, 'Mapping Iran's Online Public: Politics and Culture in the Persian Blogosphere', *Internet & Democracy Case Study Series*, 2008, Online Resource: https://cyber.harvard.edu/publications/2008/Mapping_Irans_Online_Public (accessed 18 November 2020).

Marwan Kraidy, *Reality Television and Arab Politics: Contention in Public Life* (Cambridge: Cambridge University Press, 2009).

Marwan Kraidy and Joe Khalil, *Arab Television Industries* (London: British Film Institute, 2010).

Rob Lever, 'Debate Flares on "Twitter Revolutions" Arab Spring', *Dawn*, 11 March 2013, Online Resource: https://www.dawn.com/news/791757/debate-flares-on-twitter-revolutions-arab-spring (accessed 21 January 2021).

Marc Lynch, *The Arab Uprising: The Unfinished Revolutions of the Middle East* (New York: PublicAffairs, 2013).

Jean-François Lyotard, *The Postmodern Condition: A Report on Knowledge* (Minneapolis: University of Minnesota Press, 1979).

Michael Mateas, 'Procedural Literacy: Educating the New Media Practitioner', *ETC Press*, 4 May 2007.

Middle East Journal of Culture and Communication, 5(1) (2012).

Timothy Mitchell, *Colonising Egypt* (Cambridge: Cambridge University Press, 1988).

Safiya Umoja Noble, *Algorithms of Oppression: How Search Engines Reinforce Racism* (New York: New York University Press, 2018).

Vijay Prashad, *The Death of the Nation and the Future of the Arab Revolution* (Berkeley: University of California Press, 2016).

Nicola Pratt, 'The Gender Logics of Resistance to the "War on Terror": Constructing Sex-Gender Difference Through the Erasure of Patriarchy in the Middle East', *Third World Quarterly*, 33(10).

Legacy Russell, 'Digital Dualism And The Glitch Feminism Manifesto', *Cyborgology*, 10 December 2012.

Fadi Salem and Racha Mourtada, 'Facebook Usage: Factors and Analysis', Online Resource: http://www.arabsocialmediareport.com/Facebook/LineChart.aspx (accessed 18 November 2020).

—, 'Civil Movements: The Impact of Facebook and Twitter', Online Resource: www.dsg.ae/portals/0/ASMR2.pdf (accessed 18 November 2020).

—, 'The Role of Social Media in Arab Women's Empowerment', *The Arab Social Media Report*, Dubai, UAE: Dubai School of Government, Online Resource: www.arabsocialmediareport.com/UserManagement/PDF/ASMAR%20Report%203.pdf (accessed 18 November 2020).

Laila Sakr, 'The Materiality of Virtuality: Internet Reporting On Arab Revolutions', in *Mediating the Arab Uprisings*, ed. by Adel Iskander and Bassam Haddad (Washington, DC: Tadween Publishing, 2013).

Philip Seib, *The Al Jazeera Effect: How the New Global Media Are Reshaping World Politics* (New York: Potomac Books, 2008).

Clay Shirky, 'The Political Power of Social Media', *Foreign Affairs*, January/February, 2011.

Zeynep Tufekci, *Twitter and Tear Gas: The Power and Fragility of Networked Protest* (New Haven: Yale University Press, 2017).

Barry Wellman, 'Physical Place and Cyber Place: The Rise of Networked Individualism', *International Journal of Urban and Regional Research*, 25(2) (2001), pp. 227–252.

Bassem Youssef, 'Internet and Political Mobilization in Egypt and Tunisia', *Connected in Cairo*, Online Resource: http://connectedincairo.com/2011/08/08/internet-and-political-mobilization-in-egypt-and-tunisia (accessed 18 November 2020).

About the Author

Laila Shereen Sakr is Assistant Professor of Media Theory & Practice at the University of California, Santa Barbara. At UCSB, she co-founded Wireframe, a new digital media studio that supports critical game design and digital arts practice. Sakr has shown in solo and group exhibitions and performances at galleries and museums including the San Francisco MoMA, National Gallery of Art in Jordan, Camera Austria, Cultura Digital in Brazil, Kirchner Cultural Centre in Argentina, Tahrir Cultural Center in Cairo, Fridge Art Gallery in Washington, DC, and 100 Copies in Egypt, among other venues. Over the last two decades, she has been a leading voice in the open-source movement, in particular for Arabic localization. She is Co-Editor for the open-access journal: *Media Theory*, and for *After Video* published by Open Humanities Press. She collaborates with MIT's Global Media Technologies

& Cultures Lab as a Researcher. At UCSB, she is Faculty Affiliate in the Feminist Studies Department, Media Arts and Technology Program, and the Center for Responsible Machine Learning. She serves on the advisory, executive, and steering committees for UCSB Digital Arts & Humanities Commons, Center for Middle East Studies, and Center for Information Technology & Society.

12. Where is Iran? Politics between State and Nation, Inside and Outside the Polity

Annabelle Sreberny and Gholam Khiabany

Abstract

Three powerful spatial dynamics are at work in the analysis of a country's political orbit. One is the classic remit of the international relations between states. The second is the mainstream remit of political analysis, the national dialogue – sometimes open, often constrained – between the state and its inhabitants. In addition, the third is the cross-border space between the state and its citizens who – as diaspora, exiles, and migrants – live in other countries. Too often, each is analysed in isolation, part of the intellectually unedifying division of academic work. In this chapter, we explore where contemporary Iranian politics exists and how it is played out through each of these political geographies.

Keywords: Iran, politics, nation, knowledge

The 2018 World Cup ended with France the winner, partly because its imperial past provided the country with many excellent football players of African origin. The Iranian team started at a disadvantage and did not make it beyond the first round. Nike, the Iranian team's previous shoe provider, announced that because Donald Trump reimposed sanctions against Iran, they could not support the team any longer. The players were forced to improvise and to buy their own cleats. Despite this, Team Melli[1]

1 *Melli* means 'national' in Persian, and Iranians' reference the national team as such, clearly referring to the nation and not the state. Yet the world football governing body (FIFA) referred to the team by the name of the state, Islamic Republic of Iran (IRI).

Strohmaier, A. and A. Krewani (eds.), *Media and Mapping Practices in the Middle East and North Africa: Producing Space*. Taylor & Francis Group, 2021
DOI 10.5117/9789462989092_CH12

became a viral phenomenon and was widely praised for its stylish and dapper appearance.[2] At some of the matches in Russia, Iranian women, who are banned from attending football matches at home, unfurled banners demanding access to football stadiums inside Iran; a few stadia within the country were indeed opened to women so that they could watch the Russian games relayed on-screen. An international sporting event always echoes political issues, and for Iran this World Cup echoed both international and domestic ones. Internationally, the spectre and realities of sanctions[3] were on display, and nationally, the excitement of the games was a perfect excuse for a massive public presence in the streets.

Politics is predominantly 'at a distance'. Most politicians are known to contemporary publics via media of one form or another. Now, few are on the stump anymore, talk directly to potential voters, or only briefly at rallies during elections. The television screen mediates the distance between representatives and supporters, bringing politics into the living room – as do radio, newspapers, and internet interactions, including through social media (Coleman 2007). If there is always a spatial dimension to any politics, in the contemporary world the distances have become longer while the speeds are faster. That is to say, long distance has collapsed into seconds, the time-space compression of globalization-speak (Giddens 2013). Thus, distance becomes a question of scale. This is all well known but worth repeating, to acknowledge that the claims about the deterritorialization of politics are not that new (Held et al. 2000, pp. 14–28).

A tendency within globalizing approaches and the perceived thickening of transnational relations is sometimes to ignore the ongoing significance of state-nation relations – although those remain. Indeed, state-nation relations have also become deterritorialized in fascinating ways as more and more countries have diasporic populations living outside the boundary of the state, which tries to maintain them within the political fold.

Three Geographies of Politics

Thus, we suggest that there are at least three powerful spatial dynamics currently at work in the analysis of a country's political orbit. One is the

2 Dailymail: '"Suddenly very invested in the World Cup!" Viral photo of Iran's VERY dapper national football squad leaves new fans swooning', Online Resource: http://www.dailymail.co.uk/femail/article-5820471/Suave-image-Iranian-world-cup-football-squad-social-media-swooning.html.

3 Independent, 'World Cup 2018: Nike withdraws supply of football boots to Iran national team due to new US sanctions', Online Resource: https://www.independent.co.uk/sport/football/world-cup/world-cup-2018-iran-nike-football-boots-supply-donald-trump-us-sanctions-a8396921.html.

classic remit of the international relations between states, although even here, new approaches to public diplomacy suggest that states are reaching out across geographical borders to each other's publics in unprecedented ways (Held et al. 2000, pp. 14–28) including interfering in each other's elections, most obviously Russian interference in the 2016 U.S. election.[4] The second is the classic remit of political analysis, the national dialogue – sometimes open, often constrained – between the state and its inhabitants within the country. The third is the space between the state and its citizens who live abroad – as diaspora, exiles, refugees, or migrants. Too often, each is analysed in isolation, part of the intellectually unedifying division of academic work. In addition, to complicate these geographies further, diasporas can communicate back in various ways while other states and other publics can become entangled in the conversation.

These three spaces overlap in terms of communicative dynamics, even if they describe slightly different sets of actors and agendas. These dynamics are simultaneous, perhaps affecting each other, waxing and waning as the issues change. The nation as an interpellator of people and their political imagination and participation remains strong, so distance does not fully overcome location, and space does not alter the ongoing significance of place. We cannot fully embrace the arguments about 'the death of distance' (Cairncross 2001), since proximity to the orbit of the state clearly still matters at times. Better put, we might say that the repressive reach of the state is more noticeable and harder to evade within a country – Iran is our focus here – than outside, but it is also present there (regime pressure on the families and property of BBC Persian employees is one example). We need to think of space and place in tension, even in contradiction, all the time, and we need to ask where politics is to be found: within the system or on the streets, and which streets those might be? This chapter explores these issues by examining the spatiality of contemporary Iranian politics within and outside the country.

International relations is an obvious form of politics at a distance, involving the system of internationally recognized actors at the level of a state. Two or more political systems interact in a variety of forms, from complex communications to outright war. Iran was entangled in great power politics throughout the nineteenth and 20th centuries. It is currently embroiled in a number of international conflicts, with a range of countries aligned in opposition to it. Looking merely at Iranian international relations with 'its' region, we see some Iranian military personnel on the ground in Syria. It

4 See: https://www.nytimes.com/news-event/russian-election-hacking.

is involved in the political conflict in Yemen. It enjoys some influence still inside Iraq. It has been a vocal supporter of a Palestinian state and loud critic of Israel; Prime Minister Netanyahu has ratcheted up his inflammatory language about Iranian nuclear developments at the United Nations.[5]

Beyond its geographic neighbourhood, Iran has been locked in an antagonistic relationship with the United States – often only narrowly avoiding outright military conflict – since the islamic revolution and hostage crisis of 1979. The successful negotiation of the Joint Comprehensive Plan of Action (JCPOA) with the P5+1 in July 2015 was seen as a significant breakthrough in Iran's diplomacy with the West. However, President Donald Trump withdrew the United States from the JCPOA in May 2018 and reimposed sanctions, a move condemned by the International Court of Justice in The Hague. He also used inflammatory rhetoric about Iran at the United Nations. Around 40 US military bases surround Iran.[6] A number of Iran's nuclear scientists have been murdered inside the country, state-sponsored killings that have received little publicity or analysis from Western media (Hasan 2012). Therefore, at the level of state activities, it is comparatively easy to understand the political extension of Iran into its wider neighbourhood and beyond – and the occasional hand of external states around and inside the country.

Somewhat more complicated is the dynamic nature of political communication between the state and the nation. Appadurai has quipped that 'the nation and the state are each other's project' (1990, pp. 295–310) as states try to develop a strong sense of nationhood, including a shared culture and history, while some nations wish to become states. States everywhere indeed claim to represent not only the people but also the unity of their people and wield the monopoly of violence to ensure their governmentality. It is precisely through the enforcement of such claims that the state attempts to impose a singular politics and often fails to represent the unity of the people it purports to represent. In the MENA region – and because of military interventions of imperial powers that of course were and are de-territorial – there are huge areas in which the very idea of the state or state power has ceased to exist: in Iraq, Afghanistan, Libya, Syria, Somalia, and Sudan. This idea is often referred to in the concise and somewhat crude notion of 'failed states'. The other related paradox, as Rogers Brubaker reminds us, is that the issue of membership of a national polity is not always generated 'by the movement of people over borders, but by the movement of borders over

5 Netanyahu's UN speech can be found at: https://www.youtube.com/watch?v=4xCBbPnE3oE.

6 For discussion and a map, see: https://www.aljazeera.com/indepth/interactive/2012/04/2012417131242767298.html.

people' (2010, p. 69). The collapse of the Soviet Union and the emergence of various successor states or the breakup of the former Yugoslavia are some examples. Another fact that is often ignored is that in the formation of nation states and membership of their politics, many people did not join the state but rather the state encompassed them. This is true of all empires but in particular in the case of expansion of the United States into the Caribbean and Pacific (Brubaker 2010, p. 71). In the MENA region, two countries are still known not through the representation of the unity of their people or the name of the nation but by the title of the ruling family: the Kingdom of Saudi Arabia and the Hashemite Kingdom of Jordan. The state is no longer (if it ever was) the container of the nation but often its substitute.[7]

However, we might also say that in Iran, the state and the nation are each other's nemesis, locked in fierce political confrontation. What we mean in our shorthand use of the 'nation' here is essentially a country's citizens, its publics. Here we approach Hannah Arendt's notion of the political as a contrast between a polis – or a res publica – in which citizens act together to manage their own affairs, and the state as holding the monopoly on violence and engaged in rulership over its citizens (1965, p. 280).

The Politics of the State

Clearly, the Iranian state, like all other states, is most governmental within its borders. Iranians inside are required to pay taxes, abide by all laws – including those relating to censorship, hijab, and sexual behaviour – travel on an Iranian passport, and so forth. Those individuals who question the actions of the state may feel the full force of the law through executions, imprisonments, censorship, travel bans, and more (Sreberny and Torfeh 2013). Activists in unions, in women's and minority groups (*Bahai'i*, Jews, etc.) have been joined by journalists, editors, bloggers, poets, cartoonists, musicians, and others in having experienced the repression of the Iranian state (Sreberny and Khiabany 2010; Sreberny and Torfeh 2013).

However, the space of Iran, and thus of Iranian politics, has become elongated because of the outmigration of a substantial diaspora since the 1979 islamic revolution. Approximately four million people, including some of the most highly educated, have become dispersed across many

7 The new law of nationality promulgated in Israel in July 2018 is yet another example of the lack of contiguity between people and political demarcations; the Palestinians resident inside Israel becoming increasingly disenfranchised.

different new host countries, including the United States, the United Kingdom, Canada, France, and Germany, but also Costa Rica, Malaysia, Japan, and Norway, depending on the vicissitudes of international relations, immigration regimes, and economic flux (Sreberny 2000). These diasporic communities are neither treated as postnational nor transnational but, as Moallem has suggested, as the 'excess of the nation-state' (2000, pp. 200–210), often encountering forms of racism in the host country and exclusion by 'their own state'. Over the past few years, the diaspora has acquired the right to vote in Iranian elections – yet the counting of votes often seems too fast to include that opinion (Boroujerdi and Rahimkhani 2018, pp. 101–103) – and the diaspora broadcasts and messages back to fellow countrymen and the Iranian state.

The Islamic Republic of Iran has used various approaches to persuade, cajole, and enforce its positions vis-à-vis its citizens, and these have a reach far beyond the confines of the territory of Iran. The long arm of the state can reach across borders to punish Iranians who live outside Iran. According to the Iran Human Rights Documentation Center, the Iranian state has been involved in at least 162 extrajudicial killings of political opponents in nineteen different countries around the world, including in France, Sweden, Austria, Germany, the United States, Japan, Turkey, and Switzerland.[8] High-profile killings include the assassination of Abdol-Rahman Ghassemlou, Secretary General of the Kurdish Democratic Party of Iran, in Vienna in 1989 and his successor, Sadegh Sharafkandi, in the Mykonos restaurant in Berlin in 1992. A recent victim of such an assassination by the Iranian state was Saeed Karimian (who held British citizenship), chairperson of the Dubai-based GEM TV Network. He and his Kuwaiti business partner were shot dead on 29 April 2017 in Istanbul. GEM TV runs seventeen Persian-language channels (with dubbed Turkish soap operas among its main and very popular programmes). In 2016, a revolutionary court in Iran passed a sentence of six years of imprisonment against Karimian for spreading propaganda.[9]

State violence against non-resident Iranians has also taken many forms inside the country. Recently, Iranian state activity has included the arrest of dual nationals entering the country to visit family or attend educational or

8 Iran Human Rights, Online Resource: http://www.iranhrdc.org/english/publications/reports/3152-no-safe-haven-iran-s-global-assassination-campaign.html.

9 'Iranian TV executive shot dead in Istanbul, Turkish media report', Online Resource: https://www.theguardian.com/world/2017/apr/30/iranian-gem-tv-saeed-karimian-shot-dead-in-istanbul-turkish-media-reports.

cultural events, often for these individuals to be used as a pawn in negotiations with their new host country. According to Reuters,[10] since 2015, more than 30 people with dual nationality have been arrested in Iran, including students, lawyers, academics, and charity workers; the majority detained in recent years have European citizenship. In February 2018, Kavous Seyed Emami, the Iranian Canadian founder of the Persian Wildlife Heritage Foundation, died in solitary confinement in Evin Prison, having been charged with espionage. In 2003, another Iranian Canadian, photojournalist Zahra Kazemi, was killed in the same prison, arrested for taking photographs of protesting families of political prisoners. The ongoing imprisonment without trial of the Iranian British woman, Nazanin Zaghari-Radcliffe, whose daughter has now been returned to her father in London, has raised consternation in Britain and continues in autumn 2020 despite a strong public campaign and intervention from British political figures.[11]

In addition, the Iranian state, like most other states, is very active in mobilizing politico-cultural support beyond its own borders. International media influence is important. The Islamic Republic of Iran Broadcasting (IRIB) broadcasts to foreign audiences through four international news television channels, six satellite television channels, and 30 radio channels broadcasting in 32 languages, including English, Arabic, Hebrew, Armenian, Turkish, Russian, Bengali, and Japanese. Its Jame-Jam network (comprising three channels for Europe, America, and Asia) also broadcasts specifically to the Iranian diaspora. In January 2017, the Islamic Republic of Iran parliament approved a government proposal to double the budget of IRIB. According to published reports, 'the IRIB will receive 0.7% of the government's general budget over the next five years' and is allowed to spend around $750 million annually.[12] This is despite the wide acknowledgement that IRIB has lost a significantly large slice of its domestic audience. Press TV, an English-language channel with offices in London, actively asserts its critical journalism vis-à-vis British politics. Communicative borders are fluid, and individuals can access multiple sources of information and cultural content.

10 'Exclusive – Iran's Revolutionary Guards arrest more dual nationals', Online Resource: https://uk.reuters.com/article/uk-iran-arrests-exclusive/exclusive-irans-revolutionary-guards-arrest-more-dual-nationals-idUKKBN1D90TE.

11 'Britain tells Iran it will not let matter of detained aid worker rest', Online Resource: https://uk.reuters.com/article/uk-britain-iran/britain-tells-iran-it-will-not-let-matter-of-detained-aid-worker-rest-idUKKCN1M71ZA.

12 'IRIB Budget Doubles', Online Resource: https://financialtribune.com/articles/economy-domestic-economy/57517/irib-budget-doubles.

However, while the Iranian state is active in international broadcasting, it does not like being the target of such broadcasting from others. The sensitivity of the Iranian state against what it calls 'soft war' (Sreberny 2013) is evident in its long-running hostility towards the BBC, especially BBC Persian, which was established in 2008 and is aimed at Iranians inside Iran (Sreberny and Torfeh 2014). The state had tried to block signal reception by sending up parasites, but audiences became adept at retuning their Hotbird receivers; unable to attack the station, the state has moved against its Iranian employees. The Islamic Republic of Iran has sequestered the property of BBC Persian staff and even held their families to ransom. In October 2017, the Iranian state opened a criminal investigation into what they considered to be 'conspiracy against national security' by 152 current and former BBC Persian journalists. Since the Iranian uprising in 2009, dozens of BBC journalists' families and friends as well as those contributing to BBC Persian programmes have been harassed, investigated, and put under pressure by the Iranian security forces. The National Union of Journalists (UK), International Federation of Journalists, and a number of human rights organizations have condemned the Iranian state's treatment of BBC journalists and their families. In March 2018, the BBC World Service appealed to the United Nations for support to protect its staff but the problem continues.[13]

Digital Technologies and the Possibilities for and Limits to Politics

Digital technologies may have the effect of further increasing the geographic reach of Iranian politics and culture. The internet can link a national public across geographic borders. Home and away are all the same on Instagram or Telegram, which offer instantaneous communication and can facilitate diverse and multiple publics, otherwise referred to as public sphericules or even content bubbles.

Women have used new media to considerable effect (Sreberny 2017). Yet the dynamics of such interaction reveals the complexities and different senses of place. This is particularly acute in the case of a highly publicized and successful Facebook campaign called 'My Stealthy Freedom'. Iranian journalist and activist Massih Alinejad launched the campaign in May 2014. This page was dedicated to posting images of women inside Iran who had dared to remove their hijabs. After a few days, the page had received over

13 'BBC urges UN to help protect rights of Iran staff', Online Resource: https://www.alaraby.co.uk/english/news/2018/3/12/bbc-urges-un-to-protect-its-iranian-staff.

100,000 likes, and less than a year after the launch of the campaign, the page had over 760,000 followers – a much bigger following than the winner of the 2009 presidential election, Mir-Hossein Musavi's, Facebook page at the peak of the 2009 Iranian uprising. The international success of the Stealthy Freedom campaign shows that new technologies do indeed break geographical boundaries and precipitate international sympathy and solidarity. However, these technologies are also born in specific historical societies and social relations.

The compression of time and space tends to make space seem irrelevant; however, new technologies by themselves cannot bypass assumptions, prejudices, and stereotypes. The overwhelming feeling outside Iran has been one of support, solidarity, and encouragement. Yet, the positive international coverage and support should not obscure a number of contradictions and dilemmas, especially when 'saving' Muslim women (Abu-Lughod 2013) was on the political agenda and used as an excuse for military interventions in the region. There is always a danger, as Zishad Lak reminds us, that a local struggle might be assimilated into something completely different (2014). For Lak, the Stealthy Freedom campaign is one of those instances in which 'local resistance risks being thrust into obscurity to be protected from colonial interpretations. What we should be wary of is the audience or the interlocutor that is implicit in the message around which the actor organises her actions' (2014). In every story published in mainstream British or American media are supporting statements encouraging Iranian women with phrases such as 'Go Girls!', or simply describing participants in this campaign as '[b]eautiful, smart, confident, and happy' and welcoming them 'to the 21st century' (Khiabany 2015). The mismatch between local context and concerns (internal) and the framing and reception of these concerns outside the geographical boundaries of the local (external) are rarely considered.

The digital environment in Iran has been subject to friction between government factions of the state, competition between the institutional interests of various state agencies, and tension between the state and the public. Even though the Islamic Republic of Iran has officially encouraged the use of the internet, issues of access and control over content remain controversial (Rahimi 2003, pp. 101–115). Fearful of the formation of any alternative source of social power, the state has taken a leading role in providing internet access and services. A majority of shares of the Telecommunication Company of Iran were sold to the Revolutionary Guards for $5 billion in 2009, clearly demonstrating the significance of the internet to the state. This military takeover, which was done under the banner of privatization, points to the security dimension of the internet in Iran. In addition, the Iranian state has devised a range of controls, which are implemented with greater or lesser degrees of efficiency. These controls include ISP licensing, internet filtering, censorship of sites, control of the speed of the internet, attempts to inundate the net with favourable content, the criminalization of digital activities, and the arrest of activists (Sreberny and Khiabany 2010). In short, the Iranian state, rather unsurprisingly, relies on coercion and violence (physical as well as symbolic) as its ultimate sanction.

The lack of independent media, the closure of many reformist newspapers, and the erosion of state legitimacy, coupled with the growing popular dissatisfaction in Iran, helped to increase activism in the blogosphere and develop a vigorous cyber culture – a 'blogistan', so to speak (Sreberny and Khiabany 2010). However, the collapse of the Iranian uprising in 2009 and the massive state crackdown on all forms of political activism (digital and otherwise) put an end to that phenomenon. A recent study suggested a 'high rate of blog abandonment with only around 20 percent of the prominent blogs from 2008–2009' still active in September 2013 and a general decrease in the frequency of posting: '70 percent of bloggers surveyed publish one post per month or less – in contrast to earlier years' (Giacobino et al. 2014, p. 3). However, this decline can also be attributed to the growth of social media and the migration of bloggers to new platforms.

New Social Networking Sites (SNSs) 'facilitate publication and instant promotion of any content, including short, often less analytical and re-used content and provide a consistent, seamless and cross platform experience for readers' (Giacobino et al. 2014, p. 4). While these may be the technical advantages of SNSs, they also operate in the same historical society and thus have all the hallmarks of the existing realities and constraints on expression in Iran. Significantly, the popularity of and controversy over SNSs are consistent with previous phases of new media adoption and abandonment in Iran. In 2014, Al Jazeera reported that of those who have access to the internet, 58 per cent use Facebook regularly and 37 per cent use Google+. Facebook and Twitter are still officially banned in Iran, although that has not deterred key Iranian officials, including the supreme leader, or institutions, from using such platforms.

The most interesting success stories, however, relate to the use of Viber, WhatsApp, and, more recently, Telegram. In contrast to the country's 4.5 million Facebook users, around 10 million people (12 per cent of the Iranian population) used Viber. In fact, in 2015, Iranians were the leading group of Viber users internationally, a trend similar to their earlier adoption of blogs. According to Alexa.com, Viber was ranked 6644th globally but 776th in

Iran; nearly 18 per cent of Viber users in 2015 were in Iran, ahead of Russia (8.8 per cent), the United States (8.4 per cent), India (7 per cent), and Brazil (4.9 per cent). Viber itself emerged as a top news item after the Iranian police arrested eleven people in September 2014 for circulating jokes about Ayatollah Khomeini on Viber.[14] Gholam-Hossein Mohseni-Ejei, first deputy of the judiciary branch, in a letter to Mahmood Vaezi, the communications minister, threatened that if Vaezi did not take action, the judiciary would 'take the appropriate measures' (Khosrokhavar 2012, p. 152). Mohseni-Ejei described applications such as Viber, WhatsApp, and Tango as networks 'with immoral and criminal content' (Khosrokhavar 2012, p. 152), and gave the government a month to block them. The communication minister replied that blocking such networks would not solve any problems.[15] Iranian officials and institutions were also among Viber users and the fact that it was developed by Israel had not stopped Press TV, for example, or Rafsanjani's website from using the app. Even the security officials are divided over how to tackle these networks. Ahmadi-Moghadam, the chief of the Iranian police, had admitted that the use of such apps is inevitable, but added that the state needs to find ways to monitor and control inappropriate use of these services.[16] He did urge the Supreme Council on Cyberspace to come up with appropriate guidelines regulating the use of networks such as Facebook and Viber. Prior to the Viber controversy and confusion, there was more unhappy news for users of new technologies in Iran when an Iranian court charged seven young men and women in May 2014 for posting their version of a Pharrell Williams music video on YouTube.[17]

The Viber controversy, however, had more than a political dimension; it was not simply over jokes about political leaders or irreverence towards the clergy, even though according to a report by BBC Persian, 29 per cent of all SMS in Iran are jokes.[18] Tools such as Viber and WhatsApp were and are also denting the profits of the Telecommunication Company of Iran. Viber and WhatsApp provide Iranians with what voice-over internet protocol

14 'The Most Popular Messaging App in Every Country', Online Resource: https://www.similarweb.com/blog/worldwide-messaging-apps.

15 'Iran Communication Minister: The problem will not be Solved by Closure of SNSs', Online Resource: http://www.bbc.co.uk/persian/iran/2014/10/141001_filter_communication_vaezi.

16 'Iran's Police Chief: Inevitability of use of Viber and Facebook', Online Resource: http://www.bbc.co.uk/persian/iran/2014/09/140924_l45_police_filtering_ahmadimoghaddam.

17 https://www.independent.co.uk/news/world/middle-east/young-iranians-arrested-and-made-to-confess-on-tv-for-being-too-happy-9409988.html.

18 'Nearly One Third of all SMSs are Jokes', Online Resource: http://www.bbc.co.uk/persian/iran/2015/03/150301_me_sms_iran.

had done in the early 2000s: low-cost telephone communication. TCI had reported $20 million profit in 1998, but in 2002, despite a threefold increase in the number of people with access to telephone lines in that period, TCI reported a loss of $23 million. Expectedly, private internet and telephone providers were blamed for this loss in profit and the reversal of TCI fortune and, in reaction, the government began to close many cybercafés. The Revolutionary Guards, owner of TCI, were reacting yet again with the same moral crusade and waging their own war against what certain factions of the Iranian state have labelled 'soft war', yet this struggle may actually be more about money than influence.

The controversy over SNSs in Iran continued as Telegram started to replace Viber and WhatsApp as Iranians' favourite method of communication. In October 2015, Mahmood Vaezi, still the Islamic Republic of Iran's Communications Minister, suggested that approximately 14 million Iranians use Telegram, although the social networking company claimed that the number of users was around 20 million.[19] In October 2015, Pavel Durov, the co-founder of Telegram, revealed that Iran's Ministry of Information and Communications Technology requested 'spying and censorship tools' from the service (Solomon 2015). When Telegram refused to give in to the minister's request, the ministry moved to block the service – a claim that was vehemently denied by Iranian officials. Because of their ease of use and accessibility via mobile devices, SNSs have started to replace blogs and emails. Their popularity is such that many state officials, including the supreme leader, news organizations, political parties, musicians, and various social groups have opened their own channels on Telegram to share and disseminate their works, ideas, and analysis. As an emerging company, Telegram cannot ignore such a big market in Iran, and many organizations have expressed concerns over its possible cooperation with the Iranian state. For example, on 18 November 2015, Article 19, a British human rights organization, urged Telegram to 'clarify the nature of its negotiations with Iranian officials, including the nature of any data, technology, and censorship being requested, as well as the date on which discussions began, and which elements of the Iranian regime have been involved'.[20] Such expressions of concerns about privacy and Iranian government interference with popular communication

19 'Communication Minister: 13 to 14 Million use Telegram in Iran', Online Resource: http://www.bbc.com/persian/iran/2015/10/151025_l26_iran_ict_minister_telegram.

20 'Iran: Privacy and Censorship Fears around Telegram Messaging App', Online Resource: https://www.article19.org/resources.php/resource/38196/en/iran:-privacy-and-censorship-fears-around-telegram-messaging-app.

tools and applications are not without precedent. In 2009, Nokia Siemens Networks was forced to confirm that it had provided the Telecommunication Company of Iran with equipment that allowed it to spy on its customers. This led to a campaign against the Finnish information technology company, and 'Nokia: Disconnecting People' became one of the slogans of Iranian activists in 2009.[21] Despite all of these issues, access to such tools have increased rapidly. In 2018, despite massive investment by the Iranian state to block and filter Telegram, 40 million Iranians (20 per cent of Telegram users across the world) were using it, which is around 90 per cent of Iranian adults. In the same year, Iran was among the top ten users of Instagram, with 24 million registered users. Such platforms blur the line between inside and outside and help to create a transnationally stretched Iranian politico-cultural space.

Mobile phones equipped with SNS capability are also politically relevant for other reasons; they

> not only permit the exchange of information between the individuals but also enable each person to act as an amateur photojournalist during demonstrations and their repression, sending photos and short videos to channels. In Tunisia, Al Jazeera did not have an authorized official journalist, and it broadcast the images and short sequences it received from the young amateur journalists. (Khosrokhavar 2012, p. 152)

During the 2009 Iranian uprising, Yahoo Messenger and Bluetooth technologies allowed individuals to maintain connectivity and exchange information, often with international broadcasters such as BBC Persian, which rebroadcasts this material in Iran; such technologies made it possible to either bypass the government censorship wall or remain connected when the internet was down or too slow. SNSs thus connect Iranians inside and outside Iran, but they are also heavily used by the state itself. Uses and abuses of media (social or otherwise) happen within a particular political context. Thus, while the government might wish to control, or in the last instance, close down these platforms, its own involvement with them makes such control a mixed blessing.

Uncritical accounts of the role of information technologies tend to forget what Williams has called 'uncontrolled effect' (1974, p. 18). Williams insisted on finding ways to democratize culture and cultural production and long before digital media optimism, he argued for access to cultural resources

21 'Iranian consumers boycott Nokia for "collaboration"', Online Resource: https://www.theguardian.com/world/2009/jul/14/nokia-boycott-iran-election-protests.

which would allow everyone to both produce and consume culture. Williams also pointed at alternative uses (and regulation) of technologies. In the case of television in particular, he suggested that while it is possible to discuss the social history of broadcasting on its own, 'it is unrealistic to extract from it another and perhaps more decisive process, through which, in particular economic situations, a set of scattered technical devices became an applied technology and then a social technology' (Williams 1974, p. 18). In other words, the level of one's independence as a producer is shaped by the broader material realities that structure, regulate, and circulate symbolic production. However, in the same book, we read this memorable sentence: 'For there was no way to teach a man to read the Bible which did not also enable him to read the radical press. A controlled intention became an uncontrolled effect' (Williams 1974, pp. 133–134). In the case of social media, it is not just the contradiction between the original design (controlled intention), the use of technology (uncontrollable effect), and the obvious and stark interplay between public and private that confront us. Furthermore, the guerilla political struggle and the constant shift from one software to the next, one application to another, one platform to a different one, reveals more about the nature of resistance and political engagement than technologies per se or the iron fist of state. This is, of course, not unique to Iran.

External Incursions into Iran

The crossing of territorial borders by the state to reach and influence audiences, or harass, recruit, or even kill dissident voices is not unique to the Iranian state. The obverse, the incursions into Iran by external forces, is evident and sometimes even deadlier.

During the Iranian uprising in 2009, the US government asked Twitter to postpone its scheduled maintenance service because of Twitter's alleged significance to the Iranian uprising. The US government had made every effort to appear as if it was not intervening in Iran's domestic affairs. However, Secretary of State Hillary Clinton defended the decision by saying that 'the United States believes passionately and strongly in the basic principle [...]. We promote the right of free expression'.[22]

Such attempts to promote the myths of the 'technologies of freedom' were made at the same time that the US government was suppressing

22 'Remarks With Israeli Foreign Minister Avigdor Lieberman', Online Resource: https://2009-2017.state.gov/secretary/20092013clinton/rm/2009a/06/125044.htm.

these very technologies (or alternative uses of them, particularly in the case of WikiLeaks) and, moreover, that US policy was contributing to censorship of the internet in Iran. It was not until President Obama delivered his famous message for the Iranian New Year on 20 March 2012, when he asked Iranian leaders to bring down the 'electronic curtain' of censorship, that he promised to lift the sanctions that had disconnected Iranians from the internet. The tragedy is not just that the US government had contributed to that 'electronic curtain', but rather that it was hypocritically trying to 'help' Iranians to break free of the government at a time when US economic sanctions were making the Iranian people more dependent than ever on the Iranian regime (parallel to the situation in Iraq). US sanctions threaten the livelihoods and health of the Iranian people, but at least, thanks to the US government's generosity, they could access Google+. President Obama promised that, despite objections from the Iranian government, the US would launch a Virtual Embassy that would bring 'Iranians information they cannot access from their own government-controlled media, ranging from world news to facts on US policy and culture'. As Ben Rhodes, the deputy national security adviser for strategic communications and speech-writing, commented, the next step was to make easily available 'software and services that *Iranians have told us* are essential in order to effectively use the Internet' (2015; emphasis in original).

From the coup of 1953, and perhaps even before, there have been numerous attempts to provide Iranians with what they 'need' (Gasiorowski 2012, pp. 669–678). The policy of helping the Iranian people gained considerable traction after the popular politics against the 'stolen election' of 2009 in what became known as the Green Movement. For example, in January 2010 the *Wall Street Journal* reported that the US government was searching for ways to support the Iranian opposition. In an article entitled 'U.S. Shifts Iran Focus to Support Opposition', Jay Solomon suggested that the government was considering new sanctions that would target individuals and institutions in Iran directly involved in suppressing dissent in the country. It was also reported that although Obama was not considering regime change, 'Washington remains committed to a dual-track approach of pursuing dialogue aimed at ending Iran's nuclear program while applying increasing financial pressure if the talks fail' (Solomon 2015). Eighteen months later, the hint for the 'Iranians' to tell 'us' what was essential for them came in two interviews that Hillary Clinton gave to BBC Persian and Voice of America (VoA) in October 2011. In both interviews, she explicitly suggested that the US government would have supported the uprising in

Iran if people had asked for it.[23] Around the same time, Kenneth Pollack of the Brookings Institution and Ray Takeyh, a senior fellow at the Council on Foreign Relations, argued in a joint article entitled 'Doubling Down on Iran' published in the *Washington Quarterly* that the dual-track policy of successive US administrations had failed and that the time had come to pressure Iran where it hurts most. In rethinking US strategy, they persisted in keeping the option of a 'very limited military operation' on the table, but insisted that policy-makers have to start breaking 'old taboos – such as providing direct assistance to Iranian opposition groups' (Pollack and Takeyh 2011, pp. 7–21). The strategy of 'doubling down' on Iran proposed by the authors had clear 'cultural/technological' elements. Connecting with the Green Movement (which by then had collapsed) was regarded a top priority, but the authors also pointed out that Iran is a place of labour, student, and ethnic minority protests, and while it is important for the CIA to be careful not to 'fall into traps inevitably laid by Tehran's own security forces at this point, it would be extremely productive to establish contact with Iranian opposition groups in hope of finding out what the United States could do to help them' (Pollack and Takeyh 2011, p. 16).

Such external 'help' is tied up with funding organizations and activities. Many of the initiatives involve media projects. In addition to the US government's $36 million combined budget for two official US media channels, Radio Farda and VoA, various US institutions fund other media or cultural outlets, including TV channels and websites. It is hard to determine precisely how much is invested and where, but patterns of mismanagement, abuse, and corruption have been reported.[24] The point here is that communication technologies are not simply a force for opposition from below; they also facilitate attempts to direct opposition from above and from outside. What Pollack and Takeyh suggest could help the Iranian opposition and hurt the regime 'lies in the overlapping areas of Internet freedom and information and cyber warfare' (2011, p. 17). Helping to reopen internet services in Iran or trying to keep them open was one aspect of this policy; the other was to try to wrest control over information technology from the Iranian state.

23 For a trenchant critique of the US attempt and the complicity of some Iranian actors, see: Hamid Dabashi, 'Fifth Column of the Postmodern Kind', *Al Jazeera*, http://www.aljazeera.com/indepth/opinion/2011/11/20111120132050412430.html.

24 For problems associated with VoA, see VOA PNN Watchdog, http://www. voapnnwatchdog.com/; and for reports of financial irregularity associated with the National Iranian American Council, see National Endowment for Democracy, http://www. iranian-americans.com/docs/NedReport.pdf.

However, this involved not only platform politics but also cyber warfare. Iran has been the object of some of the first moves by Western powers in this new form of warfare, presaged in the early 1990s and made real in the aughts with the deployment of Stuxnet in 2009 (Giacobino et al. 2014, p. 3). In 2012 it was ranked highest of the ten most destructive computer viruses, described as 'a cartoon villain come to life: it was the first computer virus designed specifically to cause damage in the real, as opposed to virtual, world' (Weinberger 2012, quoted in Sreberny 2013).[25] It is believed to have badly impeded Iran's nuclear enrichment programme. The story of the US–Israel collaboration in developing Stuxnet is not denied by either state, but what is usually ignored is that the product of that collaboration was not one but three viruses: Stuxnet came first, followed by Flame and DUQU. Another ignored dimension is that while such attacks occur in the virtual realm, their damage is real and physical. Such cyberattacks were part of the 'doubling down on Iran' policy that has so far involved the killing of five nuclear scientists in Iran since 2007. Technological interventions have moved from creating and supporting a fifth column in Iran to waging war in the fifth domain – the other military operational domains are land, air, sea, and space.[26]

Barack Obama expanded the War on Terror and nurtured the security state even more than his predecessor. *The Economist* reported that he 'declared America's digital infrastructure to be a "strategic national asset"' (Giacobino et al. 2014, p. 3) and appointed Howard Schmidt, the former head of security at Microsoft, as his 'cyber-security tsar' (Giacobino et al. 2014, p. 3). *The Economist* also reported that

> [i]n May [2010] the Pentagon set up its new Cyber Command (Cybercom) headed by General Keith Alexander, director of the National Security Agency (NSA). His mandate is to conduct 'full-spectrum' operations – to defend American military networks and attack other countries' systems. Precisely how, and by what rules, is secret. (Giacobino et al. 2014, p. 3)

In addition, in July 2011 *Reuters* cited Deputy Defense Secretary William Lynn as saying that the Pentagon wants to avoid militarizing cyberspace, but will treat it as an 'operational domain' (Alexander 2011). However, as already indicated, US policy towards Iran has taken a turn for the worse under Trump.

25 Sharon Weinberger, 'The top ten most-destructive computer viruses', Online Resource: https://www.smithsonianmag.com/science-nature/top-ten-most-destructive-computer-viruses-159542266/.

26 'War in the Fifth Domain: Are the Mouse and Keyboard the New Weapons of Conflict', Online Resource: http://www.economist.com/node/16478792.

2018 and the Current Political Upheaval

Nothing has shaken the regime more recently than the internal unrest that spread across numerous cities in the winter of 2017–2018, as popular concerns about the state of the economy, the environment, and indeed international affairs spilled onto the streets. This is ongoing. It also suggests that place still matters most.

During the late spring and summer 2018, there were sporadic protests around the country regarding water shortages, unpaid salaries, and severe currency depreciation. These show a new and different urban spatiality, away from major cities to smaller towns where there has historically been less investment and less economic activity. The current issues are less ideological and more economic and environmental. Iran was experiencing its worst drought in recent history. There were severe water shortages and limits to accessibility; the famous Zayanderood River in Isfahan was completely dry, mocking its famous bridges, while Lake Urumieh almost disappeared. Electricity was frequently cut, cutting the internet, halting lifts, and disrupting air conditioning in a very hot summer (Kamali Deghan 2018). These elements are the lived experience of Iranians inside the country, making them angry with an incompetent government that prefers foreign adventures to domestic investment. However, these also open up a gap with the diaspora, who do not experience such crises – or at least not in the same form.

In addition, there were some violent acts within the territory of Iran that raised the spectre of foreign-supported groups such as ISIS; however, and perhaps even more worrying for the Iranian state, an internal group, Ahwaz National Resistance, claimed responsibility.[27]

Conclusion

So where is Iran? What conclusions can we draw about the relevance of distance in contemporary national politics? We see Iran caught up in a vortex of international relations, including Putin's instructions for Iran to leave Syria as well as Trump's repudiation of JCPOA and his September 2018 Twitter spat with Rouhani. The 2020 US election that voted Trump out may change these bilateral relations for the better. Nevertheless, for many

27 'The Turbulent World of Middle East Soccer', Online Resource: http://mideastsoccer.blogspot.com/2018/09/attack-in-iran-raises-spectre-of.html.

Iranians these conflictual international relations are a diversion from the more pressing issues of domestic politics.

We see the emergence of stretched national publics, with Iranians inside and outside Iran sharing images, jokes, and political analyses. With the speed and immediacy of communications, dispersed populations can inhabit the 'same' time frame. Yet, we also see that lived experiences inside and outside the country are very different, and a politics that emerges from everyday grievances may diverge from the more abstracted, more ideological positions of those further away. Similarly, while the state might have long arms, it also has powerfully short arms, and those within its immediate purview feel a much more severe repression than those outside.

Neither the state nor the nation is a singular entity; both have profound divisions. Everyday difficulties inside Iran and vastly different diasporic lived experiences produce a psycho-cultural gap between Iranians outside and Iranians inside the territorial boundary of the country; even if for a moment, they (and the state) were all rooting for Team Melli.

References

Leila Abu-Lughod, *Do Muslim Women Need Saving?* (Cambridge: Harvard, 2013).

David Alexander, 'Pentagon to Treat Cyberspace as "Operational Domain"', Online Resource: http://www.reuters.com/article/2011/07/14/us-usa-defense-cybersecurity-idUSTRE76D5FA20110714 (accessed 18 November 2020).

Arjun Appadurai, 'Disjuncture and Difference in the Global Cultural Economy', *Theory, Culture & Society*, 7(2–3) (1990), pp. 295–310.

Hannah Arendt, *On Revolution*, revised second edition [1963] (New York: Viking Press, 1965).

Mehrzad Boroujerdi and Kourosh Rahimkhani, eds., *Postrevolutionary Iran: a Political Handbook* (Syracuse: Syracuse University Press, 2018), pp. 101–103.

Rogers Brubaker, 'Migration, Membership, and the Modern Nation-State: Internal and External Dimensions of the Politics of Belonging', *Journal of Interdisciplinary History*, 41(1) (2010), pp. 61–78.

Frances Cairncross, *The Death of Distance 2.0: How the Communications Revolution Will Change our Lives* (New York: Texere, 2001).

Stephen Coleman, 'Mediated politics and everyday life,' *International Journal of Communication*, 1 (2007), pp. 49–60.

Mark J. Gasiorowski, 'The Causes of Iran's 1953 Coup: A Critique of Darioush Bayandor's, Iran and the CIA', *Iranian Studies*, 45 (2012), pp. 669–678.

Laurent Giacobino, Arash Abadpour, Collin Anderson, Fred Petrossian, and Caroline Nellemann, 'Whither Blogestan: Evaluating Shifts in Persian Cyberspace', *Iran Media Program* (2014), p. 3.

Anthony Giddens, *The Consequences of Modernity* (New York: John Wiley & Sons, 2013).

Mehdi Hasan, 'Iran's Nuclear Scientists Are not Being Assassinated. They Are Being Murdered', *The Guardian*, Online Resource: https://www.theguardian.com/commentisfree/2012/jan/16/iran-scientists-state-sponsored-murder (accessed 18 November 2020).

David Held, Anthony McGrew, David Goldblatt, and Jonathan Perraton: 'Global Transformations: Politics, Economics and Culture', *Politics at the Edge*, ed. by Chris Pierson and Simon Tormey (London: Palgrave Macmillan, 2000), pp. 14–28.

Saeed Kamali Deghan, *The Guardian*, 20 July 2018, Online Resource: https://www.theguardian.com/world/2018/jul/20/desperate-to-find-a-way-out-iran-edges-towards-precipice (accessed 18 November 2020).

Gholam Khiabany, 'The Importance of "Social" in Social Media: The Lessons from Iran', in *The Routledge Companion to Social Media and Politics*, ed. by Axel Bruns, Gunn Enli, Eli Skogerbø, Anders Olof Larsson, and Christian Christensen (New York: Routledge, 2015), pp. 223–234.

Farhad Khosrokhavar, *The New Arab Revolutions that Shook the World* (Boulder: Paradigm Publishers, 2012).

Zishad Lak, 'Stealthy Freedoms and the Colonial Gaze', *The F Word - Feminist Media Collective*, 17 September 2014, Online Resource: http://www.feminisms.org/6640/stealthy-freedoms-and-the-colonial-gaze/index.html (accessed 21 January 2021).

Minoo Moallem, '"Foreignness" and Be/longing: Transnationalism and Immigrant Entrepreneurial Spaces', *Comparative Studies of South Asia, Africa and the Middle East*, 20(1) (2000), pp. 200–210.

Kenneth Pollack and Ray Takeyh, 'Doubling Down on Iran', *The Washington Quarterly*, 34 (2011), pp. 7-21.

Babak Rahimi, 'Cyberdissent: The Internet in Revolutionary Iran', *Middle East Review of International Affairs*, 7(3) (2003), pp. 101–115.

Ben Rhodes, 'On Nowruz, President Obama Speaks to the Iranian People', *The White House Blog*, Online Resource: http://www.whitehouse.gov/blog/2012/03/20/nowruz-president-obama-speaks-iranian-people (accessed 18 November 2020).

Jay Solomon, 'U.S. Shifts Iran Focus to Support Opposition', *The Wall Street Journal*, Online Recource: http://online.wsj.com/news/articles/SB126300060937222569 (accessed 18 November 2020).

Annabelle Sreberny, 'Media and Diasporic Consciousness: An Exploration Among Iranians in London', in *Ethnic Minorities and the Media*, ed. by Simon Cottle (London: Open University Press, 2000).

Annabelle Sreberny and Gholam Khiabany, *Blogistan* (London: I.B. Tauris, 2010).
Annabelle Sreberny, 'The New Cyber Military-Industrial Complex: International Relations Refracted through a new PRISM', plenary talk at IAMCR (Dublin: 2013).
–, 'Talking Soft About "Soft" War', *International Journal of Communication*, 7 (2013), pp. 801–804.
–, 'Women's Digital Activism: Making Change in the Middle East', *Digital Middle East: State and Society in the Information Age* (London: Hurst Publishers, 2018).
Annabelle Sreberny and Massoumeh Torfeh, *Cultural Revolution in Iran* (London: I.B. Tauris, 2013).
–, *Persian Service: The BBC and British Interests in Iran* (London: I.B. Tauris, 2014).
Raymond Williams, *Television: Technology and Cultural Form* (London: Fontana, 1974).

About the Authors

Gholam Khiabany teaches in the Department of Media, Communication and Cultural Studies at Goldsmiths, University of London. He is a member of council of management of the Institute of Race Relations, and Editorial Working Committee of Race and Class. His publications include *Iranian Media: The Paradox of Modernity* (2010); *Blogistan* co-authored with Annabelle Sreberny (2010); *Media, Democracy and Social Change* co-authored with Aeron Davis, Natalie Fenton and Des Freedman (2020); and two co-edited collections: *Liberalism in Neoliberal Times: Dimensions, Contradictions, Limits* (2017), and *After Charlie Hebdo: Terror, Racism and Free Speech* (2017).

Annabelle Sreberny is Emeritus Professor in the Centre for Global Media and Communications Studies at SOAS, University of London, was the first director of its Centre for Iranian Studies. She was president of the International Association for Media and Communication Research (IAMCR) from 2008 to 2012. Much of her research has been in the broad field of international communication and globalization, with strong foci on international news, diasporas, and feminism. Her work on the contemporary history, politics, and cultural environment of Iran includes books on the 1979 Revolution (*Small Media, Big Revolution*, 1994), the Persian blogosphere (*Blogistan*, 2010), and issues of cultural creativity and repression (*Cultural Revolution in Iran*, 2013).

13. Mapping Genocide? Giving Visual Memory to Oral Culture

Sebastian Maisel

Abstract

Yezidi religion and history had been largely transmitted orally until the late 20th century due to the closeness, isolation, and marginalization of the community in their various home countries. It was the advent of digitization that sparked a radical switch and concurrent emergence of a new class of protagonists who used social media as a tool to theorize and generalize sacred knowledge. The new actors often do not belong to the traditional class of clergy in charge of preserving and transmitting this information. In this chapter, I argue that their use of social media to spread deliberate knowledge has contributed to the development of new forms of identity and loyalty among Yezidi groups in Syria and Iraq.

Keywords: Yezidi, genocide, new media, memory

In August 2014, the Yezidi community in Iraq endured yet another genocidal attack when the radical Islamist and terror organization ISIS targeted the community in the Sinjar Mountain region. The attack included mass killings, kidnapping and enslavement of women and children, forced exile, and the destruction and looting of property and sacred buildings. Because of the magnitude of the event, the Yezidi community reacted differently to this attack than in the past. Yezidi activists, many of them women and survivors, broke with age-old customs, taboos, and traditions by publicly giving testimony to the atrocities and crimes committed against them. They did so by using their education and technological skills to promote their narratives. This story, which can be labelled 'Victims seeking Justice', reflects the transformation of Yezidi collective identity and the transmission of religious knowledge.

Yezidi religion and history have been largely transmitted orally until the late 20th century due to the closeness, isolation, and marginalization of the

North Africa: Producing Space. Taylor & Francis Group, 2021
DOI 10.5117/9789462989092_CH13

community in their various home countries. A new class of protagonists emerged since the digital era reached the Yezidi community. They use social media as a tool to theorize and generalize sacred knowledge and social conventions. The new actors often do not belong to the traditional class of clergy or patriarchal community leaders, who used to be in charge of preserving and transmitting this information. Effectively, we note an increased exposure to sacred knowledge as well as a reinterpretation of social dogmas by unorthodox scholars, activists, intellectuals, and politicians. In this chapter, I argue that their use of social media to spread deliberate knowledge has contributed to the development of new forms of identity and loyalty among Yezidi groups in Syria and Iraq. The Sinjar genocide accelerated this process.

Methodology

The main objective of this chapter is to understand how the visualization of survivors' memories spark social change. I will do this by discussing the relationship between social change, collective identity, and the production and circulation of both sacred and profane knowledge. This process traces the mapping of indigenous geographies, which is viewed as a means to establish new levels of authority and legitimacy within the Yezidi community.

Sources evaluated for this study include interviews with Yezidis from different social and religious groups, educational levels, and regional backgrounds. In addition to the oral narratives, official Iraqi, Syrian, German, and Yezidi documents from governmental and private archives were analysed to supplement the empirical data. In addition, several other mapping projects with Yezidi collaborators, to which the author contributed, will be introduced in this chapter.

Aside from ethnographic methods of inquiry such as participant observation or key informant interviews, this chapter also relies on social media practices and technologies as part of the fieldwork. In addition to the mapping project, popular social media sites such as Twitter, Facebook, and YouTube were accessed to communicate with distant informants, retrieve otherwise inaccessible data, and observe Yezidi practices of self-representation. Especially when informants and localities are located in areas of war and conflict, thus making it impossible to conduct traditional fieldwork, researchers refer to online sources. With the spread of technologies such as mobile phones and computers, the target community is more capable of sharing unfiltered and live information. For the distant analyst and observer, the raw material, documents, films, images, and other items become primary sources of high importance.

In case of the Yezidis, the number of primary sources prior to digitization was extremely small. However, over the past decade, Yezidis from all areas have gained access to various technologies and have used them vigorously to share files, photos, messages, and other information with each other and the rest of the world. Following the guidelines and recommendations of Hine (2008) as well as Poster and Pink (2012), the data was used to shape the opinion of the author and the argument of this chapter.

Introduction

To depict emerging social hierarchies, this chapter will highlight the shift in transmitting sacred and profane knowledge among the Yezidis using new tools and technologies at home and abroad in times of crisis.

The Yezidis, like other indigenous ancient religious and ethnic communities in northern Mesopotamia, have been described in hierarchical, segmented terms. Their religious structure is defined by the separation between clerics and laypeople. Most Yezidis support the idea of identifying with Kurdish origin and ethnicity, while some prefer a distinct Yezidi ancestry as a unique ethno-religious group. Spatially, the community is divided among several host countries including Iraq, Syria, Turkey, Armenia, Georgia, and the diaspora in Europe and North America.

Throughout history, they have been exposed to numerous periods of upheaval and instability. But since the time of nation-building in the Middle East after World War I, Yezidis have experienced periods of intense, large-scale population upheavals, often involuntarily due to forced displacement, ethnic and religious violence, and government interventions, some of which can be described as genocide. This study aims to investigate how recent events such as war, genocide, political change, and the promotion of specific identities have 1) affected the traditional Yezidi communities directly; 2) impacted gender roles; and 3) created new hierarchies based on the application of new digital media.

Traditional Yezidi Society

Religious authority within Yezidi society is concentrated in the hands of a few hereditary families who also represent the two main clerical castes of

Sheikh and Pir. This system of authority was reformed in the twelfth century by Sheikh Adi bin Musafir, a historic figure with strong connections to Sufi Islam. Prior to his arrival in the Hakkari Mountains, where the community lived, they already practised a proto form of Yezidism that incorporated elements of sun worship and likely Zoroastrianism. A group of religious leaders, the Pirs, stood above the general population of laypeople. Sheikh Adi added another group of clerics, the Sheikhs, and prescribed distinct marriage and interaction rules known as *sed u hed*. These rules define Yezidi society along three distinct groups, sometimes called castes, which were not permitted to intermarry and which were to guide and instruct each other. The descendants of Sheikh Adi have maintained political leadership over the community to the present day.

Little is known about the various Yezidi communities during the Ottoman Empire other that they lived secretive, reclusive lives at the periphery of society. Ottoman court records, French missionary reports, and British travellers' accounts, however, depict a community life similar to that of Sheikh Adi's time. They chose to live on the fringes away from the urban centre because of the continual discrimination by their Muslim neighbours, which resulted in mass killings, deportations, forced conversions, enslavement, and general destruction of their livelihood. The reports, however, also talk about the preservation of ancient customs (*sed u hed*) and unique religious convictions. Similar to their Muslim Kurdish neighbours, the Yezidis maintained a patriarchal, tribal family structure, which was economically based on farming and animal husbandry. In contrast to their Muslim neighbours, Yezidi communities made efforts not to integrate or assimilate with mainstream Ottoman, Kurdish, or Arab lifestyles. Instead, they favoured disassociation and avoiding cultural and/or religious exchange to a degree that the Yezidis in Sheikhan and Sinjar were known for their illiteracy. The leadership decided it was more beneficial for the community not to allow Islamic education to be spread among them. They argued that it was better to have no education than to be affected by Muslim *da'wa* ('proselytizing') and the *kuttab* (Koran schools). Thus, knowledge was concentrated in the hands of a few learned men while the rest of the community remained unexposed to religious (including Yezidism) and secular education until World War I.

With the establishment of nation states, which focused on nation-building through public education, the Yezidi communities in Turkey, Iraq, and Syria were no longer able to withhold their children from mandatory attendance at schools, where they were often forced to participate in Islamic education. Since the Yezidis were not recognized as an autonomous *millet* (an autonomous religious or ethnic community in the Ottoman Empire), they were not granted special treatment and ownership of their religious affairs and education. This also enabled governments to extend conscription laws to the Yezidis, which the community strongly opposed. Their motto of survival was still disengagement at all costs, and the community leadership in Iraq tried everything to avoid nationalistic, sectarian, or political conflicts.

The situation in Syria was less tense because the Yezidi community there was much smaller and already deeply embedded in the Kurdish sector (Maisel 2017). There was even a certain degree of tolerance towards the Yezidis, especially in the Efrin region, where Yezidi settlements can be traced back to the sixteenth century. Thus, the policies of anti-educational disengagement did not apply here. On the contrary, Yezidi community leaders such as Darwish Shamo actively pursued religious and secular education in their villages. In 1925, Shamo opened the first Yezidi School in the village of Qibare, and a Yezidi teacher from Iraq was hired to teach, among other topics, Yezidi religion. The minority-friendly attitudes of the French mandate powers allowed this to happen. When Syrian nationalists took over Syria after the country had become independent, Yezidism was no longer a permitted topic to teach. This and the physical disconnect from the main Yezidi territories in Iraq led to the formation of a diverse Yezidi identity in Syria – one that saw a decreased influence of clerics over community and identity-building measures. This void was often filled by educated laypeople, a process that has continued to the present day. Moreover, with the current political and demographic changes in Rojava (the Kurdish territories in Syria), such as the rise of the PYD (*Partiya Yekîtiya Demokrat*) and the establishment of self-autonomy, there might be greater changes for the Yezidi community in Syria and their role in shaping the future society.

New Gender Roles

In all Yezidi territories, including among the diaspora communities, gender roles have transformed. The traditional model of the patriarchal family is still widely accepted, but the recent experiences in Syria and Iraq, both traumatic and positive, pushed for a bigger role for Yezidi women in community life. The religious sphere, however, is less affected, and women are still largely excluded from it. Traditionally, they have played a minor role in the transmission of sacred knowledge. While Yezidi women have to select two female spiritual mentors (Sheikh and Pir) and a sister in the hereafter, these functions are largely passive. The only important role Yezidi women play in the religious landscape is a supportive one during weddings and funerals, where they often contribute with laments, traditional hymns, and other songs.

As such, very few historic female role models could have inspired Yezidi women to seek a more active role in social and religious life. Certainly, until the end of the 20th century, they were stuck in the traditional, patriarchal society where they had specific obligations and functions to fulfil. Seeking education, either secular or religious, was not one of those tasks. Some Yezidi women gained wider popularity throughout history mainly because of their political or social activism. They include the Grand Dame of Yezidism, Mayan Khatun, and some other female members of the Mir family. However, since the outbreak of the Kurdish national struggle and the limited involvement of Yezidi organizations, more Yezidi women from various backgrounds have used this platform to present a new image of Yezidi womanhood. In particular, the role Kurdish Yezidi women play in the PKK (*Partiya Karkerên Kurdistanê*) as politicians, fighters, activists, or supporters is noteworthy. The area of operation covered the Yezidi territories in Turkey, Iraq, and Syria, and the fallen members were always honoured and held in high esteem. A small number of Yezidi women from Turkey and Syria were among them. The number of Yezidi female fighters increased during the early years of the Syrian civil war, when many joined the Women's Protection or Defense Units YPJ (*Yekîneyên Parastina Jin*) in the battle against ISIS and other radical Islamic groups. Others joined the Asayish police force or the autonomous self-administration in high-ranking positions. The PYD has a women quota and a female co-chair, and similarly, Syrian Yezidi women became involved in the local politics of Rojava. Especially in the Efrin Canton, Yezidi women's organizations are particularly active and successful in promoting women's rights, the Yezidi religion, and the Kurdish national struggle. They use social media to connect and organize events and spread awareness among the general Kurdish population as well as the Yezidi diaspora.

The general access to the internet and availability of computers and more so of mobile phones have contributed much to enabling Yezidi women in Syria and Iraq to communicate their side of the story to a widespread and international audience. While they may be semi-analphabet, they know how to use technology to better connect and communicate with each other. The live newsfeeds from Mount Sinjar were crucial in shaping public Western opinion about getting involved and starting the air rescue campaign. Many Yezidi women were able to hide their mobile phones in ISIS captivity and convey messages and coordinates to their families. Moreover, while the past massacres against the Yezidis went almost unnoticed, the latest one was brought directly into the living rooms of the entire world by television.

The Yezidis finally caught the world's attention, but the consequences were brutal. While the Western world started a campaign to free some two hundred abducted Nigerian pupils, the number of kidnapped and enslaved Yezidi women and children reached almost ten thousand. Over the past three years, about half were eventually rescued, freed, ransomed, or managed to escape, which still leaves several thousand in ISIS captivity.

ISIS's campaign against the Yezidis has been qualified as genocide.[1] It follows a common pattern as the genocides in Armenia, Bosnia, or Rwanda: forced conversions, separation of men and women and children, mass execution of men above twelve, enslavement of young women and children, looting and destruction of property, and training and indoctrination of young Yezidi boys in ISIS camps.

Moreover, while ISIS indiscriminately and brutally attacks Sunnis, Shias, Druze, and Christians, they make special efforts to eliminate the Yezidis.[2] It must be noted, however, that this is not a new phenomenon. Intolerant and violent attitudes among the Muslim population in northern Iraq were observed long before the August 2014 attacks. ISIS is just the latest addition to the group of anti-Yezidi perpetrators. What is new is the methodology, justification, and technology of the genocide. ISIS is very transparent about their intentions and goals, one of which is the eradication of Yezidis from the area.

In their official publication, *Dabiq*, self-styled ISIS clerics try to explain the Islamic rationale that justifies the sexual enslavement of women from communities determined to be of 'pagan' or 'polytheistic' origin. Here, ISIS speaks of greater efforts to exterminate Yezidis; differentiates between People of the Book, apostates, and polytheists (Yezidis); justifies the enslavement of polytheist women as a custom of the early Muslim community; and describes the 'benefits' of having a relationship with a slave women over a prostitute (*Dabiq* 2015, pp. 14–17). A female ISIS writer providing scriptural and historic justification for the term *saby* ('female captive') further clarified this (*Dabiq* 2015, pp. 45–49). In addition, ISIS released a Q & A memorandum detailing the interaction with and treatment of the *sabaya*.[3] Finally, ISIS fighters and supporters use social media to share their experiences. At one

1 See the report 'They Came to Destroy – ISIS Crimes Against the Yezidis', published by the UN Commission of Inquiry (COI) on Syria at: http://www.ohchr.org/Documents/HRBodies/HRCouncil/CoISyria/A_HRC_32_CRP.2_en.pdf.

2 While Christians were given three options by ISIS – to convert, to leave, or to pay a special head tax – Yezidis had to choose between conversion and execution.

3 'Islamic State (ISIS) Releases Pamphlet On Female Slaves', 4 December 2014, http://www.memrijttm.org/islamic-state-isis-releases-pamphlet-on-female-slaves.html#_edn1.

point, it was reported that Yezidi women were sold online, and manuals and flyers of appropriate customer behaviour were circulated.

Considering the traditional character of Yezidi society, with their unique interpretation of honour and shame, one could assume that the crimes committed against Yezidi women would tarnish their reputation, and that they would not be allowed to return to their communities. This was the common reaction to similar incidents prior to the Sinjar massacre. Yezidi women have been kidnapped, converted to Islam, and married off to Muslim men in the past, and their ties with their Yezidi families were frequently completely severed (Kreyenbroek 2009, pp. 40–45 and pp. 50–55). However, despite the massive group and individual trauma, the community reacted differently in this case and welcomed back the women freed from ISIS captivity. Moreover, while the Yezidi religious leadership with the Mir, the Baba Sheikh, and the Spiritual Council have been rather silent or perhaps overwhelmed and paralyzed by the atrocities, they did make one important public announcement: the official and religious embracing of the Yezidi women into the community with special blessings, baptisms, and official decrees.

Yezidi women themselves, however, were the real activists for Yezidi women, refugees, and justice. Often traumatized survivors, they reached out to the world to publicize their horrific stories in an act of defiance towards traditional values and customs. Talking about rape and other sexual crimes used to be taboo; however, sharing their stories was not only an act of great personal courage but also ignited a campaign to raise awareness of the crimes and awareness of the great number of women still held captive by ISIS. Many Yezidi men have been killed, but Yezidi women continue to live with the ordeal, and some Yezidi women use social media to advocate for them. With this increased publicity, they have taken on new leadership roles in the community, certainly not in a religious capacity, but with great impact on gender roles and hierarchies within Yezidi society.

Vian Dakhil is the most prominent among these women activists. A member of the Iraqi parliament and from an upper-class family in Sinjar, she has used her elevated position to raise awareness about the genocide inside Iraq and around the world. Other women, such as Nadia Murad, Lamia Bashar, and Farida Khalaf, are survivors of ISIS sexual slavery.[4] By bravely sharing their experiences with the world, they broke taboos and the barriers of ignorance and social hierarchies. Bari Ibrahim is a Yezidi advocate from the diaspora group, and she has managed to raise awareness and aid for the

4 Murad and Khalaf both published their accounts 2017.

community in Sinjar. These five women symbolize the new Yezidi woman who represents the community from within and to the world.

Building New Hierarchies

Before August 2014, Yezidi society in Iraq was known for its strict religious observances, marriage rules, and patriarchies. As mentioned earlier, secular education did not spread among the villages until the 1960s. Furthermore, a religious education for the masses had not even started yet (with exceptions in the diaspora). It was only in the 1970s when young Yezidi academics (from religious families) started to collect oral histories and texts with the aim of releasing them and spreading the information among the Yezidi masses. Pir Khidr Suleyman and Khalil Jindy from Iraq, and Celil and Celile from Armenia, were among those pioneers who recorded the first poems, hymns, psalms, and religious chants and later transcribed and translated them (Celil and Celile 1978; Suleyman and Jindy 1979). With increasing publication opportunities, these texts eventually made it into the first textbooks on the Yezidi religion. However, they were not allowed to be used in Syrian, Turkish, or Iraqi schools, so it was left for the diaspora community to organize the first lessons in Yezidism. Yezidis in Germany were also among the first to publish newspapers, journals, and books in the 1990s when Yezidi studies was still in its infancy. The writers, publishers, organizers, and activists mainly belonged to the laypeople's caste, the *mirid*. Frequently, the clergy debunked, criticized, and rejected these works. However, in groundbreaking work, this new, second generation of Yezidi emigrants opened up the debate and included various stakeholders, including women, youth, and laypeople.

Two other facts contributed to the formation of a new leadership: first, their higher educational level and media savviness, and second, the absence of a new generation of active clergy. Thus, the new leaders filled a gap and provided the community with long sought-after answers to their identity struggle. As such and for the first time, Yezidis were reimagining their life and religion with input from the entire socio-religious spectrum. Religious instruction and intellectual debate were no longer restricted to a particular group, but allowed for the inclusion of every member of Yezidi society.

Eventually, this debate travelled back to the home countries, where new political realities helped the Yezidi communities to receive this message with open minds. Increased technological connectivity as well as increased mobility triggered a vibrant dialogue among the different geographical regions, generations, and religious castes, including attempts to 'highjack' certain religious dogmas or rituals and other attempts to disqualify nonconformist communities and opinions. While each side held their own conferences and meetings, it was in the online world where they were able to meet and discuss directly, including in forums, blogs, news sites, and social media.

Mapping the New Yezidi World

Both Arab and Kurdish nationalists have used maps as an instrument of oppression. In their discourse to instrumentalize the Yezidis, they have relied on other forms of (mis)representation such as the census and museums. Benedict Anderson singles out maps as instruments of power that profoundly shaped the way in which the colonial state imagined its dominion – the nature of the human beings it ruled, the geography of its domain, and the legitimacy of its ancestry (Anderson 2006, pp. 163–164). In the new Yezidi world – the one after genocide – Yezidis have turned the tools against the former master. Now, Yezidis use maps and other forms of representation to self-consciously control their discourse and define their domain. Like other indigenous peoples, they try to reverse histories and geographies of dispossession and discrimination, establishing what Coombes calls 'indigenous histories and geographies' to reconceptualize their participation and action (Coombes et al. 2014, pp. 845–854).

It appears that the Yezidi attempts to do that reveal several underlying trends. The current Yezidi media landscape reflects the political, spiritual, and social division among the many Yezidi communities in the homeland and in the diaspora. Although we have noted a process of reimagining Yezidi theology and a concurrent, ongoing debate over values and dogmas, this has not translated into the emergence of a unified Yezidi identity. Political calculations, strategic considerations, economic interests, and power struggles have made it impossible for the minority to rally behind a common goal. Instead, competing geographical backgrounds and party affiliations dominate the daily discourse. Each interest group has access to and is represented by influential media outlets, which diligently report one-sided stories and accusations, and claim the sole status of representation and victimhood.

Various influential outlets are omnipresent in today's Yezidi media landscape; among the most popular are two sites that operate from within the Yezidi homeland: Bahzani[5] and Lalishduhok.[6] Both sites are, however, extremely partial to Yezidi events inside Iraq and only rarely report on affairs

5 www.bahzani.net.
6 www.lalishduhok.com.

from Syria or the diaspora, and never from Armenia. Lalishduhok is also the mother company for well-known Yezidi journals and newspapers, including *Majallat Lalish*, *Denge Lalish*, and some other publications by local branches of the Lalish Cultural Centre, which receives most of its funding from the KDP (*Partiya Demokrat a Kurdistanê*). The small number of printed hard copies is now substituted by a complete and free online version.

A recent addition to the online news sites is *Ezidipress*, which was established after the August 2014 attacks and which established itself in the field due to its in-depth, less partisan reporting from all geographical and social angles of Yezidism. The people working at *Ezidipress* are mostly young professionals with little religious background but a sincere interest in those topics. The fact that *Ezidipress* has offices and staff in all Yezidi territories contributes to the positive popular opinion of the agency.

The diaspora community is largely at odds with itself, or to be more precise, the diaspora organizations are divided and have failed to reach a consensus on religious, political, and social issues. They have also been unable to unify under one central organization. In Germany, for example, Yezidis congregate in small, local associations and try to trump the others in political activism, aide campaigns, and media representation. However, their activities mostly cancel each other out. When each organization tries to organize a meeting (and picture) with the German chancellor Angela Merkel and gain political and financial support from the German government, the results are negligible.

However, efforts are made by Yezidi organizations in the homeland and abroad to use digital media to their advantage because they recognize the broad audience they can reach. In the following, I would like to introduce three mapping projects in which apps, social media, GIS data, and other solutions are used to enhance the understanding of Yezidi history and religiosity.

Mapping Historic Genocides

The first project started on 15 August 2014, immediately following the first attacks on Sinjar. A team of journalists, academics, and activists described the long-lasting history of anti-Yezidi genocides and the fact that the current one was a continuation of these repeated attacks. Through interactive maps and infographics, the visualization traces those attacks back to the twelfth century. It also shows the usual Yezidi response to the attacks, which is to migrate to nearby or distant areas of safety. By doing so, they have often established permanent homes in new territories and had to abandon their former, ancestral homeland.

The collaboration with a major US newspaper allowed for more visibility of the Yezidi discourse of discrimination and migration. The Yezidi community in the United States is still small: perhaps two thousand individuals. However, as Krajeski (2018) shows, reaching out to the strongest political and military power in the world created positive connections and responses. The decision by the Obama administration to rescue the Yezidis from Sinjar Mountain and to bomb ISIS positions is one of these immediate benefits.

The other outcome of this project was for the Yezidi community to contextualize their multilayered history and identity. One of the main obstacles to the development of a strong, unified Yezidi identity is the distribution of Yezidi communities across a number of countries. Here, the community reflects on the notion of a sacred homeland and the loss of it due to the continual impositions of their enemies. It also helps to establish a geographical depository of former and current Yezidi locations, which allows the community to appreciate their geographical diversity and historic depth.

Mapping Yezidi Territory

The second project is a collaboration between Yezidi activists and scholars from Turkey, Syria, and Iraq with the goal of mapping local Yezidi migration in eastern Anatolia and northern Mesopotamia. This multilayered outcome illustrates where Yezidi communities used to live in Turkey, Syria, and Iraq at specific points in history: 1832, 1892, 1918, 1961, 1975, 1992, 2007, and 2014. These dates coincide with major attacks on the Yezidi communities and subsequent population movements. The final graph shows which communities moved into the former Yezidi territories.

Since the days of independence of their nation states, the ancient communities in the borderland between Turkey, Syria, and Iraq have experienced periods of large-scale population movement, often involuntarily due to displacement, ethnic and religious violence, and government interventions. Under the new circumstances, with the rise of radical Islamist movements, the Yezidi community and other minority groups will be forced out of their traditional spaces, which ultimately will lead to a remodelling of the region's identity landscape.

The mapping of Yezidi migration is significant because it identifies the former and current population development in an understudied area, and it will ultimately increase our understanding of the region's ethno-religious diversity. The project aims to trace former and recent Yezidi communities in their homeland and the diaspora by mapping routes of migration and identifying causes and consequences of the movement. It highlights the ongoing process of de-territorialization and identity transformation of the Yezidi community.

Mapping Mass Graves

This project run by Yazda, a Yezidi NGO in Iraq (unaffiliated with the religious and political leaders), uses satellite and GIS data in addition to local oral sources to locate and identify mass graves around Mount Sinjar.

It was estimated that 3100 Yezidis were killed during the attacks, with nearly half of them executed and the other half dying on Mount Sinjar because of the ISIS siege (Cetorelli et al. 2017). Those who were executed (shot, beheaded, or burned alive) were largely males, but many females – especially older females – were among the victims too. Survivors reported mass executions in many Yezidi villages on both the northern and southern sides of the mountain. After the areas were liberated from ISIS occupation, those reports helped in locating the grave sites. However, technology such as drones and satellite images were needed to physically find the exact locations, because the graves have been ploughed over. They were left for three years, and some areas were further damaged during the military campaigns.

Yazda composed a detailed, interactive map with the location of the sites and information about the victims in an effort to document the genocide against the Yezidi community. Their focus included the collection of survivor testimonies, the identification of mass graves in the Sinjar region, and the recording of the physical destruction of Yezidi material culture and towns to try to prevent them from returning to their homelands and effectively erase their presence from the region. The stated objective was to preserve the memory of what befell the Yezidi community, preserve knowledge about Yezidi culture as it existed in Sinjar prior to the attempted genocide, provide the International community with a detailed and accurate account of the genocide, and collect evidence of crimes against the Yezidis in the effort to secure genocide recognition for the Yezidi case (Yazda 2016, p. 3).

The project is ongoing as more grave sites are discovered. As of now, 68 mass graves have been identified. Efforts to secure and investigate the sites are underway, and survivor accounts from all Yezidi groups and castes are being collected and analysed.

ISIS Usage of Technology to Enslave Yezidis

It should be mentioned here that ISIS, too, is very competent in using up-to-date media technologies. They have used devices, technology, and social media to recruit new members, to reach out to their constituency and the larger world, and to sell and trade Yezidi sex slaves. The latter have been the focus of this chapter: ISIS members are thought to have used WhatsApp or Telegram to post portfolios of Yezidis girls, women, and families in order to trade or sell them to other ISIS members. The Associated Press reported on these accounts: a girl for sale is 'Virgin. Beautiful. 12 years old [...] Her price has reached $12,500 and she will be sold soon' (Hinhant et al. 2016). They also viewed an advertisement on WhatsApp for a mother with a three-year-old and a seven-month-old baby, with a price of $3700. 'She wants her owner to sell her', read the posting, followed by a photo (Hinnant et al. 2017). Apparently, ISIS maintains a thorough database with a photograph of the Yezidi women and information about their owners. This knowledge can be shared among other ISIS members who want to purchase a Yezidi slave or who work at the checkpoints, preventing the Yezidis from escaping. The usage of encrypted apps cannot be prevented; Yezidi activists trying to liberate the women can only monitor these transactions. However, sometimes they manage to 'win' an auction and pay the ransom to free their women.

Conclusion

Technology has had a massive impact on the life and culture of the Yezidis. It has affected their communication, representation, education, and interaction with the non-Yezidi world. As shown in these examples and the arguments brought forward in this chapter, the approach to seeking knowledge in Yezidism has changed dramatically away from a discriminative, elitist process towards a mass culture of learners and contributors. Concurrently, the oral nature of the sacred knowledge has been altered and scripturized, reflecting a higher standard of education and higher awareness of religious topics. This supports the process of reimagining religiosity and historiographies, as it gives voice to previously ignored or silenced groups such as youth, women, and laypeople. Ultimately, this will lead to a transformation of social and religious structures. Allison claims that 'Yezidism will change from an oral religion of orthopraxy to a scriptural religion of orthodoxy' (2001, p. 50). However, the current process of transformation will extend beyond the religious realm.

New media and technology have triggered the changing of gender and generation roles as well as larger processes of identity building within the

Yezidi community. Apps and maps help the Yezid community to recover from trauma and the disruption of their livelihood. They recreate an image of the past, provide assistance, increase communication, and develop new norms and standards.

Additionally, as a by-product, tools have been created to better safeguard the community from future harm, as these mapping tools have increased our knowledge of how previous attacks were orchestrated. These tools may also help prevent future attacks as they visualize the spatial conditions and circumstances on the ground that have allowed the attacks to occur.

References

Christine Allison, *The Yezidi Oral Tradition in Iraqi Kurdistan* (Richmond: Psychology Press, 2001).

Benedict Anderson, *Imagined Communities: Reflections on the Origin and Spread of Nationalism* (London/New York: Verso Publishers, 2006).

Ordixane Celil and Celile Celile, *Zargotina Kurda* (Moscow: Nauka Publishers, 1978).

Valeria Cetorelli, Isaac Sasson, Nazar Shabila, and Gilbert Burnham G, 'Mortality and Kidnapping Estimates for the Yazidi Population in the Area of Mount Sinjar, Iraq, in August 2014: A retrospective household survey', *PLoS Medicine*, 14(5) (2017), pp. 1–15.

Brad Coombes et al., 'Indigenous Geographies III – Methodological Innovation and the Unsettling of Participatory Research', *Progress in Human Geography*, 38(6) (2014), pp. 845–854.

Dabiq Magazine, 7, 2015, Online Resource: https://azelin.files.wordpress.com/2015/02/the-islamic-state-e2809cdc481biq-magazine-722.pdf (accessed 18 November 2020).

Christine Hine, 'Overview: Virtual Ethnography: Modes, Varieties, Affordances', in *Handbook of Online Research Methods*, ed. by Nigel G. Fielding, Raymond M. Lee, and Grant Blank (London: Sage, 2008).

Lori Hinnant, Maya Alleuzzo, and Balint Szlanko, 'How ISIS Uses Smart Phone Apps to Sell and Register its Sex Slave Captives', *Associated Press*, Online Resource: https://www.pbs.org/newshour/world/how-isis-uses-smart-phone-apps-to-sell-and-register-its-sex-slave-captives (accessed 18 November 2020).

—, 'Islamic State Tightens Grip on Captives Held as Sex Slaves', *Associated Press*, Online Resource: https://apnews.com/article/7685e20b2883496abc3d0911de6c67ab (accessed 20 November 2020).

Farida Khalaf, *The Girl who Escaped ISIS* (New Yok: Atria Books, 2017).

Jenna Krajeski, 'The Daring Plan to Save a Religious Minority from ISIS', *The New Yorker*, 26 (2018).

Philip Kreyenbroek, *Yezidism in Europe – Different Generations Speak about their Religion* (Wiesbaden: Harrassowitz, 2009).
Sebastian Maisel, *Yezidis in Syria: Identity Building among a Double Minority* (Lanham: Lexington Books, 2017).
Memri. Jihad and Terrorism Threat Monitor, 'Islamic State (ISIS) Releases Pamphlet On Female Slaves', 2014, Online Resource: http://www.memrijttm.org/islamic-state-isis-releases-pamphlet-on-female-slaves.html#_edn1 (accessed 18 November 2020).
Nadia Murad, *The Last Girl – My Story of Captivity, and my Fight against the Islamic State* (New York: Tim Duggan Books, 2017).
John Postill and Sarah Pink, 'Social Media Ethnography: The Digital Researcher in a Messy', *Web Media International Australia*, 145(1) (2012), pp. 123–134.
Raoul Ranoa and Jon Schleuss, 'Yazidi persecution spans centuries', *Los Angeles Times*, 2014, Online Resource: http://beta.latimes.com/world/middleeast/la-fg-g-yazidi-a-history-of-persecution-20140815-htmlstory.html (accessed 18 November 2020).
Khidir Suleyman and Khalil Jindy, *Êzdiyatî: liber Roṣnaya Hindek Têkstêd Aîne Êzdiyan* (Baghdad, 1979).
UN Human Rights Council, 'They Came to Destroy – ISIS Crimes Against the Yezidis', 2016, Online Resource: http://www.ohchr.org/Documents/HRBodies/HRCouncil/CoISyria/A_HRC_32_CRP.2_en.pdf (accessed 18 November 2020).
Yazda Organization, *A Report from the Yazda Documentation Project on Mass Graves of Yazidis Killed by the Islamic State Organization or Local Affiliates On or After August 3, 2014*, 2016, Online Resource: https://www.yazda.org/wp-content/uploads/2017/07/Yazda-Report-on-Mass-Graves-Jan-28-2016-2-Copy-1.pdf (accessed 18 November 2020).

About the Author

Sebastian Maisel is a full professor of Arabic Language and Translations at Leipzig University. His research focus is on social transformation among rural communities and minority groups, for which he has conducted fieldwork among the Bedouin tribes in Saudi Arabia, the Yezidis in Syria and Iraq, as well as Dinka slave soldiers from Sudan. He primarily uses sources that include oral narratives, material culture, and virtual representations. His publications include *The Customary Law of the Bedouins in Northern Arabia* (2006), *The Kingdom of Saudi Arabia* with D. Long (2010), *Yezidis in Syria – Identity Building among a Double Minority* (2016), and *The Kurds: An Encyclopedia of Life, Culture and Society*, (2018).

14. Reconfiguring the Kurdish Nation on YouTube: Spatial Imaginations, Revolutionary Lyrics, and Colonial Knowledge

Andrea Fischer-Tahir

Abstract

Around the globe, colonized nations and indigenous peoples make use of colonial photography of the nineteenth and early 20th century to provide documentation of oppression, slavery, and extinction, as well as of resistance and survival. In Iraqi Kurdish nationalist discourse, for example, this mode of representation can be observed from the 1960s and the first Kurdish modern guerrilla war. The digital age has multiplied the possibilities of translocating knowledge and has promoted Kurdish national aspirations for a nation state. Yet, this chapter will show that colonial images of the Kurdish bodies and of Kurdistan's geography still have use value in identity discourses shaped by the aid of social media and mobile media, which bears witness to the simultaneity of modes of representation.

Keywords: Youtube, Kurdistan, cultural memory, nation

Marx wrote in 1852 that women and men 'make their own history, but they do not make it as they please; [...] but under circumstances existing already, given and transmitted from the past' (1972, p. 115). As Marx is still popular in Kurdish nationalist discourse, at least among its left wing,[1] I will take him as a point of departure to discuss aspects of nationalist space production in the Kurdistan Region of Iraq (*Herêm-î Kurdistan*), or as the Kurds alternatively call it, South Kurdistan (*Kurdistan-î başûr*). However, this

1 Recently, *Capital* has been translated from Persian into Kurdish Sorani; to this day, Kurdish leftists, at least in Iraq, have relied on the Arabic or Persian translations.

Strohmaier, A. and A. Krewani (eds.), *Media and Mapping Practices in the Middle East and North Africa: Producing Space*. Taylor & Francis Group, 2021
DOI 10.5117/9789462989092_CH14

chapter more or less leaves aside the conflicts over new borders established by the Kurdish movement, as well as over resources (oil, gas, and water, for example) and political power that usually preoccupy studies on Kurdistan in Iraq (Stansfield 2003; Natali 2010; Zedalis 2012). Instead, this chapter predominantly deals with the discursive side of space-making and with representations of history and geography in discourses on Kurdish identity.

This chapter will first expound the problems of Kurdistan in geography – that is, 'the Kurds' place' in metageographies (Lewis and Wigen 1997), on the one hand, and Kurdish representations challenging the hegemonic geographical order, on the other. The second section will provide some notes on the issue of colonialism in the context of Kurdistan, and it will shed light on the colonial heritage in imaginations of Kurdistan and the Kurds, shifting the analytical perspective from the 'perceived' to the 'conceived space' (Lefebvre 1991). In the third section, I will discuss a YouTube video representing a Kurdish anthem as well as photographic documents of Kurdish revolts between 1918 and 1991. The discussion aims to identify and contextualize imaginations of geography and history in this particular discourse. The question is to what extent cultural heritage and colonial knowledge are constructed as opposition or to what extent these different corpora of 'substantive assertions and ideas' to 'act on the world' (Barth 2002, p. 2) become harmonized to conceive of Kurdistan as a geographical entity of historical legitimacy. The discussion will exemplify how those constraints resulting from pre-existing circumstances structure practices of representation. It will become clear that the specific experience of European as well as of internal colonialism in the context of Kurdistan resulted in an ambivalent relationship to European colonialism and its respective discourses 'to tame' the unfamiliar Oriental (Said 2003, p. 54).

Metageography and Dissident Geography

The number of Kurds in the MENA (Middle East and North Africa) region, Caucasus, Central Asia,[2] and the diaspora in Europe varies between 36 and 45 million, depending on sources such as official state censuses, assessments of human rights groups, or international organizations such as the European Commission. In their homeland, Kurds make up approximately 85 per cent of the population.[3] The problems of assessments in terms of

2 For Kurds in the Caucasus and for deportations to Central Asia, see Müller 2000.

3 See: http://www.institutkurde.org/en/info/the-kurdish-population-1232551004 (last accessed 17 November 2020).

numbers predominantly result from politics of marginalization, oppression, and cultural assimilation pursued by the Iranian, Turkish, Iraqi, and Syrian states since their respective independence. In Turkey, for example, the existence of Kurds has been denied for many decades, whereas state politics in Iraq involved the vague acceptance of cultural difference to the Arab main population as well as genocidal persecution.[4] In contrast to the statistical numbers and the relatively compact territory of settlement, Kurds are labelled 'minority' not only in the administrative logic of the states where they live but also in international political and academic imaginations of the MENA region. The term 'minority' is deeply embedded in legal discourses on the protection and rights of non-dominant groups within states, defined in terms of nation, ethnicity, religion, language, and/or indigeneity. Yet, the definition of majorities and minorities depends on scale (Derichs 2017, p. 35) as well as on the system of reference; namely, the geographical order of the world. Martin W. Lewis and Kären E. Wigen argue that there is a 'set of spatial structures through which people order their knowledge of the world: the often unconscious frameworks that organize studies of history, sociology, anthropology, economics, political science, or even natural history' (1997, p. ix). They call these imaginations about the segmentation of the world into continents, subcontinents, world regions, and states 'metageographies'. These frameworks, however, are the result of historical processes – namely, capitalism, colonialism, and nation-building – and at the same time these frames persist due to geopolitical power relations, normalization, and education; the prevailing Eurocentrism inherent to knowledge production; and the authoritative 'myth of the nation state' (Lewis and Wigen 1997, p. 7). From my perspective, it is striking that despite the constructivist turn in studies on nations and nationalism and the canonic status of Benedict Anderson's *Imagined Communities*, inquiry in social science and the humanities continues to rely on internationally recognized nation states as a system of reference. Consequently, when 'minorities' manage to establish command over areas of their settlement, these territories are still conceived of as anomalous, if not as a security problem or threat to the regional order (Nolutshungu 1996; Caspersen and Stansfield 2011).

One such 'anomaly' is the Kurdistan Region of Iraq. This administratively and politically semi-independent unit is first and foremost the outcome of a decades-long struggle of the Kurdish movement, a movement that is in itself heterogeneous and conflictual, and at the same time transnationally entangled with strands of the Kurdish movement in Turkey, Syria, and

4 For an overview of Kurdish history, see McDowall 1996.

Iran. Governed by the Kurdish Regional Government (KRG) since 1992, the territory comprises 73,000 spare kilometres, making up 20 per cent of Iraqi territory. The borders of the region are disputed, under constant pressure, and subject to military conflict. These borders, however, are the result of the Kurdish uprising in March 1991 and the events that followed in that spring: brutal repression of the revolt by the Iraqi regime, international humanitarian intervention, establishment of a no-flight zone under predominantly US protection, and negotiations between the Kurdish parties and the Iraqi regime (Cook 1995). In May 1992, the Kurds elected their own parliament and constituted the Kurdistan Region, whose borders coincide more or less with the borders of the governorates of Dohuk, Erbil, and Sulaimaniya. During the war of 2003, which led to the overthrow of the Iraqi regime, and during the war against the Islamic State since 2014, the Kurds have expanded their territory. In September 2017, the Kurdistan Regional Government held a referendum for independence with more than 90 per cent of the votes cast agreeing to separation from Iraq. The Iraqi central government, as well as Iran and Turkey reacted promptly with embargo, political isolation of the KRG and military reconquest of Kirkuk and other parts of South Kurdistan.

Indeed, Kurds in Iraq speak about the area in question as 'Kurdistan', 'region' (*herêm*), 'homeland' (*welat*), or 'South Kurdistan' (*Kurdistan-î başûr*). This latter term partly results from British and French colonial partition of the Middle East. Kurdish geography describes the homeland of the Kurdish people as situated in West Asia (*xorawe-î asiya*) and segmented into North, West, East and South Kurdistan, with each of the single parts 'under the command' (*bindestî*) of Turkey, Syria, Iran, and Iraq, respectively. Alternative views add to Greater Kurdistan a Northeastern part; namely, the Kurdish settlements in the Caucasus. The territory comprises between 475,000 and 506,000 square kilometres, depending on the inclusion of exclaves in the Caucasus, Central Asia, and Central Anatolia (Zend 2001, pp. 9–10; Izady 1992, pp. 1–10). In fact, this specific geography hardly has an equivalent in Kurdish history before 1920. Instead, under Ottoman and Persian rule, borders cutting across Kurdistan were perceived in terms of provinces established by the empires or of the Kurdish emirates, such as the emirates of Bahdinan, Soran, and Baban in the south (Bruinessen 1989). As 'phantom borders' (Hirschhausen et al. 2015), these principalities recurred in the structuration of the Kurdish movement and the violence-loaded competition between the leading parties, the Kurdistan Democratic Party (KDP) and the Patriotic Union of Kurdistan (PUK).

Without a doubt, the Kurdish nationalist movement has been challenging the order of the MENA region from the beginning of its emergence. However, at the heart of the Kurdish nationalist agenda lies the idea of national self-determination with the nation state as model of reference. Yet as a 'majority' in their homeland, Kurds and Kurdish nationalist discourse themselves tend to construct minorities such as Christians of all denominations, Yezidis, Shabak, and Turkmens.[5] Thus, reconfiguring geography in the context of Kurdistan and the Kurdish nationalist movement is both resistance to metageography and obedience to its logic.

Colonialism and Colonial Knowledge

British colonial politics in South Kurdistan during and following World War I ranged between, on the one hand, attempts to make representatives of the local elite allies against the Ottoman Empire and later the Turkish Republic (Mosul debate in 1926) and, on the other hand, the brutal oppression of insurgencies led by more or less the same elites. The Kurds, for their part, were motivated to rebellion by the regional wave of nationalist movements that claimed independence and the creation of nation states of their own, as well as by Lenin's call for the right to self-determination and the Wilsonian doctrine of self-determination of peoples. From 1918 (complete British occupation of Mesopotamia) to 1920 (League of Nations confers British Mandate on Iraq), up to 1932 (formal acceptance of Iraq as a member of the League of Nations), the colonial power faced several local revolts; namely, the various insurrections led by Sheikh Mehmud Berzinci in Sulaimaniya (Edmonds 1957). However, British colonial rule over Iraq did not end until the Iraqi monarchy was abolished in 1958.

In Kurdish national historiography, British colonialism, along with the French, is identified as the principle reason for the partition of Kurdistan in the early 20th century and the missed historical chance to create a Kurdish nation state; the arbitrary colonial decisions and injustice are symbolized in the treaties of Sykes-Picot (1916), Sèvres (1920), and Lausanne (1923). However, in Kurdish historiography, colonialism started much earlier with the Battle of Chaldiran (1514) and the Treaty of Zohab (1639), events that established the territorial partition of Kurdistan between the Ottoman Empire and the Persians/Safavids (Baxewan 1999, pp. 56–63; Izady 1992, p. 51). In addition, Kurdish historiography and political discourse – under the given historical

5 In the Kurdistan Parliament, eleven seats are reserved for minorities; in constitutional documents, others rank after the Kurds. See: Constitution Draft of the Kurdistan Region of Iraq 2009, Articles 5–6; Charter of the Social Contract 2013. See also the contribution of Maisel in this volume.

and present-day circumstances – are more preoccupied with later forms of colonialism imposed by the Iraqi, Turkish, Iranian, and Syrian regimes. In all parts of Kurdistan, the Kurds faced deprivation of political, social, cultural, and economic rights, denial of their cultural identity and even of their existence, and in addition, the appropriation of natural resources located in their territory. In this context, Ismail Beşikçi coined the notion of 'Kurdistan as international colony', a notion in line with Lenin's concept of internal colonialism to describe regionally manifested uneven development, structural dependency, and political injustice within states, a concept developed further for states based on settler colonialism and postcolonial nation states (Marquard 1957). For the South Kurdish context, collective suffering, especially under the dictatorship of the Baath regime (1968–2003), is the central feature of national identity. The key events of reference here are the annihilation of 8000 male members of the Barzani tribe in 1983, the chemical gas attack on Halabja with 5000 people killed, and the Anfal Campaign in 1988 with 182,000 men and women disappeared or murdered (Mlodoch 2015; Fischer-Tahir 2012).

Constantly faced with marginalization and oppression, the search for historical meaning is rendered in Kurdish nationalist discourse as a matter of cultural survival. At the same time, history and geography are constructed according to the hope for national self-determination while negotiating the 'horizon of expectation' with the 'space of experience' (Koselleck 2010). Thus, Kurdish historians, geographers, and other symbol-handlers are in search of the historically proven spatial control over their territorial domain. In doing so, Islamic and Kurdish sources from past centuries are explored, as well as the works of British, French, Russian, German, or Italian scholars, travellers, or agents of the colonial apparatus. The most prominent narrative established by historiography and political discourse is the view of the Kurds as descended ethnically and linguistically from the Medes, a theory discussed by Orientalists in the 1920s (Minorski 1986; Nikitin 1956). When Kurdish intellectuals introduced a Kurdish calendar in the 1950s, they drew on that theory insofar as they took the destruction of the Assyrian capital Niniveh in 612 B.C. by the nomadic Medes as the beginning of the Kurdish time (Bohrmann 1991).[6] Orientalists, however, also suggested that the Kurds

6 The calendar is of political meaning for the Kurds, especially as it places – just like the Iranian calendar – Newroz as the beginning of the year. The KRG also experimented with the calendar in official affairs. However, when I worked in the Ministry of Higher Education in Sulaimaniya in 2005–2006, I could observe the confusion of my Kurdish colleagues when official documents came in for bureaucratic purposes, which state, apart from the date according to the Gregorian calendar, an Islamic and a Kurdish date.

descended from autochthonous peoples such as the Corduene and Cyrti, as they appear in the reports of ancient Greek and Roman historians and geographers, and that the Kurds belong to the indigenous peoples of Asia Minor (Marr 1923). These theories were embraced very early by Kurdish historiography (Talabani 1971, pp. 23–49; Izady 1992, pp. 28–34), and inspire academic studies to the present (Bābān 1989; Pîrbal 2006, pp. 75–94).[7]

Deeply embedded in Kurdish nationalist discourse is the narrative that the Kurdish woman has a much better social status than the women of the ethnic 'other' (Turks, Arabs, Iranians). This narrative is often proved in reference to publications of British military officers, who described Kurdish women as 'comparatively free' (Noel 1919, p. 67) or 'practically as free as in any European country' (Soane 2007, p. 268). These are statements based merely on the encounter with individual women of the notables and on reports by Kurdish men, but useful in arguments of distinction (e.g. Basīr 1972, p. 108; Xeznedar 2012, pp. 5–6). Recent Kurdish feminist discourse hesitates to deconstruct these narratives. It tends, however, to correct these views in referring to later female European researchers (e.g. Hansen 1961) who indeed address in their work the gender inequality inherent to the Kurdish patriarchal society. Moreover, the feminist discourse emphasizes today's problems of gender-based violence (Hardi 2013; Fischer-Tahir 2009).

From my own observations in Kurdistan over the last 20 years and based on the reading of Kurdish academic and other intellectual work, I argue that the identity discourse makes a sharp distinction between European colonial politics that led to the partition of Kurdistan on the one hand, and those colonial imaginations serving to prove the Kurds' place in history and among the civilizations, on the other hand. The genuine problem with colonialism is not so much 'the West' but (the regimes of) Iraq, Turkey, Iran, and Syria.

Imagined Geography in Lyrics, Photographs, and Revolutionary Marches

This section exemplifies geographical imaginations with the aid of a YouTube video uploaded under the title 'We always will be Kurds', or '*Her kurd ebîn*'. Based on the number of users and the fact that it has been reproduced several times, the video seems to be highly popular among Kurds in South

7 For a critical discussion of Kurdish Studies, see Scalbert-Yücel and Le Ray 2006; for studies on the Kurds in Tsarist Russia, see Alsancakli 2016; for a discussion on whether Soviet Kurdish studies pursued a view named 'Orientalism' by Edward Said, see Leezenberg 2011.

Kurdistan, other parts of Kurdistan, and in the diaspora. From a social anthropologist perspective, I will contextualize, describe, and interpret this discourse fragment, paying particular attention to word and image to identify narratives and images as well as the strands of various and overlapping discourses on the past, present, and future of Kurdistan.

Encountering 'We always will be Kurds'

I remember the days before the elections to the Iraqi Parliament on 30 January 2005. These elections were the first to be held under the Transitional Law imposed by the US occupation. The dominant Kurdish parties had formed the Democratic Patriotic Alliance of Kurdistan, including the Kurdistan Democratic Party (KDP) and Patriotic Union of Kurdistan (PUK), the Kurdistan Communist Party, the Islamic Union of Kurdistan, and others. On the same day, elections were also to be held in the Kurdistan Region for reorganizing the Kurdish Regional Parliament, which was dissolved during the Kurdish militia war (1994–1997); former war enemies now worked to unify the administration after years of division. From a Kurdish nationalist perspective, the election day was important in two related directions: to prove to 'the Arab' political entities (be they Sunni or Shiite, religious or secular, partners or competitors) the power of the Kurds on the Iraqi national level; to prove the legitimacy of the Kurdish parties as they indeed governed the Iraqi north.

As elsewhere in the world, in Kurdish everyday life television is omnipresent, depending on the provision of electricity. You watch and do not watch TV wherever you go: in the household, when visiting relatives and friends, in offices of the administration, in political and cultural institutions, and of course in shops and restaurants. I remember one day I sat with some Kurdish colleagues in a restaurant in Sulaimaniya for lunch when '*Her kurd ebîn*' was shown on TV. For a moment, my colleagues stopped talking, and one of them said reverentially 'this is very beautiful' (*zor ciwan-e*). Then the talk turned to the coming elections, and it was clear to my colleagues that they would vote 'for Kurdistan'. The reason why I remember this scene so well is that I started feeling what they felt. I had watched the video several times at my brother-in-law's house in Sulaimaniya, but I still found it exciting and touching, especially in the way it combined images of collective suffering with the folklorist images my colleagues had deemed 'beautiful'. I also remember that moment because I envied my colleagues a bit for having the right to vote, and thus the chance to change something.

The video 'We always will be Kurds' offers a journey through time: it strings together photographs documenting the struggle for the national liberation of the Kurds roughly between 1918 and 1991. The lyrics accompanying the photographs are based on a Kurdish poem and are displayed in Kurdish subtitles, accompanied by the blaring sound of a revolutionary military march. When the PUK Satellite channel KurdSat aired this video in 2004–2005, it provided the following information: lyrics by Ibrahim Ahmed, composed by Shanaz Ibrahim Ahmed. KurdSat had started to operate in January 2000 as the official TV of the PUK, which, since semi-independence in 1992, had already operated a terrestrial TV station called *Gel-î Kurd.* After the KDP's satellite channel KurdistanTV and the PKK-operated Med-TV, KurdSat was the third Kurdish satellite television station, and many followed in the next years. KurdSat was based in the city of Sulaimaniya, the stronghold of the PUK.[8]

The poem's author, Ibrahim Ahmed, was a well-known Kurdish politician, guerrilla fighter, and intellectual from Sulaimaniya (1914–2000). He studied law at Baghdad University and started a career as a judge in Kurdistan, but soon turned to the political field. Over the next four decades, he significantly influenced the structuration and development of the Kurdish movement in Iraq. Inspired by Marxist-Leninist world views and theories of political practice, he became a leading figure of the KDP, established in 1946 (analogue to the KDP-Iran) as a socially and politically heterogeneous alliance. Left-wing ideas of progress and political organization brought him and like-minded leading figures such as Jalal Talabani (1933–2017), who became his son-in-law, into conflict with Mulla Mustafa Barzani (1903–1979), the tribal and conservative leader of the KDP. This conflict eventually resulted in the founding of the PUK in 1975. However, conflicts between the Ibrahim Ahmed/ Jalal Talabani group and the Barzanis arose not only from different views in terms of political decision-making, the role and function of the party and the Peshmerga forces. There were also different views of the relationship between Kurds and Arabs on the level of the Iraqi nation state (Brentjes 1964; McDowall 1996). Several decades later, Ahmed's views on the necessity of an Iraq-wide democratization found their echo in the politics of negotiation and balance as pursued by Jalal Talabani, who acted as the president of Iraq between 2005 and 2014. At the same time, Mulla Mustafa's sons, above all Masud Barzani, continued to prefer a rather radical nationalist stance towards Baghdad, visible in the referendum for independence in 2017. Apart from

8 On Kurdish media and especially media in the Kurdistan Region, see Sheyholislami 2011; Reporters without Borders 2010; Fischer-Tahir 2017.

his political influence, Ibrahim Ahmed is also well known (at least among Iraqi Kurds) for his work in the fields of journalism and literature, having established several journals and published novels and short stories. He wrote his poem *'Her kurd ebîn'* in 1981, during his exile in London (Berzincî 2004, p. 69). However, the poem did not become popular until his daughter Shanaz, herself a leading figure of the PUK, directed the production of the KurdSat video, which has today gained the status as the unofficial Kurdish anthem.

Upon my request, a journalist from KurdSat provided me with the video in 2007, and I saved it on my computer in Berlin. Some months ago, I found my Kurdish husband at home cleaning our flat with headphones in his ears and singing *'Her Kurd ebîn'*. His voice became especially loud and committed when he sang the refrain: 'I am not an Arab, not an Iranian and not a mountain Turk' – a rhyme which I had heard relatives, friends, or colleagues in Sulaimaniya say whenever this song on TV flushed into an everyday-life situation: 'it has great power' (*zor be qewet-e*). Searching for the video on the internet, I realized that multiple versions of the KurdSat video *'Her Kurd ebîn'* were uploaded between 2006 and 2017, some slightly or more distinctively modified in the representation of images, but still all with a duration of four minutes and a few seconds, as long as the anthem lasts. In some videos, the subtitles are in Sorani, Sorani and English, and Sorani using Latin letters, Kurmanci, or Turkish. What happened to the KurdSat production is not spectacular in our times but still requires some reflection. If I accept the theoretical view in terms of 'the "content" of any medium is always another medium' (McLuhan 2001, p. 8), then it can be said that YouTube as a medium remediated 'other, mostly older forms of mediation' (Grusin 2009, p. 65) such as TV, radio, photography, and writing. If I tentatively draw on the idea that YouTube is 'structured by a competition for attention' and the 'logic [...] of exchange value' (Schröter 2009, pp. 342–343), the material can be investigated under the question of how this exchange value comes into existence and why.

Geography of Disobedience and Distinction

Let me start with a representation of Ibrahim Ahmed's poem as it most often appears in the video,[9] translated into English and in Kurdish-Sorani

9 For the video production, the original text was slightly restructured; see Berzincî 2004, pp. 68–69.

(for simplicity, the reading is transliterated according to the Bedr-Khan-Alphabet; the English translation is roughly taken from a translation circulating the web).[10]

We always will be Kurds	Her Kurd ebîn
Hate filled invaders	Dagîrkerî dil pir le qîn
Savages without conscience	Dirinde-î bê wîjdan-û dîn
You can't force us not to be Kurds	Na tiwanî wa ekey Kurd nebîn
We have always been Kurds and always will be Kurds	Her Kurd buyn-û her Kurd ebîn
Before Islam	Le pêş Musilmanêtî-da
Before Fire worshipping	Le pêş agirperestî-da
During imprisonment and in victory	Be dîlî-u be serbestîda
We have always been Kurds and always will be Kurds	Her Kurd buyn-u her Kurd ebîn
I am not an Arab, not an Iranian and not a Mountain Turk	Ne Arabim ne Îranîm ne Turkêkî şaxistanîm
History will sing with me	Legel xom mêjûş her eleyt.
That I am a Kurd, a Kurdistani	Ke Kurdim-u Kurdistanim
[…]	
All peoples are my brothers	Gelan hemû bira-î minin
Only my oppressors are my enemy	Her zordaranim dujminin
I am not demanding anyone's land	Daway welat-î kes nakem
[…]	
Even if you flatten the Mountains Qandil	Eger Qendil-u Agirî-u Şirîn
Ararat and Shirin to the ground	yeksan ekey legel ruy zemin
You can't force us not to be Kurds	Natiwanî wa ekey Kurd nebîn
We have always been Kurds and always will be Kurds	Her Kurd buyn-u her Kurd ebîn

The rhymes clearly address anger, pride, and disobedience, and at the same time represent a vision of Greater Kurdistan. In the context of 1981, the lyrics originally echoed the defeat of the Kurdish movement in the war imposed on them by the Iraqi regime in 1974–1975. It can also be seen as a statement to the reorganized resistance, or 'new revolution' (*şorş-î nwê*), started by

10 See: http://lyricstranslate.com/de/her-kurd-eb%C3%AEn-we-always-will-be-kurds.html#ixzz4zAA8S1qP (last accessed 17 November 2020).

the PUK as an alliance of three more or less leftist groups, including those of Talabani and Ahmed in 1975. I would also interpret the original lyrics as a comment on the cruelty Kurdish parties faced in Iran when the religious regime of Tehran intensified pressure on the opposition, as well as a comment on the terror imposed on the Kurds (and Turkish leftists) after the military putsch in Turkey in 1980 (McDowall 1996). However, given that repression by the respective regimes did not end and that the struggle for liberation started much earlier and still continues, the remediated text flows through space and time.

These rhymes clearly address the colonialism under Arab, Iranian, and Turkish regimes, these invaders represented in metaphors of 'hate' (literally: 'the heart filled with hate'), 'savagery' (*derinde*), and 'absence of conscience' or 'unscrupulousness' (*bêwîjdan*). Without being explicitly stated, Kurds are imagined as the opposite of all that: they are kind-hearted, civilized, and have good consciences. Interestingly, the argument turns to the historical legitimacy of the Kurds and Kurdistan: Kurds inhabited the land before the Muslim invaders and even before the Medes – who knew the fire cult – settled down there. It is this historical rootedness of the Kurds in their homeland that dooms all invaders' attempts of assimilation to defeat, even if they deny the Kurds in their existence, stigmatizing them as 'Mountain Turks'. The invaders, with their politics of destruction and deportation, may raze the highest of the Kurdish mountains to the ground, but they will not destroy the Kurdish identity. The mountains Qandil, Shirin, and Ararat are addressed not only as topographical markers but also as places that played and still play a crucial role in Kurdish resistance, as refuge of the armed forces for survival and reorganization, and as spaces of desire and hope. Shirin (Barzan territory in Dohuk province) and Qandil (Iraq–Iran border, Erbil province) in the territory of Kurdistan in Iraq are connected to Mount Ararat in Turkey, so imagining Greater Kurdistan across national borders. To distinguish between (Arab, Turkish, and Iranian) 'tyrants' and the respective people ('all peoples are my brothers') might be interpreted as a merely rhetorical manoeuvre. It reflects, however, Ibrahim Ahmed's political views on the future of the Kurds within an Iraq based on solidarity between the popular masses and forces of progress.

Clearly, the lyrics are an expression of identity, resistance, and survival. In this way, they recall an older, prominent text of Kurdish revolution poetry, *Ey reqîb* ('Oh Enemy') by Younes Ra'uf, written in 1938. Ra'uf's poem was composed to serve as the anthem of the Kurdish Republic of Mahabad in 1946, the Kurdistan Region as the official anthem adopted it in 1992, and recently it was taken up by the Rojava cantons (West Kurdistan) in North Syria. *Ey reqîb* begins as follows:

> Oh enemy! The Kurdish people live on
> They have not been crushed by the weapons of any time
> Let no one say Kurds are dead, they are living
> They live and never shall we lower our flag.[11]

Later, the rhymes state 'we are the descendants of the red banner of the revolution' and 'we are the descendants of the Medes and of Cyaxares [king of the Medes]'. *Ey reqîb* positions the Kurds against the enemies more radically, without the socialist touch of international solidarity, even though the 'red banner' refers to communist revolutions. Both poems explore theories about the origin of the Kurds to highlight the denial of rights deriving from history and to legitimate the national liberation struggle. One reason for the popularity of 'We always will be Kurds' is surely that the audiences in both South Kurdistan and Greater Kurdistan were prepared for the song, as *Ey reqîb* is a long-standing and well-established representation in Kurdish national culture. What mainly constitutes the exchange value of the video within the market of attention is the refrain 'I am not an Arab, not an Iranian and not a Mountain Turk' – read by audiences as a powerful statement of distinction and disobedience.

Display the Nation

The original KurdSat video 'We always will be Kurds' includes more than 50 photographs taken between 1900 and 2000 by a wide range of people. Many of these photographs originate from Susan Meiselas's voluminous illustrated book *Kurdistan: In the Shadow of History* (1997). Thus, the KurdSat video reproduces that illustrated book, which itself is based on a selection of pictures from various sources: personal archives or public collections in South Kurdistan, museums and libraries in Western states, archives of Western photographers, and the international press. However, what is not 'old' in Meiselas's book has been 'made old' for the video production, and so the whole journey through Kurdistan appears in black and white with a touch of green. According to the rhythm of the song, the photographs

11 See: http://cabinet.gov.krd/a/d.aspx?l=16&s=03010100&r=200&a=16878&s=010000 (last accessed 17 November 2020).

appear in quick succession or as stills, with the camera moving upward and downward on a single picture, or zooming in and out.

Interestingly, the journey starts with a photograph taken by a French journalist in 1974 during the war of the Iraqi regime against the rebellious Kurds after negotiations for autonomy had failed. The picture shows a pupil standing at a blackboard, turning attention to the words written on the board: 'Kurdistan, Mort, Kurdistan'. A man in the background holds a kerosene lamp above the girl (Meiselas 1997, p. 271): 'Mort'– death, a picture arranged for the French journalist to turn international attention to the war in Kurdistan. In the KurdSat video, with an expected Kurdish audience not likely proficient in French, the image serves instead to remind of experiences such as learning under conditions of war, discrimination against the Kurdish language, and punishment for using the word 'Kurdistan'. At the same time, this representation echoes the developmental narrative of girls as both the target and the symbol of progress, a well-known narrative to Kurdish symbol-handlers. In addition, this picture as a starting point highlights children as the most vulnerable part of society.

What follows is a photographic narration of the Kurdish struggle for survival and the nation state, and thus of collective suffering and resistance. In doing so, the journey crosses borders established by colonialism. Thus, we see Sheikh Mehmud Bernzinci in his stronghold of Sulaimaniya in 1919 and Ismael Agha Simko, who led a rebellion in Urmia/East Kurdistan in 1919–1920. Several pictures tell the story of the Republic of Mahabad in 1946, from its foundation, its main figures such as the president Qazi Mohammed, and his execution in March 1947 by the Iranian regime. The journey also recalls the memory of Iraqi Kurdish military officers executed in 1947, revolutionaries executed by the Iranian army in 1979, and Kurdish civilians who were the target of the Turkish military in the 1970s. Other pictures document the genocidal attack on Halabja on 16 March 1988 and the dramatic mass escape of Kurds into the border regions of Turkey and Iran in March–April 1991, as the people feared new gas attacks to suppress their uprising in 1991. The journey also takes the viewer to the guerrilla war in South Kurdistan between 1961 and 1970 and its leading figure Mulla Mustafa Barzani. However, as the video is a product of his rival, the PUK, the focus is mostly on the PUK Peshmerga and Talabani in 1975, including a scene showing a few women as part of the armed forces. At the end, the viewer can see – in slow motion – a Peshmerga talking to an (invisible) audience.

Several of the photographs from 1900 to the 1940s are taken by British, American, or German military officers, diplomats, journalists, and professional photographers – among them the British Major E.B. Soane and his wife Lynette Soane, political officer C.J. Edmonds, the British administrator A.T.

Wilson, US intelligence officer A.B. Roosevelt, and the missionary Augusta Guthart. The latter is shown in a photograph from 1923 in which she – as a modern Western woman – is sitting in the middle of Kurdish men in tribal clothes; the picture was originally taken for a missionary newspaper (Meiselas 1997, p. 113). The video also re-establishes the narrative of the Kurdish woman as freer than that of the neighbours. Thus, the journey brings the viewer back to Adila Xanum Caf Begzade, who had temporarily acted as tribal leader after the death of her husband. The journey also brings us to Hafsa Xan Naqib, a cousin and sister-in-law of Sheikh Mehmud Berzinci. Both women represent exceptional figures among the tribal and religious elites of Halabja and Sulaimaniya at that time.

To sum up, the video can be taken as a fragment of overlapping discourses: Kurdish identity discourse, political discourse within the Iraqi Kurdish political field, Kurdish historiographical discourse, Orientalist discourse, and Kurdish feminist discourse. At the same time, the symbol-handlers re-establish a geographical imagination of connectivity, belonging, and history.

However, the geography of Kurdistan is imagined not only with the aid of historical events but also through photographs representing what in Kurdish is called *kurdewarî*, or 'Kurdish folklore'. This ranges from historical as well as present-day pictures of rural landscapes with people representing imaginations of Kurdish nomadic life to recent portraits of rurality taken by a photographer in Kurdistan of Turkey, and again the various parts of Kurdistan are rendered as one inseparable territory. At the same time, the pictures suggest knowledge about the essence of Kurdish cultural heritage. Interestingly, these latter portraits were coloured in the original (Meiselas 1997, pp. 372–373). What is striking is that the journey through Greater Kurdistan leaves out the western part of the homeland, North Syria. This could be explained by the fact that in Meiselas's illustrated book, Kurdistan in Syria has been left aside. However, according to my observations, in particular in Sulaimaniya, what happened in West Kurdistan did not gain attention before the uprising of Qamishli in 2004, whereas greater public interest in that far part of the Kurdish homeland started with the battle for Kobane in 2014. In this sense, the colonial partition of Kurdistan prevails even if the Kurdish symbol-handlers attempt to overcome those borders established by European colonialism and prolonged by the Turkish, Syrian, Iranian, and the Iraqi governments (the latter of which the Kurds have been a part of since 2005). Yet, the vision of Greater Kurdistan becomes more complete in some of the recent reproductions of the anthem. In response to the developments of the Rojava cantons, and in support of the Democratic

Union Party (PYD), YouTube users or producers highlight in their picture gallery, for example, photographs of armed women fighting the Islamic State (such as Women's Protection Units or the Yezidi women's battalion).

Reclaiming Spatial Control

The makers of 'We always will be Kurds' act on the world under given circumstances, adjusting their practice to the technological possibilities developed in the meantime. The KurdSat anthem is a cultural representation of Kurdish identity, converging different discourses about Kurdistan's past, present, and future. References to knowledge suitable to prove ancient history is part of the nation-building project. A coherent history of continued rebellion and suffering since the beginning of the 20th century is needed to provide the group members with meaning and a sense of responsibility. Representations of this kind also serve to legitimate the ruling political class of the Kurdistan Region. The 'emotional geography' (Derichs 2017, p. 154) re-established by the video works without an abstract map. Rather, the space production involves narratives and images that transgress the borders established by colonialism. Even though this geography only has validity within discourses of the marginalized, it challenges established metageographies.

The use of Ibrahim Ahmed's poem in the KurdSat video, along with the use of photographs taken from different sources, also gives rise to reflection on the travel of representations and some more functions of the video. It ensures that these pictures not be forgotten in place-based archives and museums; now, the video itself functions as a de-territorialized archive and museum. Moreover, it accomplishes the function of a schoolbook: the educational mandate is played out in the lyrics, the photography-based lecture on history and society, scenes such as a school under fire, and, at the end, the Peshmerga as the person who has the last word. In terms of representational aesthetics, the black-and-white design is striking. Yet, it perfectly fits the archival character and supports the argument of a distinguished culture as testified long ago even by outsiders. In this way, the video can also be seen as a part of a politics of cultural heritage – a concept which itself is a heritage of colonialism with its modes of representation (Daher and Maffi 2007).

However, if disobedience and the struggle for the nation are the topic of 'We always will be Kurds', why the references to Orientalist knowledge? At first glance, the simultaneity of positive references to colonial photography and at the same time to anti-colonial struggles may appear contradictory. Yet, using colonial photography within an agenda of resistance is a widespread practice among indigenous peoples around the world. Actors make use of respective representations from the nineteenth and early 20th century to provide documentation of oppression, slavery, and extinction, and they transform the colonial gaze into statements about survival, cultural achievements, and glorious histories, as well as into political claims. In addition, reconfiguring the landscapes of oppression into geographies of disobedience is a practice of exercising symbolically spatial control as well as a way of reclaiming spatial control over a defined territory. As for the Kurdish context, the experience with European colonialism has interfered with the colonial experience since the establishment of the national states of Iraq, Syria, Iran, and Turkey. Therefore, it is even easier to harmonize contradictions arising from politics of heritage.

References

Sascha Alsancakli, 'The Early History of Kurdish Studies (1787–1901)', *Die Welt des Islams*, 56 (2016), pp. 55–88.

Benedict Anderson, *Imagined Communities.Reflections on the Origin and Spread of Nationalism* (London: Verso, 1991).

Jamāl Bābān, *Usūl asmā' mudun wa'l-mawāqi' al-'irāqiyya* [*The origin and names of the Iraqi cities and places*] (Baghdad: Al-Ajiyāl, 1989).

Fredrik Barth, 'An Anthropology of Knowledge', *Current Anthropology*, 43(1) (2002), pp. 1–11.

Kāmil Hasan al-Basīr, 'Al-Mar'a al-kurdistāniyya fī't-tā'īkh' ['The Kurdish woman in the history'], in *Masīrat Ittihādunā* [*The Path of our Union*], ed. by Women's Union of Kurdistan (Baghdad: Dār al-Jāhiz, 1972), pp. 99–140.

Hawrê Baxewan, *Hawrêname bo mêjû-î Kurdistan-u kurd* [*Hawrêname of the History of Kurdistan and the Kurds*] (Sulaimaniya: Serdem z Rûn, 1999).

'Omer Me'ruf Berzincî, *Brayim Ehmed: Koy berhem-e şi'riyekan* [*Ibrahim Ehmed: Collected poems*] (Sulaimaniya: Ministry of Culture, 2007).

Ismail Beşikçi, *International Colony Kurdistan* (London: Taderon Press, 2004).

Lothar Bohrmann, 'Der "Ursprung der Kurden" in der westlichen Orientalistik und die Entwicklung des kurdischen Geschichtsbewusstseins', *Ethnographisch-Archäologische Zeitschrift*, 32 (1991), pp. 316–331.

Burkhardt Brentjes, 'Zu einigen Problemen der Geschichte der kurdischen Nationalbewegung', *Wissenschaftliche Zeitschrift der Martin-Luther-Universität*, 9(10) (1964), pp. 679–694.

Martin van Bruinessen, *Agha, Sheich und Staat* (Berlin: Edition Parabolis, 1989).

Nina Caspersen and Gareth Stansfield, *Unrecognized States in the International System* (London and New York: Routledge, 2011).
Helena Cook, *The Safe Haven in Northern Iraq. International Responsibility for Iraqi Kurdistan* (London: University of Essex, 1995).
Rami Daher and Irene Maffi, *The Politics and Practices of Cultural Heritage in the Middle East: Positioning the Past in Contemporary Societies* (New York: I.B. Tauris, 2007).
Claudia Derichs, 'Shifting Epistemologies in Area Studies: From Space to Scale', *Middle East – Topics & Arguments*, 4(2) (2015), pp. 29–36.
---, *Knowledge Production, Area Studies, and Global Cooperation* (New York: Routledge, 2017).
Cecil J. Edmonds, *Kurds, Turks and Arabs. Politics, Travel and Research in North-Eastern-Iraq. 1919–1925* (Oxford: Oxford University Press, 1957).
Andrea Fischer-Tahir, *Brave Men, Pretty Women? Gender and Symbolic Violence in Iraqi-Kurdish Urban Society* (Berlin: EZKS, 2009).
---, 'The Concept of Genocide as Part of Knowledge Production in Iraqi Kurdistan', in *Writing the History of Iraq: Historiographical and Political Challenges*, ed. by Jordi Tejel, Peter Sluglett, Riccardo Bocco, and Hamit Bozarslan (London: World Scientific Publishing, 2012), pp. 226–243.
---, 'Media, Political Culture, and the Shadows of the Militia War in Iraqi Kurdistan', in *The Kurdish Question Revisted*, ed. by Gareth Stansfield and Mohammed Shareef (London: Hurst, 2017), pp. 497–508.
Richard Grusin, 'YouTube at the End of New Media', in *The YouTube Reader*, ed. by Pelle Snickars and Patrick Vonderau (Stockholm: National Library of Sweden, 2009), pp. 24–39.
Henny Harald Hansen, *The Kurdish Woman's Life* (Copenhagen: Nationalmuseet, 1961).
Choman Hardi, 'Women's Activism in Iraqi Kurdistan: Achievements, Shortcoming and Obstacles', *Kurdish Studies*, 1 (2013), pp. 44–64.
Béatrice von Hirschhausen, Hannes Grandits Claudia Kraft, Dietmar Müller, and Thomas Serrier, eds., *Phantomgrenzen: Räume und Akteure in der Zeit neu denken, Phantomgrenzen im östlichen Europa* (Göttingen: Wallstein Verlag, 2015).
Mehdad R. Izady, *The Kurds. A Concise Handbook* (Washington, DC: Taylor & Francis, 1992).
Reinhart Koselleck, *Vergangene Zukunft. Zur Semantik geschichtlicherZeiten* (Frankfurt: Suhrkamp, 2010).
Michiel Leezenberg, 'Soviet Kurdology and Kurdish Orientalism', *The Heritage of Soviet Oriental Studies*, ed. by Michael Kemper and Stephan Conermann (London and New York: Routledge, 2011), pp. 86–102.
Henri Lefebvre, *The Production of Space* (Oxford: Basil Blackwell, 1991).

Martin Lewis and Kären E. Wigen, *The Myth of Continents: A Critique of Metageography* (Berkeley: University of California Press, 1997).

Leo Marquard, *South Africa's Colonial Policy* (Johannesburg: Institute of Race Relations, 1957).

Nikolaj Marr, *Der japhetitische Kaukasus und das dritte ethnische Element im Prozess mittelländischer Kultur* (Berlin, Stuttgart, and Leipzig: Kohlhammer, 1923).

Karl Marx, *Der 18. Brumaire des Louis Bonaparte* (Berlin: Karl Dietz Verlag, (1972), pp. 111–207.

David McDowall, *A Modern History of the Kurds* (London and New York: I.B. Tauris, 1996).

Marshall McLuhan, *Understanding Media* (London and New York Routledge, 2001).

Susan Meiselas, *Kurdistan: In the Shadow of History* (New York: Random House, 1997).

Vladimir F. Minorski, 'Kurds, Kurdistan in History', *Encyclopedia of Islam*, V (1986), pp. 447–464.

Karin Mlodoch, *The Limits of Trauma Discourse. Women Anfal Survivors in Kurdistan-Iraq* (Berlin: Klaus Schwarz, 2015).

Daniel Müller, 'The Kurds of Soviet Azerbaijan, 1920–91', *Central Asian Survey*, 19(1) (2000), pp. 41–77.

Denise Natali, *The Kurdish Quasi-State: Development and Dependency in Post-Gulf War Iraq* (New York: Syracuse, 2010).

Basile Nikitin, *Les Kurdes, étude sociologique et historique* (Paris: Klincksieck, 1956).

Edward M. Noel, *Diary of Major E. M. Noel on Special Duty in Kurdistan from 14 June to 21 September 1919* (Basra: Superintendent Government Press, 1919).

Sam C. Nolutshungu, *Margins of Insecurity: Minorities and International Security* (New York: University of Rochester Press, 1996).

Ferhad Pîrbal, *Kurd le dîd-î rojhelatnasekanewe* [The Kurds from the perspective of Orientalists] (Erbil: Aras, 2006).

Reporters without Borders, *Between Freedom and Abuses: the Media Paradox in Iraqi Kurdistan* (Paris: Reporters without Borders, 2010).

Edward Said, *Orientalism* (London: Penguin Books, 2003).

Clemence Scalbert-Yücel and Marie Le Ray, 'Knowledge, Ideology, and Power: Deconstructing Kurdish Studies', *European Journal of Turkish Studies*, 5 (2006).

Jens Schröter, 'On the Logic of the Digital Archive', in *The YouTube Reader*, ed. by Pelle Snickars and Patrick Vonderau (Stockholm: National Library of Sweden, 2009), pp. 330–346.

Jaffer Sheyholislami, *Kurdish Identity, Discourse and New Media* (New York: Palgrave MacMillan, 2011).

Ely Banister Soane, *To Mesopotamia and Kurdistan in Disguise. With Historical Note on the Kurdish Tribes and Chaldeans* (London: Murray, 2007).

Gareth Stansfield, *Iraqi Kurdistan: Political Development and Emergent Democracy* (London: Routledge, 2003).
Jalāl Tālabānī, *Kurdistān wa'l-haraka al-qawmiyya al-kurdiyya* [*Kurdistan and the Kurdish national movement*] (Beirut: Dār at-Talī'a, 1971).
'Elî Xeznedar, *Afrete le nawdarekan-î kurd. Hêz-î betiwana le bergrîkirdin-î nîştîmanekeyan* [*Famous Kurdish women: A strong force for the defense of their nation*] (Erbil: Karox, 2012).
Rex J. Zedalis, *Oil and Gas in the Disputed Kurdish Territories: Jurisprudence, Regional Minorities and Natural Resources in a Federal System* (London: Routledge, 2012).
Karîm Zend, *Cugraf-î Kurdistan* [*Geography of Kurdistan*] (Sulaimaniya: Xak, 2001).

About the Author

Andrea Fischer-Tahir studied Arabic and Oriental Studies, anthropology and the history of religion. In Kurdistan-Iraq, she conducted extensive field research on political history, memory, genocide in Kurdish historiography, gender, and media. She worked at Koye University Kurdistan, the ZMO Berlin, as well as at Philipps-Universität Marburg (Germany). Her dissertation discussed martyrdom, mobilization and memory in Kurdistan. Her book *Gender and Symbolic Violence in Iraqi Kurdish Urban Society* (2009) has been translated into Kurdish. She is co-editor of *Peripheralization: Spatial Differentiation and Social Injustice* (2013) and of *Disciplinary Spaces: Spatial control, forced assimilation, and narratives of progress since the 19th century* (2017).

Index

For Product Safety Concerns and Information please contact our EU representative GPSR@taylorandfrancis.com Taylor & Francis Verlag GmbH, Kaufingerstraße 24, 80331 München, Germany

Batch number: 10725973

Printed by Printforce, the Netherlands